Section 14

Equipment for Distillation, Gas Absorption, Phase Dispersion, and Phase Separation

Henry Z. Kister, M.E., C.Eng., C.Sc. *Senior Fellow and Director of Fractionation Technology, Fluor Corporation; Fellow, American Institute of Chemical Engineers; Fellow, Institution of Chemical Engineers (UK); Member, Institute of Energy (Section Editor, Equipment for Distillation and Gas Absorption)*

Paul M. Mathias, Ph.D. *Technical Director, Fluor Corporation; Member, American Institute of Chemical Engineers (Design of Gas Absorption Systems)*

D. E. Steinmeyer, P.E., M.A., M.S. *Distinguished Fellow, Monsanto Company (retired); Fellow, American Institute of Chemical Engineers; Member, American Chemical Society (Phase Dispersion)*

W. R. Penney, Ph.D., P.E. *Professor of Chemical Engineering, University of Arkansas; Member, American Institute of Chemical Engineers (Gas-in-Liquid Dispersions)*

B. B. Crocker, P.E., S.M. *Consulting Chemical Engineer; Fellow, American Institute of Chemical Engineers; Member, Air Pollution Control Association (Phase Separation)*

James R. Fair, Ph.D., P.E. *Professor of Chemical Engineering, University of Texas; Fellow, American Institute of Chemical Engineers; Member, American Chemical Society, American Society for Engineering Education, National Society of Professional Engineers (Section Editor of the 7th edition and major contributor to the 5th, 6th, and 7th editions)*

INTRODUCTION

DESIGN OF GAS ABSORPTION SYSTEMS

EQUIPMENT FOR DISTILLATION AND GAS ABSORPTION: TRAY COLUMNS

EQUIPMENT FOR DISTILLATION AND GAS ABSORPTION: PACKED COLUMNS

OTHER TOPICS FOR DISTILLATION AND GAS ABSORPTION EQUIPMENT

PHASE DISPERSION

PHASE SEPARATION

Nomenclature

a, a_e	Effective interfacial area	m^2/m^3	ft^2/ft^3
a_p	Packing surface area per unit volume	m^2/m^3	ft^2/ft^3
A	Absorption factor $L_M/(mG_M)$	-/-	-/-
A	Cross-sectional area	m^2	ft^2
A_a	Active area, same as bubbling area	m^2	ft^2
A_B	Bubbling (active) area	m^2	ft^2
A_D	Downcomer area (straight vertical downcomer)	m^2	ft^2
A_{da}	Downcomer apron area	m^2	ft^2
A_{DB}	Area at bottom of downcomer	m^2	ft^2
A_{DT}	Area at top of downcomer	m^2	ft^2
A_e, A'	Effective absorption factor (Edmister)	-/-	-/-
A_f	Fractional hole area	-/-	-/-
A_h	Hole area	m^2	ft^2
A_N	Net (free) area	m^2	ft^2
A_S	Slot area	m^2	ft^2
A_{SO}	Open slot area	m^2	ft^2
A_T	Tower cross-section area	m^2	ft^2
c	Concentration	kg-mol/m^3	lb-mol/ft^3
c'	Stokes-Cunningham correction factor for terminal settling velocity	-/-	-/-
C	C-factor for gas loading, Eq. (14-77)	m/s	ft/s
C_l	Coefficient in regime transition correlation, Eq. (14-129)	-/-	-/-
C_1, C_2	Parameters in system limit equation	m/s	ft/s
C_3, C_4	Constants in Robbins' packing pressure drop correlation	-/-	-/-
CAF	Flood C-factor, Eq. (14-88)	m/s	ft/s
CAF_0	Uncorrected flood C-factor, Fig. 14-30	—	ft/s
C_d	Coefficient in clear liquid height correlation, Eq. (14-116)	-/-	-/-
C_G	Gas C-factor; same as C	m/s	ft/s
C_L	Liquid loading factor, Eq. (14-144)	m/s	ft/s
C_{LG}	A constant in packing pressure drop correlation, Eq. (14-143)	$(m/s)^{0.5}$	$(ft/s)^{0.5}$
CP	Capacity parameter (packed towers), Eq. (14-140)		
C_{SB}, C_{sb}	C-factor at entrainment flood, Eq. (14-80)	m/s	ft/s
C_{sbf}	Capacity parameter corrected for surface tension	m/s	ft/s
C_v, C_V	Discharge coefficient, Fig. 14-35	-/-	-/-
C_w	A constant in weep rate equation, Eq. (14-123)	-/-	-/-
C_{XY}	Coefficient in Eq. (14-159) reflecting angle of inclination	-/-	-/-
d	Diameter	m	ft
d_b	Bubble diameter	m	ft
d_h, d_H	Hole diameter	mm	in
d_o	Orifice diameter	m	ft
d_{pc}	Cut size of a particle collected in a device, 50% mass efficiency	μm	ft
d_{psd}	Mass median size particle in the pollutant gas	μm	ft
d_{pa50}	Aerodynamic diameter of a real median size particle	μm	ft
d_w	Weir diameter, circular weirs	mm	in
D	Diffusion coefficient	m^2/s	ft^2/s
D	Tube diameter (wetted-wall columns)	m	ft
D_{32}	Sauter mean diameter	m	ft
D_g	Diffusion coefficient	m^2/s	ft^2/h
D_p	Packing particle diameter	m	ft
D_T	Tower diameter	m	ft
D_{tube}	Tube inside diameter	m	ft
D_{vm}	Volume mean diameter	m	ft
e	Absolute entrainment of liquid	kg-mol/h	lb-mol/h
e	Entrainment, mass liquid/mass gas	kg/kg	lb/lb
E	Plate or stage efficiency, fractional	-/-	-/-
E	Power dissipation per mass	W	Btu/lb
E_a	Murphree tray efficiency, with entrainment, gas concentrations, fractional	-/-	-/-
E_g	Point efficiency, gas phase only, fractional	-/-	-/-
E_{oc}	Overall column efficiency, fractional	-/-	-/-
E_{OG}	Overall point efficiency, gas concentrations, fractional	-/-	-/-
E_{mv}, E_{MV}	Murphree tray efficiency, gas concentrations, fractional	-/-	-/-
E_s	Entrainment, kg entrained liquid per kg gas upflow	kg/kg	lb/lb
f	Fractional approach to flood	-/-	-/-
f	Liquid maldistribution fraction	-/-	-/-
f_{max}	Maximum value of f above which separation cannot be achieved	-/-	-/-
f_w	Weep fraction, Eq. (14–121)	-/-	-/-
F	Fraction of volume occupied by liquid phase, system limit correlation, Eq. (14-170)	-/-	-/-
F	F-factor for gas loading Eq. (14-76)	$m/s(kg/m^3)^{0.5}$	$ft/s(lb/ft^3)^{0.5}$
F_{LG}	Flow parameter, Eq. (14-89) and Eq. (14-141)	-/-	-/-
F_p	Packing factor	m^{-1}	ft^{-1}
F_{pd}	Dry packing factor	m^{-1}	ft^{-1}
FPL	Flow path length	m	ft
Fr	Froude number, clear liquid height correlation, Eq. (14-120)	-/-	-/-
Fr_h	Hole Froude number, Eq. (14-114)	-/-	-/-
F_w	Weir constriction correction factor, Fig. 14-38	-/-	-/-
g	Gravitational constant	m/s^2	ft/s^2
g_c	Conversion factor	1.0 kg·m/ $(N \cdot s^2)$	32.2 lb·ft/ $(lbf \cdot s^2)$
G	Gas phase mass velocity	$kg/(s.m^2)$	$lb/(hr \cdot ft^2)$
G_f	Gas loading factor in Robbins' packing pressure drop correlation	$kg/(s \cdot m^2)$	$lb/(h \cdot ft^2)$
G_M	Gas phase molar velocity	kg·mol/ $(s.m^2)$	lb·mol/ $(h.ft^2)$
GPM	Liquid flow rate	—	gpm
h	Pressure head	mm	in
h'_{dc}	Froth height in downcomer	mm	in
h'_L	Pressure drop through aerated mass on tray	mm	in
h_c	Clear liquid height on tray	mm	in
h_{cl}	Clearance under downcomer	mm	in
h_{ct}	Clear liquid height at spray to froth transition	mm	in
h_d	Dry pressure drop across tray	mm	in
h_{da}	Head loss due to liquid flow under downcomer apron	mm	in
h_{dc}	Clear liquid height in downcomer	mm	in
h_{ds}	Calculated clear liquid height, Eq. (14-108)	mm	in
h_f	Height of froth	mm	in
h_{fow}	Froth height over the weir, Eq. (14-117)	mm	in
h_{hg}	Hydraulic gradient	mm	in
h_{Lo}	Packing holdup in preloading regime, fractional	-/-	-/-
h_{Lt}	Clear liquid height at froth to spray transition, corrected for effect of weir height, Eq. (14-96)	mm	in
h_{ow}	Height of crest over weir	mm	in
h_T	Height of contacting	m	ft
h_t	Total pressure drop across tray	mm	in
h_w	Weir height	mm	in
H	Height of a transfer unit	m	ft
H	Henry's law constant	kPa/mol fraction	atm /mol fraction
H'	Henry's law constant	$kPa/(kmol \cdot m^3)$	$psi/(lb \cdot mol.ft^3)$
H_G	Height of a gas phase transfer unit	m	ft
H_L	Height of a liquid phase transfer unit	m	ft
H_{OG}	Height of an overall transfer unit, gas phase concentrations	m	ft
H_{OL}	Height of an overall transfer unit, liquid phase concentrations	m	ft

Nomenclature *(Continued)*

H'	Henry's law coefficient	kPa/mol frac	atm/mol frac
HETP	Height equivalent to a theoretical plate or stage	m	ft
J_G^*	Dimensionless gas velocity, weep correlation, Eq. (14-124)	-/-	-/-
J_L^*	Dimensionless liquid velocity, weep correlation, Eq. (14-125)	-/-	-/-
k	Individual phase mass transfer coefficient	kmol /(s·m²·mol frac)	lb·mol/(s·ft²·mol frac)
k_1	First order reaction velocity constant	1/s	1/s
k_2	Second order reaction velocity constant	m³/(s·kmol)	ft³/(h·lb·mol)
k_g	Gas mass-transfer coefficient, wetted-wall columns [see Eq. (14-171) for unique units]		
k_G	gas phase mass transfer coefficient	kmol /(s·m²·mol frac)	lb.mol/(s·ft²·mol frac)
k_L	liquid phase mass transfer coefficient	kmol /(s·m²·mol frac)	lb·mol/(s·ft²·mol frac)
K	Constant in trays dry pressure drop equation	mm·s²/m²	in·s²/ft²
K	Vapor-liquid equilibrium ratio	-/-	-/-
K_C	Dry pressure drop constant, all valves closed	mm·s²/m²	in·s²/ft²
K_D	Orifice discharge coefficient, liquid distributor	-/-	-/-
K_g	Overall mass-transfer coefficient	kg·mol/(s·m²·atm)	lb·mol/(h·ft²·atm)
K_O	Dry pressure drop constant, all valves open	mm·s²/m²	in·s²/ft²
K_{OG}, K_G	Overall mass transfer coefficient, gas concentrations	kmol /(s·m²·mol frac)	lb·mol/(s·ft²·mol frac)
K_{OL}	Overall mass transfer coefficient, liquid concentrations	kmol/(s·m²·mol frac)	lb.mol/(s·ft²·mol frac)
L	Liquid mass velocity	kg/(m²·s)	lb/ft²·h
L_f	Liquid loading factor in Robbins' packing pressure drop correlation	kg/(s·m²)	lb/(h·ft²)
L_m	Molar liquid downflow rate	kg·mol/h	lb·mol/h
L_M	Liquid molar mass velocity	kmol/(m²·s)	lb·mol/(ft²·h)
L_S	Liquid velocity, based on superficial tower area	m/s	ft/s
L_w	Weir length	m	in
m	An empirical constant based on Wallis' countercurrent flow limitation equation, Eqs. (14-123) and (14-143)	-/-	-/-
m	Slope of equilibrium curve = dy^*/dx	-/-	-/-
M	Molecular weight	kg/kmol	lb/(lb·mol)
n	Parameter in spray regime clear liquid height correlation, Eq. (14-84)	mm	in
n_A	Rate of solute transfer	kmol/s	lb·mol/s
n_D	Number of holes in orifice distributor	-/-	-/-
N_a	Number of actual trays	-/-	-/-
N_A, N_t	Number of theoretical stages	-/-	-/-
N_{OG}	Number of overall gas-transfer units	-/-	-/-
N_p	Number of tray passes	-/-	-/-
p	Hole pitch (center-to-center hole spacing)	mm	in
p	Partial pressure	kPa	atm
p_{BM}	Logarithmic mean partial pressure of inert gas	kPa	atm
P, p_T	Total pressure	kPa	atm
P^0	Vapor pressure	kpa	atm
Q, q	Volumetric flow rate of liquid	m³/s	ft³/s
Q'	Liquid flow per serration of serrated weir	m·³/s	ft³/s
Q_D	Downcomer liquid load, Eq. (14-79)	m/s	ft/s
Q_L	Weir load, Eq. (14-78)	m³/(h·m)	gpm/in
Q_{MW}	Minimum wetting rate	m³/(h·m²)	gpm/ft²
R	Reflux flow rate	kg·mol/h	lb·mol/h
R	Gas constant		
R_h	Hydraulic radius	m	ft
R_{vw}	Ratio of valve weight with legs to valve weight without legs, Table (14-11)	-/-	-/-

S	Length of corrugation side, structured packing	m	ft
S	Stripping factor mG_M/L_M	-/-	-/-
S	Tray spacing	mm	in
S_e, S'	Effective stripping factor (Edmister)	-/-	-/-
SF	Derating (system) factor, Table 14-9	-/-	-/-
t_t	Tray thickness	mm	in
t_v	Valve thickness	mm	in
T	Absolute temperature	K	°R
TS	Tray spacing; same as S	mm	in
U,u	Linear velocity of gas	m/s	ft/s
U_a	Velocity of gas through active area	m/s	ft/s
U_a^*	Gas velocity through active area at froth to spray transition	m/s	ft/s
U_h, u_h	Gas hole velocity	m/s	ft/s
U_L, u_L	Liquid superficial velocity based on tower cross-sectional area	m/s	ft/s
U_n	Velocity of gas through net area	m/s	ft/s
U_{nf}	Gas velocity through net area at flood	-/-	-/-
U_t	Superficial velocity of gas	m/s	ft/s
v_H	Horizontal velocity in trough	m/s	ft/s
V	Linear velocity	m/s	ft/s
V	Molar vapor flow rate	kg·mol/s	lb·mol/h
W	Weep rate	m³/s	gpm
x	Mole fraction, liquid phase (note 1)	-/-	-/-
x'	Mole fraction, liquid phase, column 1 (note 1)		
x''	Mole fraction, liquid phase, column 2 (note 1)		
x^*, x°	Liquid mole fraction at equilibrium (note 1)	-/-	-/-
y	Mole fraction, gas or vapor phase (note 1)	-/-	-/-
y'	Mole fraction, vapor phase, column 1 (note 1)		
y''	Mole fraction, vapor phase, column 2 (note 1)		
y^*, y°	Gas mole fraction at equilibrium (note 1)		
Z	Characteristic length in weep rate equation, Eq. (14-126)	m	ft
Z_p	Total packed height	m	ft

Greek Symbols

α	Relative volatility	-/-	-/-
β	Tray aeration factor, Fig. (14-37)	-/-	-/-
ε	Void fraction	-/-	-/-
ϕ	Contact angle	deg	deg
ϕ	Relative froth density	-/-	-/-
γ	Activity coefficient	-/-	-/-
Γ	Flow rate per length	kg/(s·m)	lb/(s·ft)
δ	Effective film thickness	m	ft
η	Collection eficiency, fractional	-/-	-/-
η	Factor used in froth density correlation, Eq. (14-118)	-/-	-/-
λ	Stripping factor = $m/(L_M/G_M)$	-/-	-/-
μ	Absolute viscosity	Pa·s	cP or lb/(ft·s)
μm	Micrometers	m	-/-
ν	Kinematic viscosity	m²/s	cS
π	3.1416. . . .	-/-	-/-
θ	Residence time	s	s
θ	Angle of serration in serrated weir	deg	deg
ρ	Density	kg/m³	lb/ft³
ρ_M	Valve metal density	kg/m³	lb/ft³
σ	Surface tension	mN/m	dyn/cm
χ	Parameter used in entrainment correlation, Eq. (14-95)	-/-	-/-
ψ	Fractional entrainment, moles liquid entrained per mole liquid downflow	k·mol/k·mol	lb·mol/lb·mol
Φ	Fractional approach to entrainment flood	-/-	-/-
ΔP	Pressure drop per length of packed bed	mmH$_2$O/m	inH$_2$O/ft
$\Delta\rho$	$\rho_L - \rho_G$	kg/m³	lb/ft³

Subscripts

A	Species A
AB	Species A diffusing through species B

Nomenclature (*Concluded*)

Subscripts	
B	Species B
B	Based on the bubbling area
d	Dry
da	Downcomer apron
dc	Downcomer
dry	Uncorrected for entrainment and weeping
e	Effective value
f	Froth
Fl	Flood
flood	At flood
G, g	Gas or vapor
h	Based on hole area (or slot area)
H_2O	Water
i	Interface value
L, l	Liquid
m	Mean
min	Minimum
MOC	At maximum operational capacity

Subscripts	
n, N	On stage n
N	At the inlet nozzle
NF, nf	Based on net area at flood
p	Particle
S	Superficial
t	Total
ult	At system limit (ultimate capacity)
V	Vapor
w	Water
1	Tower bottom
2	Tower top

Dimensionless Groups	
N_{Fr}	Froude number = $(U_L^2)/(Sg)$,
N_{Re}	Reynolds number = $(D_{tube} U_{ge} \rho_G)/(\mu_G)$
N_{Sc}	Schmidt number = $\mu/(\rho D)$
N_{We}	Weber number = $(U_L^2 \rho_L S)/(\sigma g_c)$

NOTE: 1. Unless otherwise specified, refers to concentration of more volatile component (distillation) or solute (absorption).

GENERAL REFERENCES: Astarita, G., *Mass Transfer with Chemical Reaction,* Elsevier, New York, 1967. Astarita, G., D. W. Savage and A. Bisio, *Gas Treating with Chemical Solvents,* Wiley, New York, 1983. Billet, R., *Distillation Engineering,* Chemical Publishing Co., New York, 1979. Billet, R., *Packed Column Analysis and Design,* Ruhr University, Bochum, Germany, 1989. Danckwerts, P. V., *Gas-Liquid Reactions,* McGraw-Hill, New York, 1970. *Distillation and Absorption 1987,* Rugby, U.K., Institution of Chemical Engineers. *Distillation and Absorption 1992,* Rugby, U.K., Institution of Chemical Engineers. *Distillation and Absorption 1997,* Rugby, U.K., Institution of Chemical Engineers. *Distillation and Absorption 2002,* Rugby, U.K., Institution of Chemical Engineers. *Distillation and Absorption 2006,* Rugby, U.K., Institution of Chemical Engineers. *Distillation Topical Conference Proceedings,* AIChE Spring Meetings (separate Proceedings Book for each Topical Conference): Houston, Texas, March 1999; Houston, Texas, April 22–26, 2001; New Orleans, La., March 10–14, 2002; New Orleans, La., March 30–April 3, 2003; Atlanta, Ga., April 10–13, 2005. Hines, A. L., and R. N. Maddox, *Mass Transfer—Fundamentals and Applications,* Prentice Hall, Englewood Cliffs, New Jersey, 1985. Hobler, T., *Mass Transfer and Absorbers,* Pergamon Press, Oxford, 1966. Kister, H. Z., *Distillation Operation,* McGraw-Hill, New York, 1990. Kister, H. Z., *Distillation Design,* McGraw-Hill, New York, 1992. Kister, H. Z., and G. Nalven (eds.), *Distillation and Other Industrial Separations,* Reprints from CEP, AIChE, 1998. Kister, H. Z., *Distillation Troubleshooting,* Wiley, 2006. Kohl, A. L., and R. B. Nielsen, *Gas Purification,* 5th ed., Gulf, Houston, 1997. Lockett, M.J., *Distillation Tray Fundamentals,* Cambridge, U.K., Cambridge University Press, 1986. Mačkowiak, J., "Fluiddynamik von Kolonnen mit Modernen Füllkorpern und Packungen für Gas/Flussigkeitssysteme," Otto Salle Verlag, Frankfurt am Main und Verlag Sauerländer Aarau, Frankfurt am Main, 1991. Schweitzer, P. A. (ed.), *Handbook of Separation Techniques for Chemical Engineers,* 3d. ed., McGraw-Hill, New York, 1997. Sherwood, T. K., R. L. Pigford, C. R. Wilke, *Mass Transfer,* McGraw-Hill, New York, 1975. Stichlmair, J., and J. R. Fair, *Distillation Principles and Practices,* Wiley, New York, 1998. Strigle, R. F., Jr., *Packed Tower Design and Applications,* 2d ed., Gulf Publishing, Houston, 1994. Treybal, R. E., *Mass Transfer Operations,* McGraw-Hill, New York, 1980.

INTRODUCTION

Definitions Gas absorption is a unit operation in which soluble components of a gas mixture are dissolved in a liquid. The inverse operation, called stripping or desorption, is employed when it is desired to transfer volatile components from a liquid mixture into a gas. Both absorption and stripping, in common with distillation (Sec. 13), make use of special equipment for bringing gas and liquid phases into intimate contact. This section is concerned with the design of gas-liquid contacting equipment, as well as with the design of absorption and stripping processes.

Equipment Absorption, stripping, and distillation operations are usually carried out in vertical, cylindrical columns or towers in which devices such as plates or packing elements are placed. The gas and liquid normally flow countercurrently, and the devices serve to provide the contacting and development of interfacial surface through which mass transfer takes place. Background material on this mass transfer process is given in Sec. 5.

Design Procedures The procedures to be followed in specifying the principal dimensions of gas absorption and distillation equipment are described in this section and are supported by several worked-out examples. The experimental data required for executing the designs are keyed to appropriate references or to other sections of the handbook.

For absorption, stripping, and distillation, there are three main steps involved in design:

1. *Data on the gas-liquid or vapor-liquid equilibrium for the system at hand.* If absorption, stripping, and distillation operations are considered equilibrium-limited processes, which is the usual approach, these data are critical for determining the maximum possible separation. In some cases, the operations are considered rate-based (see Sec. 13) but require knowledge of equilibrium at the phase interface. Other data required include physical properties such as viscosity and density and thermodynamic properties such as enthalpy. Section 2 deals with sources of such data.

2. *Information on the liquid- and gas-handling capacity of the contacting device chosen for the particular separation problem.* Such information includes pressure drop characteristics of the device, in order that an optimum balance between capital cost (column cross section) and energy requirements might be achieved. Capacity and pressure drop characteristics of the available devices are covered later in this Sec. 14.

3. *Determination of the required height of contacting zone for the separation to be made as a function of properties of the fluid mixtures and mass-transfer efficiency of the contacting device.* This determination involves the calculation of mass-transfer parameters such as heights of transfer units and plate efficiencies as well as equilibrium or rate parameters such as theoretical stages or numbers of transfer units. An additional consideration for systems in which chemical reaction occurs is the provision of adequate residence time for desired reactions to occur, or minimal residence time to prevent undesired reactions from occurring. For equilibrium-based operations, the parameters for required height are covered in the present section.

Data Sources in the Handbook Sources of data for the analysis or design of absorbers, strippers, and distillation columns are manifold, and a detailed listing of them is outside the scope of the presentation in this section. Some key sources within the handbook are shown in Table 14-1.

Equilibrium Data Finding reliable gas-liquid and vapor-liquid equilibrium data may be the most time-consuming task associated with the design of absorbers and other gas-liquid contactors, and yet it may be the most important task at hand. For gas solubility, an important data source is the set of volumes edited by Kertes et al., *Solubility Data Series,* published by Pergamon Press (1979 ff.). In the introduction to each volume, there is an excellent discussion and definition of the various methods by which gas solubility data have been reported, such as the Bunsen coefficient, the Kuenen coefficient, the Ostwalt coefficient, the absorption coefficient, and the Henry's law coefficient. The fifth edition of *The Properties of Gases and Liquids* by Poling, Prausnitz, and O'Connell (McGraw-Hill, New York, 2000) provides data and recommended estimation methods for gas solubility as well as the broader area of vapor-liquid equilibrium. Finally, the Chemistry Data Series by Gmehling et al., especially the title *Vapor-Liquid Equilibrium Collection* (DECHEMA, Frankfurt, Germany, 1979 ff.), is a rich source of data evaluated against the various models used for interpolation and extrapolation. Section 13 of this handbook presents a good discussion of equilibrium *K* values.

TABLE 14-1 Directory to Key Data for Absorption and Gas-Liquid Contactor Design

Type of data	Section
Phase equilibrium data	
Gas solubilities	2
Pure component vapor pressures	2
Equilibrium *K* values	13
Thermal data	
Heats of solution	2
Specific heats	2
Latent heats of vaporization	2
Transport property data	
Diffusion coefficients	
Liquids	2
Gases	2
Viscosities	
Liquids	2
Gases	2
Densities	
Liquids	2
Gases	2
Surface tensions	2
Packed tower data	
Pressure drop and flooding	14
Mass transfer coefficients	5
HTU, physical absorption	5
HTU with chemical reaction	14
Height equivalent to a theoretical plate (HETP)	
Plate tower data	
Pressure drop and flooding	14
Plate efficiencies	14
Costs of gas-liquid contacting equipment	14

DESIGN OF GAS ABSORPTION SYSTEMS

General Design Procedure The design engineer usually is required to determine (1) the best solvent; (2) the best gas velocity through the absorber, or, equivalently, the vessel diameter; (3) the height of the vessel and its internal members, which is the height and type of packing or the number of contacting trays; (4) the optimum solvent circulation rate through the absorber and stripper; (5) temperatures of streams entering and leaving the absorber and stripper, and the quantity of heat to be removed to account for the heat of solution and other thermal effects; (6) pressures at which the absorber and stripper will operate; and (7) mechanical design of the absorber and stripper vessels (predominantly columns or towers), including flow distributors and packing supports. This section covers these aspects.

The problem presented to the designer of a gas absorption system usually specifies the following quantities: (1) gas flow rate; (2) gas composition of the component or components to be absorbed; (3) operating pressure and allowable pressure drop across the absorber; (4) minimum recovery of one or more of the solutes; and, possibly, (5) the solvent to be employed. Items 3, 4, and 5 may be subject to economic considerations and therefore are left to the designer. For determination of the number of variables that must be specified to fix a unique solution for the absorber design, one may use the same phase-rule approach described in Sec. 13 for distillation systems.

Recovery of the solvent, occasionally by chemical means but more often by distillation, is almost always required and is considered an integral part of the absorption system process design. A more complete solvent-stripping operation normally will result in a less costly absorber because of a lower concentration of residual solute in the regenerated (lean) solvent, but this may increase the overall cost of the entire absorption system. A more detailed discussion of these and other economical considerations is presented later in this section.

The design calculations presented in this section are relatively simple and usually can be done by using a calculator or spreadsheet. In many cases, the calculations are explained through design diagrams. It is recognized that most engineers today will perform rigorous, detailed calculations using process simulators. The design procedures presented in this section are intended to be complementary to the rigorous computerized calculations by presenting approximate estimates and insight into the essential elements of absorption and stripping operations.

Selection of Solvent and Nature of Solvents When a choice is possible, preference is given to solvents with high solubilities for the target solute and high selectivity for the target solute over the other species in the gas mixture. A high solubility reduces the amount of liquid to be circulated. The solvent should have the advantages of low volatility, low cost, low corrosive tendencies, high stability, low viscosity, low tendency to foam, and low flammability. Since the exit gas normally leaves saturated with solvent, solvent loss can be costly and can cause environmental problems. The choice of the solvent is a key part of the process economic analysis and compliance with environmental regulations.

Typically, a solvent that is chemically similar to the target solute or that reacts with it will provide high solubility. Water is often used for polar and acidic solutes (e.g., HCl), oils for light hydrocarbons, and special chemical solvents for acid gases such as CO_2, SO_2, and H_2S. Solvents are classified as physical and chemical. A chemical solvent forms complexes or chemical compounds with the solute, while physical solvents have only weaker interactions with the solute. Physical and chemical solvents are compared and contrasted by examining the solubility of CO_2 in propylene carbonate (representative physical solvent) and aqueous monoethanolamine (MEA; representative chemical solvent).

Figures 14-1 and 14-2 present data for the solubility of CO_2 in the two representative solvents, each at two temperatures: 40 and 100°C.

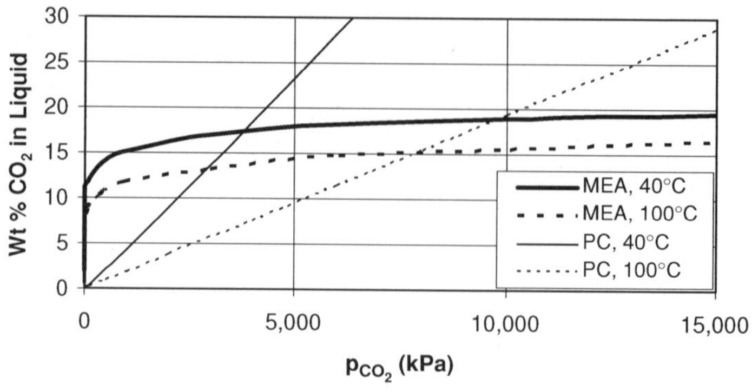

FIG. 14-1 Solubility of CO_2 in 30 wt% MEA and propylene carbonate. Linear scale.

The propylene carbonate data are from Zubchenko et al. [*Zhur. Priklad. Khim.*, **44**, 2044–2047 (1971)], and the MEA data are from Jou, Mather, and Otto [*Can. J. Chem. Eng.*, **73**, 140–147 (1995)]. The two figures have the same content, but Fig. 14-2 focuses on the low-pressure region by converting both composition and pressure to the logarithm scale. Examination of the two sets of data reveals the following characteristics and differences of physical and chemical solvents, which are summarized in the following table:

Characteristic	Physical solvent	Chemical solvent
Solubility variation with pressure	Relatively linear	Highly nonlinear
Low-pressure solubility	Low	High
High-pressure solubility	Continues to increase	Levels off
Heat of solution—related to variation of solubility with temperature at fixed pressure	Relatively low and approximately constant with loading	Relatively high and decreases somewhat with increased solute loading

Chemical solvents are usually preferred when the solute must be reduced to very low levels, when high selectivity is needed, and when the solute partial pressure is low. However, the strong absorption at low solute partial pressures and the high heat of solution are disadvantages for stripping. For chemical solvents, the strong nonlinearity of the absorption makes it necessary that accurate absorption data for the conditions of interest be available.

Selection of Solubility Data Solubility values are necessary for design because they determine the liquid rate necessary for complete or economic solute recovery. Equilibrium data generally will be found in one of three forms: (1) solubility data expressed either as weight or mole percent or as Henry's law coefficients; (2) pure-component vapor pressures; or (3) equilibrium distribution coefficients (K values). Data for specific systems may be found in Sec. 2; additional references to sources of data are presented in this section.

To define completely the solubility of gas in a liquid, it is generally necessary to state the temperature, equilibrium partial pressure of the solute gas in the gas phase, and the concentration of the solute gas in the liquid phase. Strictly speaking, the total pressure of the system should also be identified, but for low pressures (less than about 507 kPa or 5 atm), the solubility for a particular partial pressure of the solute will be relatively independent of the total pressure.

For many physical systems, the equilibrium relationship between solute partial pressure and liquid-phase concentration is given by Henry's law:

$$p_A = Hx_A \tag{14-1}$$

or

$$p_A = H'c_A \tag{14-2}$$

where H is Henry's law coefficient expressed in kPa per mole fraction solute in liquid and H' is Henry's law coefficient expressed in kPa·m^3/kmol.

Figure 14-1 indicates that Henry's law is valid to a good approximation for the solubility CO_2 in propylene carbonate. In general, Henry's law is a reasonable approximation for physical solvents. If Henry's law holds, the solubility is defined by knowing (or estimating) the value of the constant H (or H').

Note that the assumption of Henry's law will lead to incorrect results for solubility of chemical systems such as CO_2-MEA (Figs. 14-1 and 14-2) and HCl-H_2O. Solubility modeling for chemical systems requires the use of a *speciation model,* as described later in this section.

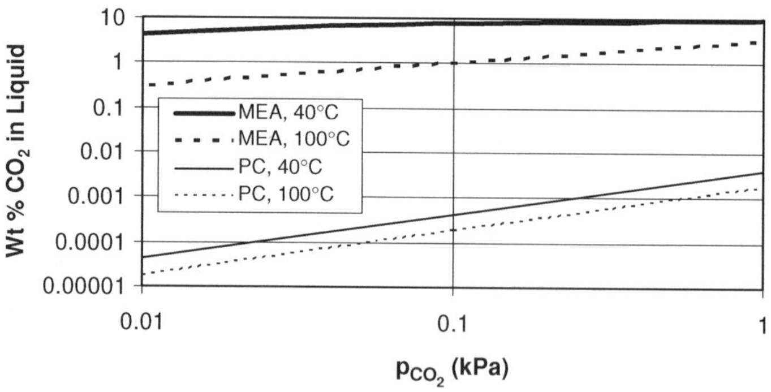

FIG. 14-2 Solubility of CO_2 in 30 wt% MEA and propylene carbonate. Logarithm scale and focus on low-pressure region.

For quite a number of physically absorbed gases, Henry's law holds very well when the partial pressure of the solute is less than about 101 kPa (1 atm). For partial pressures above 101 kPa, H may be independent of the partial pressure (Fig. 14-1), but this needs to be verified for the particular system of interest. The variation of H with temperature is a strongly nonlinear function of temperature as discussed by Poling, Prausnitz, and O'Connell (*The Properties of Gases and Liquids*, 5th ed., McGraw-Hill, New York, 2000). Consultation of this reference is recommended when temperature and pressure extrapolations of Henry's law data are needed.

The use of Henry's law constants is illustrated by the following example.

Example 1: Gas Solubility It is desired to find out how much hydrogen can be dissolved in 100 weights of water from a gas mixture when the total pressure is 101.3 kPa (760 torr; 1 atm), the partial pressure of the H_2 is 26.7 kPa (200 torr), and the temperature is 20°C. For partial pressures up to about 100 kPa the value of H is given in Sec. 3 as 6.92×10^6 kPa (6.83×10^4 atm) at 20°C. According to Henry's law,

$$x_{H_2} = p_{H_2}/H_{H_2} = 26.7/6.92 \times 10^6 = 3.86 \times 10^{-6}$$

The mole fraction x is the ratio of the number of moles of H_2 in solution to the total moles of all constituents contained. To calculate the weights of H_2 per 100 weights of H_2O, one can use the following formula, where the subscripts A and w correspond to the solute (hydrogen) and solvent (water):

$$\left(\frac{x_A}{1-x_A}\right)\frac{M_A}{M_W}100 = \left(\frac{3.86 \times 10^{-6}}{1-3.86 \times 10^{-6}}\right)\frac{2.02}{18.02}100$$

$$= 4.33 \times 10^{-5} \text{ weights } H_2/100 \text{ weights } H_2O$$

$$= 0.43 \text{ parts per million weight}$$

Pure-component vapor pressure can be used for predicting solubilities for systems in which **Raoult's law** is valid. For such systems $p_A = p^0{}_A x_A$, where $p^0{}_A$ is the pure-component vapor pressure of the solute and p_A is its partial pressure. Extreme care should be exercised when using pure-component vapor pressures to predict gas absorption behavior. Both vapor-phase and liquid-phase nonidealities can cause significant deviations from Raoult's law, and this is often the reason particular solvents are used, i.e., because they have special affinity for particular solutes. The book by Poling, Prausnitz, and O'Connell (op. cit.) provides an excellent discussion of the conditions where Raoult's law is valid. Vapor-pressure data are available in Sec. 3 for a variety of materials.

Whenever data are available for a given system under similar conditions of temperature, pressure, and composition, **equilibrium distribution coefficients** ($K = y/x$) provide a much more reliable tool for predicting vapor-liquid distributions. A detailed discussion of equilibrium K values is presented in Sec. 13.

Calculation of Liquid-to-Gas Ratio The minimum possible liquid rate is readily calculated from the composition of the entering gas and the solubility of the solute in the exit liquor, with equilibrium being assumed. It may be necessary to estimate the temperature of the exit liquid based upon the heat of solution of the solute gas. Values of latent heat and specific heat and values of heats of solution (at infinite dilution) are given in Sec. 2.

The actual liquid-to-gas ratio (solvent circulation rate) normally will be greater than the minimum by as much as 25 to 100 percent, and the estimated factor may be arrived at by economic considerations as well as judgment and experience. For example, in some packed-tower applications involving very soluble gases or vacuum operation, the minimum quantity of solvent needed to dissolve the solute may be insufficient to keep the packing surface thoroughly wet, leading to poor distribution of the liquid stream.

When the solvent concentration in the inlet gas is low and when a significant fraction of the solute is absorbed (this often the case), the approximation

$$y_1 G_M = x_1 L_M = (y_1^o/m) L_M \tag{14-3}$$

leads to the conclusion that the ratio mG_M/L_M represents the fractional approach of the exit liquid to saturation with the inlet gas, i.e.,

$$mG_M/L_M = y_1^o/y_1 \tag{14-4}$$

Optimization of the liquid-to-gas ratio in terms of total annual costs often suggests that the molar liquid-to-gas ratio L_M/G_M should be about 1.2 to 1.5 times the theoretical minimum corresponding to equilibrium at the rich end of the tower (infinite height or number of trays), provided flooding is not a problem. This, for example, would be an alternative to assuming that $L_M/G_M \approx m/0.7$.

When the exit-liquor temperature rises because of the heat of absorption of the solute, the value of m changes through the tower, and the liquid-to-gas ratio must be chosen to give reasonable values of m_1G_M/L_M and m_2G_M/L_M, where the subscripts 1 and 2 refer to the bottom and top of the absorber, respectively. For this case, the value of m_2G_M/L_M will be taken to be somewhat less than 0.7, so that the value of m_1G_M/L_M will not approach unity too closely. This rule-of-thumb approach is useful only when the solute concentration is low and heat effects are negligible.

When the solute has a large heat of solution or when the feed gas contains high concentrations of the solute, one should consider the use of internal cooling coils or intermediate liquid withdrawal and cooling to remove the heat of absorption.

Selection of Equipment Trays and random packings have been extensively used for gas absorption; structured packings are less common. Compared to trays, random packings have the advantages of availability in low-cost, corrosion-resistant materials (such as plastics and ceramics), low pressure drop (which can be an advantage when the tower is in the suction of a fan or compressor), easy and economic adaptability to small-diameter (less than 0.6-m or 2-ft) columns, and excellent handling of foams. Trays are much better for handling solids and fouling applications, offer greater residence time for slow absorption reactions, can better handle high L/G ratios and intermediate cooling, give better liquid turndown, and are more robust and less prone to reliability issues such as those resulting from poor distribution. Details on the operating characteristics of tray and packed towers are given later in this section.

Column Diameter and Pressure Drop Flooding determines the minimum possible diameter of the absorber column, and the usual design is for 60 to 80 percent of the flooding velocity. In near-atmospheric applications, pressure drop usually needs to be minimized to reduce the cost of energy for compression of the feed gas. For systems having a significant tendency to foam, the maximum allowable velocity will be lower than the estimated flooding velocity. Methods for predicting flooding velocities and pressure drops are given later in this section.

Computation of Tower Height The required height of a gas absorption or stripping tower for physical solvents depends on (1) the phase equilibria involved; (2) the specified degree of removal of the solute from the gas; and (3) the mass-transfer efficiency of the device. These three considerations apply to both tray and packed towers. Items 1 and 2 dictate the required number of theoretical stages (tray tower) or transfer units (packed tower). Item 3 is derived from the tray efficiency and spacing (tray tower) or from the height of one transfer unit (packed tower). Solute removal specifications are usually derived from economic considerations.

For tray towers, the approximate design methods described below may be used in estimating the number of theoretical stages, and the tray efficiencies and spacings for the tower can be specified on the basis of the information given later. Considerations involved in the rigorous design of theoretical stages for tray towers are treated in Sec. 13.

For packed towers, the continuous differential nature of the contact between gas and liquid leads to a design procedure involving the solution of differential equations, as described in the next subsection. Note that the design procedures discussed in this section are not applicable to reboiled absorbers, which should be designed according to the procedures described in Sec. 13.

Caution is advised in distinguishing between systems involving pure physical absorption and those in which chemical reactions can significantly affect design procedures. Chemical systems require additional procedures, as described later in this section.

Selection of Stripper Operating Conditions Stripping involves the removal of one or more components from the solvent through the application of heat or contacting it with a gas such as steam, nitrogen, or air. The operating conditions chosen for stripping normally result in a low solubility of solute (i.e., high value of m), so that the ratio mG_M/L_M will be larger than unity. A value of 1.4 may be used for rule-of-thumb calculations involving pure physical absorption. For tray-tower calculations, the stripping factor $S = KG_M/L_M$, where $K = y^0/x$ usually is specified for each tray.

When the solvent from an absorption operation must be regenerated for recycling to the absorber, one may employ a "pressure-swing" or "temperature-swing" concept, or a combination of the two, in specifying the stripping operation. In pressure-swing operation, the temperature of the stripper is about the same as that of the absorber, but the stripping pressure is much lower. In temperature-swing operation, the pressures are about equal, but the stripping temperature is much higher than the absorption temperature.

In pressure-swing operation, a portion of the gas may be "sprung" from the liquid by the use of a flash drum upstream of the stripper feed point. This type of operation has been discussed by Burrows and Preece [*Trans. Inst. Chem. Eng.*, **32,** 99 (1954)] and by Langley and Haselden [*Inst. Chem. Eng. Symp. Ser. (London)*, no. 28 (1968)]. If the flashing of the liquid takes place inside the stripping tower, this effect must be accounted for in the design of the upper section in order to avoid overloading and flooding near the top of the tower.

Often the rate at which residual absorbed gas can be driven from the liquid in a stripping tower is limited by the rate of a chemical reaction, in which case the liquid-phase residence time (and hence the tower liquid holdup) becomes the most important design factor. Thus, many stripper regenerators are designed on the basis of liquid holdup rather than on the basis of mass-transfer rate.

Approximate design equations applicable only to the case of pure physical desorption are developed later in this section for both packed and tray stripping towers. A more rigorous approach using distillation concepts may be found in Sec. 13. A brief discussion of desorption with chemical reaction is given in the subsection "Absorption with Chemical Reaction."

Design of Absorber-Stripper Systems The solute-rich liquor leaving a gas absorber normally is distilled or stripped to regenerate the solvent for recirculation back to the absorber, as depicted in Fig. 14-3. It is apparent that the conditions selected for the absorption step (e.g., temperature, pressure, L_M/G_M) will affect the design of the stripping tower, and conversely, a selection of stripping conditions will affect the absorber design. The choice of optimum operating conditions for an absorber-stripper system therefore involves a combination of economic factors and practical judgments as to the operability of the system within the context of the overall process flow sheet. In Fig. 14-3, the stripping vapor is provided by a reboiler; alternately, an extraneous stripping gas may be used.

An appropriate procedure for executing the design of an absorber-stripper system is to set up a carefully selected series of design cases and then evaluate the investment costs, the operating costs, and the operability of each case. Some of the economic factors that need to be considered in selecting the optimum absorber-stripper design are discussed later in the subsection "Economic Design of Absorption Systems."

Importance of Design Diagrams One of the first things a designer should do is to lay out a carefully constructed equilibrium curve $y^0 = F(x)$ on an xy diagram, as shown in Fig. 14-4. A horizontal line corresponding to the inlet-gas composition y_1 is then the locus of feasible outlet-liquor compositions, and a vertical line corresponding to the inlet-solvent-liquor composition x_2 is the locus of outlet-gas compositions. These lines are indicated as $y = y_1$ and $x = x_2$, respectively on Fig. 14-4.

For gas absorption, the region of feasible operating lines lies above the equilibrium curve; for stripping, the feasible region for operating lines lies below the equilibrium curve. These feasible regions are bounded by the equilibrium curve and by the lines $x = x_2$ and $y = y_1$. By inspection, one should be able to visualize those operating lines that are feasible and those that would lead to "pinch points" within the tower. Also, it is possible to determine if a particular proposed design for solute recovery falls within the feasible envelope.

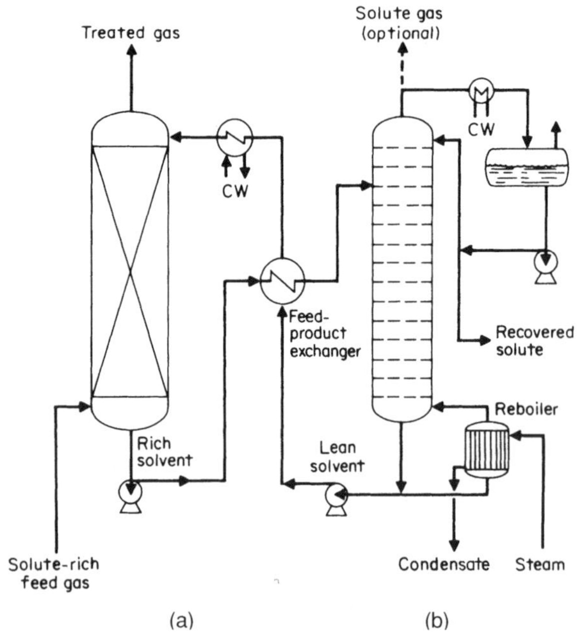

FIG. 14-3 Gas absorber using a solvent regenerated by stripping. (*a*) Absorber. (*b*) Stripper.

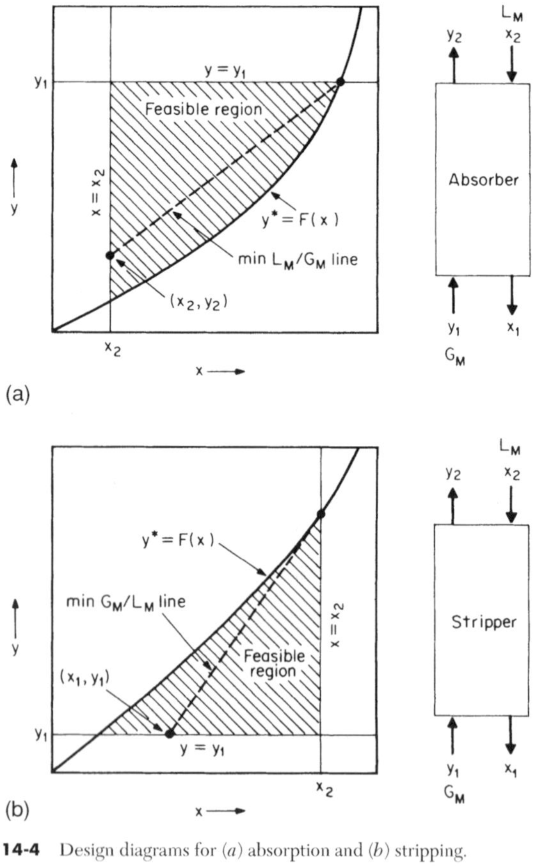

FIG. 14-4 Design diagrams for (*a*) absorption and (*b*) stripping.

Once the design recovery for an absorber has been established, the operating line can be constructed by first locating the point x_2, y_2 on the diagram. The intersection of the horizontal line corresponding to the inlet gas composition y_1 with the equilibrium curve $y^0 = F(x)$ defines the theoretical minimum liquid-to-gas ratio for systems in which there are no intermediate pinch points. This operating line which connects this point with the point x_2, y_2 corresponds to the minimum value of L_M/G_M. The actual design value of L_M/G_M should normally be around 1.2 to 1.5 times this minimum value. Thus, the actual design operating line for a gas absorber will pass through the point x_2, y_2 and will intersect the line $y = y_1$ to the left of the equilibrium curve.

For stripping one begins by using the design specification to locate the point x_1, y_1; then the intersection of the vertical line $x = x_2$ with the equilibrium curve $y^0 = F(x)$ defines the theoretical minimum gas-to-liquid ratio. The actual value of G_M/L_M is chosen to be about 20 to 50 percent higher than this minimum, so the actual design operating line will intersect the line $x = x_2$ at a point somewhat below the equilibrium curve.

PACKED-TOWER DESIGN

Methods for estimating the height of the active section of counterflow differential contactors such as packed towers, spray towers, and falling-film absorbers are based on rate expressions representing mass transfer at a point on the gas-liquid interface and on material balances representing the changes in bulk composition in the two phases that flow past each other. The rate expressions are based on the interphase mass-transfer principles described in Sec. 5. Combination of such expressions leads to an integral expression for the number of transfer units or to equations related closely to the number of theoretical stages. The paragraphs which follow set forth convenient methods for using such equations, first in a general case and then for cases in which simplifying assumptions are valid.

Use of Mass-Transfer-Rate Expression Figure 14-5 shows a section of a packed absorption tower together with the nomenclature that will be used in developing the equations that follow. In a differential section dh, we can equate the rate at which solute is lost from the gas phase to the rate at which it is transferred through the gas phase to the interface as follows:

$$-d(G_M y) = -G_M\,dy - y\,dG_M = N_A a\,dh \qquad (14\text{-}5)$$

In Eq. (14-5), G_M is the gas-phase molar velocity [kmol/(s·m²)], N_A is the mass-transfer flux [kmol/(s·m²)], and a is the effective interfacial area (m²/m³).

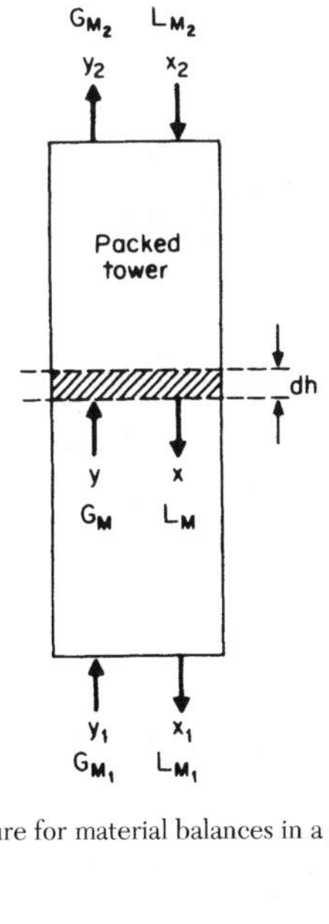

FIG. 14-5 Nomenclature for material balances in a packed-tower absorber or stripper.

When only one component is transferred,

$$dG_M = -N_A a\,dh \qquad (14\text{-}6)$$

Substitution of this relation into Eq. (14-5) and rearranging yield

$$dh = -\frac{G_M\,dy}{N_A a\,(1-y)} \qquad (14\text{-}7)$$

For this derivation we use the gas-phase rate expression $N_A = k_G(y - y_i)$ and integrate over the tower to obtain

$$h_T = \int_{y_2}^{y_1} \frac{G_M\,dy}{k_G a(1-y)(y-y_i)} \qquad (14\text{-}8)$$

Multiplying and dividing by y_{BM} place Eq. (14-8) into the $H_G N_G$ format

$$h_T = \int_{y_2}^{y_1} \left(\frac{G_M}{k_G\,a y_{BM}}\right) \frac{y_{BM}\,dy}{(1-y)(y-y_i)}$$

$$= H_{G,\text{av}} \int_{y_2}^{y_1} \frac{y_{BM}\,dy}{(1-y)(y-y_i)} = H_{G,\text{av}} N_G \qquad (14\text{-}9)$$

The general expression given by Eq. (14-8) is more complex than normally is required, but it must be used when the mass-transfer coefficient varies from point to point, as may be the case when the gas is not dilute or when the gas velocity varies as the gas dissolves. The values of y_i to be used in Eq. (14-8) depend on the local liquid composition x_i and on the temperature. This dependency is best represented by using the operating and equilibrium lines as discussed later.

Example 2 illustrates the use of Eq. (14-8) for scrubbing chlorine from air with aqueous caustic solution. For this case one can make the simplifying assumption that y_i, the interfacial partial pressure of chlorine over the caustic solution, is zero due to the rapid and complete reaction of the chlorine after it dissolves. We note that the feed gas is not dilute.

Example 2: Packed Height Requirement Let us compute the height of packing needed to reduce the chlorine concentration of 0.537 kg/(s·m²), or 396 lb/(h·ft²), of a chlorine-air mixture containing 0.503 mole-fraction chlorine to 0.0403 mole fraction. On the basis of test data described by Sherwood and Pigford (*Absorption and Extraction*, McGraw-Hill, 1952, p. 121) the value of $k_G a y_{BM}$ at a gas velocity equal to that at the bottom of the packing is equal to 0.1175 kmol/(s·m³), or 26.4 lb·mol/(h·ft³). The equilibrium back pressure y_i can be assumed to be negligible.

Solution. By assuming that the mass-transfer coefficient varies as the 0.8 power of the local gas mass velocity, we can derive the following relation:

$$\hat{K}_G a = k_G a y_{BM} = 0.1175 \left[\frac{71y + 29(1-y)}{71y_1 + 29(1-y_1)}\left(\frac{1-y_1}{1-y}\right)\right]^{0.8}$$

where 71 and 29 are the molecular weights of chlorine and air respectively. Noting that the inert-gas (air) mass velocity is given by $G'_M = G_M(1 - y) = 5.34 \times 10^{-3}$ kmol/(s·m²), or 3.94 lb·mol/(h·ft²), and introducing these expressions into the integral gives

$$h_T = 1.82\int_{0.0403}^{0.503} \left[\frac{1-y}{29 + 42y}\right]^{0.8} \frac{dy}{(1-y)^2 \ln\,[1/(1-y)]}$$

This definite integral can be evaluated numerically by the use of Simpson's rule to obtain $h_T = 0.305$ m (1 ft).

Use of Operating Curve Frequently, it is not possible to assume that $y_i = 0$ as in Example 2, due to diffusional resistance in the liquid phase or to the accumulation of solute in the liquid stream. When the backpressure cannot be neglected, it is necessary to supplement the equations with a material balance representing the operating line or curve. In view of the countercurrent flows into and from the differential section of packing shown in Fig. 14-5, a steady-state material balance leads to the following equivalent relations:

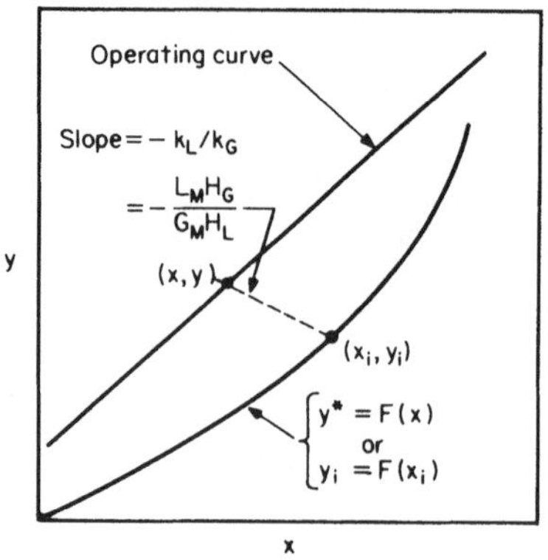

FIG. 14-6 Relationship between equilibrium curve and operating curve in a packed absorber; computation of interfacial compositions.

$$d(G_M y) = d(L_M x) \qquad (14\text{-}10)$$

$$G'_M \frac{dy}{(1-y)^2} = L'_M \frac{dx}{(1-x)^2} \qquad (14\text{-}11)$$

where L'_M = molar mass velocity of the inert-liquid component and G'_M = molar mass velocity of the inert gas; L_M, L'_M, G_M, and G'_M are superficial velocities based upon the total tower cross section.

Equation (14-11) is the differential equation of the operating curve, and its integral around the upper portion of the packing is the equation for the operating curve.

$$G'_M\left[\frac{y}{1-y} - \frac{y_2}{1-y_2}\right] = L'_M\left[\frac{x}{1-x} - \frac{x_2}{1-x_2}\right] \qquad (14\text{-}12)$$

For dilute solutions in which the mole fractions of x and y are small, the total molar flows G_M and L_M will be nearly constant, and the operating-curve equation is

$$G_M(y - y_2) = L_M(x - x_2) \qquad (14\text{-}13)$$

This equation gives the relation between the bulk compositions of the gas and liquid streams at each height in the tower for conditions in which the operating curve can be approximated as a straight line.

Figure 14-6 shows the relationship between the operating curve and the equilibrium curve $y_i = F(x_i)$ for a typical example involving solvent recovery, where y_i and x_i are the interfacial compositions (assumed to be in equilibrium). Once y is known as a function of x along the operating curve, y_i can be found at corresponding points on the equilibrium curve by

$$(y - y_i)/(x_i - x) = k_L/k_G = L_M H_G / G_M H_L \qquad (14\text{-}14)$$

where L_M = molar liquid mass velocity, G_M = molar gas mass velocity, H_L = height of one transfer unit based upon liquid-phase resistance, and H_G = height of one transfer unit based upon gas-phase resistance. Using this equation, the integral in Eq. (14-8) can be evaluated.

Calculation of Transfer Units In the general case, the equations described above must be employed in calculating the height of packing required for a given separation. However, if the local mass-transfer coefficient $k_G a y_{BM}$ is approximately proportional to the first power of the local gas velocity G_M, then the height of one gas-phase transfer unit, defined as $H_G = G_M/k_G a y_{BM}$, will be constant in Eq. (14-9). Similar considerations lead to an assumption that the height of one overall gas-phase transfer unit H_{OG} may be taken as constant. The height of packing required is then calculated according to the relation

$$h_T = H_G N_G = H_{OG} N_{OG} \qquad (14\text{-}15)$$

where N_G = number of gas-phase transfer units and N_{OG} = number of overall gas-phase transfer units. When H_G and H_{OG} are not constant, it may be valid to employ averaged values between the top and bottom of the tower and the relation

$$h_T = H_{G,\text{av}} N_G = H_{OG,\text{av}} N_{OG} \qquad (14\text{-}16)$$

In these equations, the terms N_G and N_{OG} are defined by Eqs. (14-17) and (14-18).

$$N_G = \int_{y_2}^{y_1} \frac{y_{BM}\,dy}{(1-y)(y-y_i)} \qquad (14\text{-}17)$$

$$N_{OG} = \int_{y_2}^{y_1} \frac{y^{o}_{BM}\,dy}{(1-y)(y-y^{o})} \qquad (14\text{-}18)$$

Equation (14-18) is the more useful one in practice. It requires either actual experimental H_{OG} data or values estimated by combining individual measurements of H_G and H_L by Eq. (14-19). Correlations for H_G, H_L, and H_{OG} in nonreacting systems are presented in Sec. 5.

$$H_{OG} = \frac{y_{BM}}{y^{o}_{BM}} H_G + \frac{mG_M}{L_M}\frac{x_{BM}}{y^{o}_{BM}} H_L \qquad (14\text{-}19a)$$

$$H_{OL} = \frac{x_{BM}}{x^{o}_{BM}} H_L + \frac{L_M}{mG_M}\frac{y_{BM}}{x^{o}_{BM}} H_G \qquad (14\text{-}19b)$$

On occasion, the changes in gas flow and in the mole fraction of inert gas can be neglected so that inclusion of terms such as $1 - y$ and y^{o}_{BM} can be approximated, as is shown below.

One such simplification was suggested by Wiegand [*Trans. Am. Inst. Chem. Eng.*, **36,** 679 (1940)], who pointed out that the logarithmic-mean mole fraction of inert gas y^{o}_{BM} (or y_{BM}) is often very nearly equal to the arithmetic mean. Thus, substitution of the relation

$$\frac{y^{o}_{BM}}{(1-y)} = \frac{(1-y^{o}) + (1-y)}{2(1-y)} = \frac{y-y^{o}}{2(1-y)} + 1 \qquad (14\text{-}20)$$

into the equations presented above leads to the simplified forms

$$N_G = \frac{1}{2}\ln\frac{1-y_2}{1-y_1} + \int_{y2}^{y1}\frac{dy}{y-y_i} \qquad (14\text{-}21)$$

$$N_{OG} = \frac{1}{2}\ln\frac{1-y_2}{1-y_1} + \int_{y2}^{y1}\frac{dy}{y-y^{o}} \qquad (14\text{-}22)$$

The second (integral) terms represent the numbers of transfer units for an infinitely dilute gas. The first terms, usually only a small correction, give the effect of a finite level of gas concentration.

The procedure for applying Eqs. (14-21) and (14-22) involves two steps: (1) evaluation of the integrals and (2) addition of the correction corresponding to the first (logarithmic) term. The discussion that follows deals only with the evaluation of the integral term (first step).

The simplest possible case occurs when (1) both the operating and equilibrium lines are straight (i.e., the solutions are dilute); (2) Henry's law is valid ($y^0/x = y_i/x_i = m$); and (3) absorption heat effects are negligible. Under these conditions, the integral term in Eq. (14-21) may be computed by Colburn's equation [*Trans. Am. Inst. Chem. Eng.*, **35,** 211 (1939)]:

$$N_{OG} = \frac{1}{1 - mG_M/L_M}\ln\left[\left(1 - \frac{mG_M}{L_M}\right)\left(\frac{y_1 - mx_2}{y_2 - mx_2}\right) + \frac{mG_M}{L_M}\right] \qquad (14\text{-}23)$$

Figure 14-7 is a plot of Eq. (14-23) from which the value of N_{OG} can be read directly as a function of mG_M/L_M and the ratio of concentrations. This plot and Eq. (14-23) are equivalent to the use of a logarithmic mean of terminal driving forces, but they are more convenient because one does not need to compute the exit-liquor concentration x_1.

In many practical situations involving nearly complete cleanup of the gas, an approximate result can be obtained from the equations just presented even when the simplifications are not valid, i.e., solutions are concentrated and heat effects occur. In such cases the driving forces in the upper part of the tower are very much smaller than those at the bottom, and the value of mG_M/L_M used in the equations should be the ratio of the operating line L_M/G_M in the low-concentration region near the top of the tower.

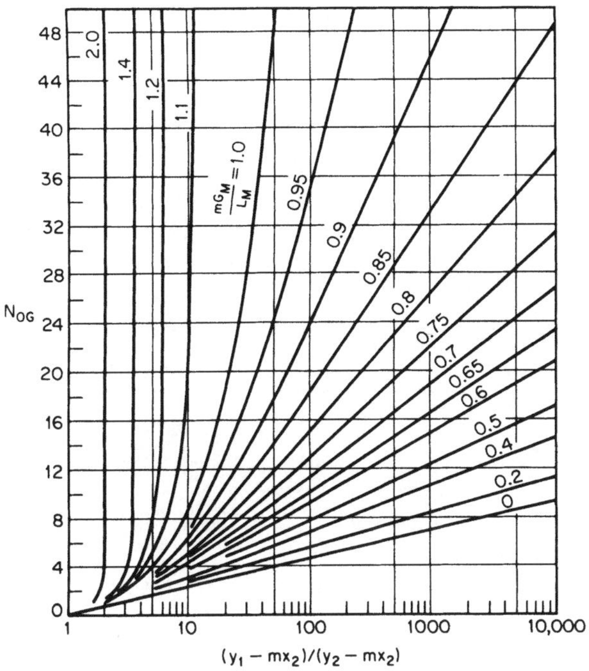

FIG. 14-7 Number of overall gas-phase mass-transfer units in a packed absorption tower for constant mG_M/L_M; solution of Eq. (14-23). (*From Sherwood and Pigford,* Absorption and Extraction, *McGraw-Hill, New York, 1952.*)

Another approach is to divide the tower arbitrarily into a lean section (near the top) where approximate methods are valid, and to deal with the rich section separately. If the heat effects in the rich section are appreciable, consideration should be given to installing cooling units near the bottom of the tower. In any event, a design diagram showing the operating and equilibrium curves should be prepared to check the applicability of any simplified procedure. Figure 14-10, presented in Example 6, is one such diagram for an adiabatic absorption tower.

Stripping Equations Stripping or desorption involves the removal of a volatile component from the liquid stream by contact with an inert gas such as nitrogen or steam or the application of heat. Here the change in concentration of the liquid stream is of prime importance, and it is more convenient to formulate the rate equation analogous to Eq. (14-6) in terms of the liquid composition x. This leads to the following equations defining the number of transfer units and height of transfer units based on liquid-phase resistance:

$$h_T = H_L \int_{x_2}^{x_1} \frac{x_{BM}dx}{(1-x)(x_i-x)} = H_L N_L \qquad (14\text{-}24)$$

$$h_T H_{OL} \int_{x_2}^{x_1} \frac{x^o_{BM}dx}{(1-x)(x^o-x)} = H_{OL} N_{OL} \qquad (14\text{-}25)$$

where, as before, subscripts 1 and 2 refer to the bottom and top of the tower, respectively (see Fig. 14-5).

In situations where one cannot assume that H_L and H_{OL} are constant, these terms need to be incorporated inside the integrals in Eqs. (14-24) and (14-25), and the integrals must be evaluated numerically (using Simpson's rule, for example). In the normal case involving stripping without chemical reactions, the liquid-phase resistance will dominate, making it preferable to use Eq. (14-25) together with the approximation $H_L \approx H_{OL}$.

The Weigand approximations of the above integrals, in which arithmetic means are substituted for the logarithmic means (x_{BM} and x^0_{BM}), are

$$N_L = \frac{1}{2} \ln \frac{1-x_1}{1-x_2} + \int_{x_1}^{x_2} \frac{dx}{x-x_i} \qquad (14\text{-}26)$$

$$N_{OL} = \frac{1}{2} \ln \frac{1-x_1}{1-x_2} + \int_{x_1}^{x_2} \frac{dx}{x-x^o} \qquad (14\text{-}27)$$

In these equations, the first term is a correction for finite liquid-phase concentrations, and the integral term represents the numbers of transfer units required for dilute solutions. In most practical stripper applications, the first (logarithmic) term is relatively small.

For dilute solutions in which both the operating and the equilibrium lines are straight and in which heat effects can be neglected, the integral term in Eq. (14-27) is

$$N_{OL} = \frac{1}{1 - L_M/mG_M} \ln \left[\left(1 - \frac{L_M}{mG_M}\right)\left(\frac{x_2 - y_1/m}{x_1 - y_1/m}\right) + \frac{L_M}{mG_M} \right] \qquad (14\text{-}28)$$

This equation is analogous to Eq. (14-23). Thus, Fig. 14-7 is applicable if the concentration ratio $(x_2 - y_1/m)/(x_1 - y_1/m)$ is substituted for the abscissa and the parameter on the curves is identified as L_M/mG_M.

Example 3: Air Stripping of VOCs from Water A 0.45-m diameter packed column was used by Dvorack et al. [*Environ. Sci. Tech.* **20,** 945 (1996)] for removing trichloroethylene (TCE) from wastewater by stripping with atmospheric air. The column was packed with 25-mm Pall rings, fabricated from polypropylene, to a height of 3.0 m. The TCE concentration in the entering water was 38 parts per million by weight (ppmw). A molar ratio of entering water to entering air was kept at 23.7. What degree of removal was to be expected? The temperatures of water and air were 20°C. Pressure was atmospheric.

Solution. For TCE in water, the Henry's law coefficient may be taken as 417 atm/mf at 20°C. In this low-concentration region, the coefficient is constant and equal to the slope of the equilibrium line m. The solubility of TCE in water, based on H = 417 atm, is 2390 ppm. Because of this low solubility, the entire resistance to mass transfer resides in the liquid phase. Thus, Eq. (14-25) may be used to obtain N_{OL}, the number of overall liquid phase transfer units.

In the equation, the ratio $x_{BM^\circ}/(1 - x)$ is unity because of the very dilute solution. It is necessary to have a value of H_L for the packing used, at given flow rates of liquid and gas. Methods for estimating H_L may be found in Sec. 5. Dvorack et al. found H_{OL} = 0.8 m. Then, for h_T = 3.0 m, $N_L = N_{OL}$ = 3.0/0.8 = 3.75 transfer units.

Transfer units may be calculated from Eq. 14-25, replacing mole fractions with ppm concentrations, and since the operating and equilibrium lines are straight,

$$N_{OL} = \frac{38 - (\text{ppm})_{\text{exit}}}{\ln 38/(\text{ppm})_{\text{exit}}} = 3.75$$

Solving, $(\text{ppm})_{\text{exit}}$ = 0.00151. Thus, the stripped water would contain 1.51 parts per billion of TCE.

Use of HTU and K_Ga Data In estimating the size of a commercial gas absorber or liquid stripper it is desirable to have data on the overall mass-transfer coefficients (or heights of transfer units) for the system of interest, and at the desired conditions of temperature, pressure, solute concentration, and fluid velocities. Such data should best be obtained in an apparatus of pilot-plant or semiworks size to avoid the abnormalities of scale-up. Within the packing category, there are both random and ordered (structured) packing elements. Physical characteristics of these devices will be described later.

When no K_Ga or HTU data are available, their values may be estimated by means of a generalized model. A summary of useful models is given in Sec. 5. The values obtained may then be combined by use of Eq. (14-19) to obtain values of H_{OG} and H_{OL}. This simple procedure is not valid when the rate of absorption is limited by chemical reaction.

Use of HETP Data for Absorber Design Distillation design methods (see Sec. 13) normally involve determination of the number of theoretical equilibrium stages N. Thus, when packed towers are employed in distillation applications, it is common practice to rate the efficiency of tower packings in terms of the height of packing equivalent to one theoretical stage (HETP).

The HETP of a packed-tower section, valid for either distillation or dilute-gas absorption and stripping systems in which constant molal overflow can be assumed and in which no chemical reactions occur, is related to the height of one overall gas-phase mass-transfer unit H_{OG} by the equation

$$\text{HETP} = H_{OG}\frac{\ln (mG_M/L_M)}{(mG_M/L_M - 1)} \tag{14-29}$$

For gas absorption systems in which the inlet gas is concentrated, the corrected equation is

$$\text{HETP} = \left(\frac{y^{o}_{BM}}{1-y}\right)_{\text{av}} H_{OG}\frac{\ln (mG_M/L_M)}{mG_M/L_M - 1} \tag{14-30}$$

where the correction term $y^{o}_{BM}/(1-y)$ is averaged over each individual theoretical stage. The equilibrium compositions corresponding to each theoretical stage may be estimated by the methods described in the next subsection, "Tray-Tower Design." These compositions are used in conjunction with the local values of the gas and liquid flow rates and the equilibrium slope m to obtain values for H_G, H_L, and H_{OG} corresponding to the conditions on each theoretical stage, and the local values of the HETP are then computed by Eq. (14-30). The total height of packing required for the separation is the summation of the individual HETPs computed for each theoretical stage.

TRAY-TOWER DESIGN

The design of a tray tower for gas absorption and gas-stripping operations involves many of the same principles employed in distillation calculations, such as the determination of the number of theoretical trays needed to achieve a specified composition change (see Sec. 13). Distillation differs from absorption because it involves the separation of components based upon the distribution of the various substances between a vapor phase and a liquid phase when all components are present in both phases. In distillation, the new phase is generated from the original phase by the vaporization or condensation of the volatile components, and the separation is achieved by introducing reflux to the top of the tower.

In gas absorption, the new phase consists of a relatively nonvolatile solvent (absorption) or a relatively insoluble gas (stripping), and normally no reflux is involved. This section discusses some of the considerations peculiar to gas absorption calculations for tray towers and some of the approximate design methods that can be applied (when simplifying assumptions are valid).

Graphical Design Procedure Construction of design diagrams (xy curves showing the equilibrium and operating curves) should be an integral part of any design involving the distribution of a single solute between an inert solvent and an inert gas. The number of theoretical trays can be stepped off rigorously, provided the curvatures of the operating and equilibrium lines are correctly represented in the diagram. The procedure is valid even though an inert solvent is present in the liquid phase and an inert gas is present in the vapor phase.

Figure 14-8 illustrates the graphical method for a three theoretical stage system. Note that in gas absorption the operating line is above the equilibrium curve, whereas in distillation this does not happen. In gas stripping, the operating line will be below the equilibrium curve.

On Fig. 14-8, note that the stepping-off procedure begins on the operating line. The starting point x_f, y_3 represents the compositions of the entering lean wash liquor and of the gas exiting from the top of the tower, as defined by the design specifications. After three steps one reaches the point x_1, y_f representing the compositions of the solute-rich feed gas y_f and of the solute-rich liquor leaving the bottom of the tower x_1.

Algebraic Method for Dilute Gases By assuming that the operating and equilibrium curves are straight lines and that heat effects are negligible, Souders and Brown [*Ind. Eng. Chem.*, **24**, 519 (1932)] developed the following equation:

$$(y_1 - y_2)/(y_1 - y^{o}_2) = (A^{N+1} - A)/(A^{N+1} - 1) \tag{14-31}$$

where N = number of theoretical trays, y_1 = mole fraction of solute in the entering gas, y_2 = mole fraction of solute in the leaving gas, $y^{o}_2 = mx_2$ = mole fraction of solute in equilibrium with the incoming solvent

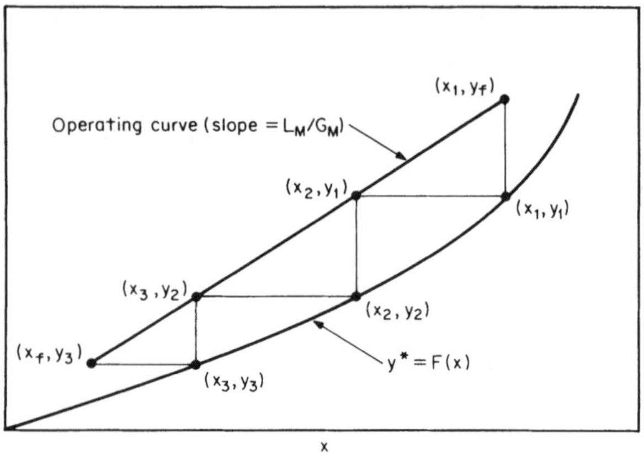

FIG. 14-8 Graphical method for a three-theoretical-plate gas-absorption tower with inlet-liquor composition x_f and inlet-gas composition y_f.

(zero for a pure solvent), and A = absorption factor = L_M/mG_M. Note that the absorption factor is the reciprocal of the expression given in Eq. (14-4) for packed columns.

Note that for the limiting case of $A = 1$, the solution is given by

$$(y_1 - y_2)/(y_1 - y^{o}_2) = N/(N + 1) \tag{14-32}$$

Although Eq. (14-31) is convenient for computing the composition of the exit gas as a function of the number of theoretical stages, an alternative equation derived by Colburn [*Trans. Am. Inst. Chem. Eng.*, **35**, 211 (1939)] is more useful when the number of theoretical plates is the unknown:

$$N = \frac{\ln [(1 - A^{-1})(y_1 - y^{o}_2)/(y_2 - y^{o}_2) + A^{-1}]}{\ln (A)} \tag{14-33}$$

The numerical results obtained by using either Eq. (14-31) or Eq. (14-33) are identical. Thus, the two equations may be used interchangeably as the need arises.

Comparison of Eqs. (14-33) and (14-23) shows that

$$N_{OG}/N = \ln (A)/(1 - A^{-1}) \tag{14-34}$$

thus revealing the close relationship between theoretical stages in a plate tower and mass-transfer units in a packed tower. Equations (14-23) and (14-33) are related to each other by virtue of the relation

$$h_T = H_{OG}N_{OG} = (\text{HETP})N \tag{14-35}$$

Algebraic Method for Concentrated Gases When the feed gas is concentrated, the absorption factor, which is defined in general as $A = L_M/KG_M$ and where $K = y^0/x$, can vary throughout the tower due to changes in temperature and composition. An approximate solution to this problem can be obtained by substituting the "effective" adsorption factors A_e and A' derived by Edmister [*Ind. Eng. Chem.* **35**, 837 (1943)] into the equation

$$\frac{y_1 - y_2}{y_1} = \left[1 - \frac{1}{A'}\frac{(L_M x)_2}{(G_M y)_1}\right]\frac{A_e^{N+1} - A_e}{A_e^{N+1} - 1} \tag{14-36}$$

where subscripts 1 and 2 refer to the bottom and top of the tower, respectively, and the absorption factors are defined by the equations

$$A_e = \sqrt{A_1(A_2 + 1) + 0.25} - 0.5 \tag{14-37}$$

$$A' = A_1(A_2 + 1)/(A_1 + 1) \tag{14-38}$$

This procedure has been applied to the absorption of C_5 and lighter hydrocarbon vapors into a lean oil, for example.

Stripping Equations When the liquid feed is dilute and the operating and equilibrium curves are straight lines, the stripping equations analogous to Eqs. (14-31) and (14-33) are

$$(x_2 - x_1)/(x_2 - x^{0}_1) = (S^{N+1} - S)/(S^{N+1} - 1) \tag{14-39}$$

where $x_1^0 = y_1/m$; $S = mG_M/L_M = A^{-1}$; and

$$N = \frac{\ln [(1-A)(x_2 - x_1^0)/(x_1 - x_1^0) + A]}{\ln (S)} \tag{14-40}$$

For systems in which the concentrations are large and the stripping factor S may vary along the tower, the following Edmister equations [*Ind. Eng. Chem.*, **35,** 837 (1943)] are applicable:

$$\frac{x_2 - x_1}{x_2} = \left[1 - \frac{1}{S'}\frac{(G_My)_1}{(L_Mx)_2}\right]\frac{S_e^{N+1} - S_e}{S_e^{N+1} - 1} \tag{14-41}$$

where

$$S_e = \sqrt{S_2(S_1 + 1) + 0.25} - 0.5 \tag{14-42}$$

$$S' = S_2(S_1 + 1)/(S_2 + 1) \tag{14-43}$$

and the subscripts 1 and 2 refer to the bottom and top of the tower respectively.

Equations (14-37) and (14-42) represent two different ways of obtaining an effective factor, and a value of A_e obtained by taking the reciprocal of S_e from Eq. (14-42) will not check exactly with a value of A_e derived by substituting $A_1 = 1/S_1$ and $A_2 = 1/S_2$ into Eq. (14-37). Regardless of this fact, the equations generally give reasonable results for approximate design calculations.

It should be noted that throughout this section the subscripts 1 and 2 refer to the bottom and to the top of the apparatus respectively regardless of whether it is an absorber or a stripper. This has been done to maintain internal consistency among all the equations and to prevent the confusion created in some derivations in which the numbering system for an absorber is different from the numbering system for a stripper.

Tray Efficiencies in Tray Absorbers and Strippers Computations of the theoretical trays N assume that the liquid on each tray is completely mixed and that the vapor leaving the tray is in equilibrium with the liquid. In practice, complete equilibrium cannot exist since interphase mass transfer requires a finite driving force. This leads to the definition of an overall tray efficiency

$$E = N_{\text{theoretical}}/N_{\text{actual}} \tag{14-44}$$

which can be correlated with the system design variables.

Mass-transfer theory indicates that for trays of a given design, the factors that have the biggest influence on E in absorption and stripping towers are the physical properties of the fluids and the dimensionless ratio mG_M/L_M. Systems in which mass transfer is gas-film-controlled may be expected to have efficiencies as high as 50 to 100 percent, whereas tray efficiencies as low as 1 percent have been reported for the absorption of low-solubility (large-m) gases into solvents of high viscosity.

The fluid properties of interest are represented by the Schmidt numbers of the gas and liquid phases. For gases, the Schmidt numbers are normally close to unity and independent of temperature and pressure. Thus, gas-phase mass-transfer coefficients are relatively independent of the system.

By contrast, the liquid-phase Schmidt numbers range from about 10^2 to 10^4 and depend strongly on temperature. The temperature dependence of the liquid-phase Schmidt number derives primarily from the strong dependence of the liquid viscosity on temperature.

Consideration of the preceding discussion in connection with the relationship between mass-transfer coefficients (see Sec. 5)

$$1/K_G = 1/k_G + m/k_L \tag{14-45}$$

indicates that the variations in the overall resistance to mass transfer in absorbers and strippers are related primarily to variations in the liquid-phase viscosity μ and the slope m. O'Connell [*Trans. Am. Inst. Chem. Eng.*, **42,** 741 (1946)] used the above findings and correlated the tray efficiency in terms of the liquid viscosity and the gas solubility. The O'Connell correlation for absorbers (Fig. 14-9) has Henry's law constant in lb·mol/(atm·ft³), the pressure in atmospheres, and the liquid viscosity in centipoise.

The best procedure for making tray efficiency corrections (which can be quite significant, as seen in Fig. 14-9) is to use experimental data from a prototype system that is large enough to be representative of the actual commercial tower.

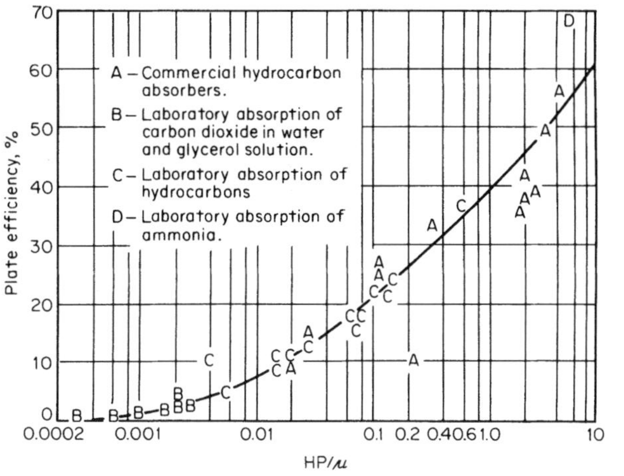

FIG. 14-9 O'Connell correlation for overall column efficiency E_{oc} for absorption. H is in lb·mol/(atm·ft³), P is in atm, and μ is in cP. To convert HP/μ in pound-moles per cubic foot-centipoise to kilogram-moles per cubic meter-pascal-second, multiply by 1.60×10^4. [*O'Connell,* Trans. Am. Inst. Chem. Eng., ***42,*** 741 (*1946*).]

Example 4: Actual Trays for Steam Stripping The number of actual trays required for steam-stripping an acetone-rich liquor containing 0.573 mole percent acetone in water is to be estimated. The design overhead recovery of acetone is 99.9 percent, leaving 18.5 ppm weight of acetone in the stripper bottoms. The design operating temperature and pressure are 101.3 kPa and 94°C respectively, the average liquid-phase viscosity is 0.30 cP, and the average value of $K = y°/x$ for these conditions is 33.

By choosing a value of $mG_M/L_M = S = A^{-1} = 1.4$ and noting that the stripping medium is pure steam (i.e., $x_1^0 = 0$), the number of theoretical trays according to Eq. (14-40) is

$$N = \frac{\ln [(1 - 0.714)(1000) + 0.714]}{\ln (1.4)} = 16.8$$

The O'Connell parameter for gas absorbers is $\rho_L/KM\mu_L$, where ρ_L is the liquid density, lb/ft³; μ_L is the liquid viscosity, cP; M is the molecular weight of the liquid; and $K = y°/x$. For the present design

$$\rho_L/KM\mu_L = 60.1/(33 \times 18 \times 0.30) = 0.337$$

and according to the O'Connell graph for absorbers (Fig. 14-7) the overall tray efficiency for this case is estimated to be 30 percent. Thus, the required number of actual trays is 16.8/0.3 = 56 trays.

HEAT EFFECTS IN GAS ABSORPTION

Overview One of the most important considerations involved in designing gas absorption towers is to determine whether temperatures will vary along the height of the tower due to heat effects; note that the solute solubility usually depends strongly on temperature. The simplified design procedures described earlier in this section become more complicated when heat effects cannot be neglected. The role of this section is to enable understanding and design of gas absorption towers where heat effects are important and cannot be ignored.

Heat effects that cause temperatures to vary from point to point in a gas absorber are (1) the heat of solution (including heat of condensation, heat of mixing, and heat of reaction); (2) the heat of vaporization or condensation of the solvent; (3) the exchange of sensible heat between the gas and liquid phases; and (4) the loss of sensible heat from the fluids to internal or external coils.

There are a number of systems where heat effects definitely cannot be ignored. Examples include the absorption of ammonia in water, dehumidification of air using concentrated H_2SO_4, absorption of HCl in water, absorption of SO_3 in H_2SO_4, and absorption of CO_2 in alkanolamines. Even for systems where the heat effects are mild, they may not be negligible; an example is the absorption of acetone in water.

Thorough and knowledgeable discussions of the problems involved in gas absorption with significant heat effects have been presented by Coggan and Bourne [*Trans. Inst. Chem. Eng.*, **47,** T96, T160 (1969)]; Bourn, von Stockar, and Coggan [*Ind. Eng. Chem. Proc. Des. Dev.*, **13,** 115, 124 (1974)]; and von Stockar and Wilke [*Ind. Eng. Chem. Fundam.*, **16,** 89 (1977)]. The first two of these references discuss tray-tower absorbers and include experimental studies of the absorption of ammonia in water. The third reference discusses the design of packed-tower absorbers and includes a shortcut design method based on a semitheoretical correlation of rigorous design calculations. All these authors demonstrate that when the solvent is volatile, the temperature inside an absorber can go through a maximum. They note that the least expensive and most common of solvents—water—is capable of exhibiting this "hot-spot" behavior.

Several approaches may be used in modeling absorption with heat effects, depending on the job at hand: (1) treat the process as isothermal by assuming a particular temperature, then add a safety factor; (2) employ the classical adiabatic method, which assumes that the heat of solution manifests itself only as sensible heat in the liquid phase and that the solvent vaporization is negligible; (3) use semitheoretical shortcut methods derived from rigorous calculations; and (4) employ rigorous methods available from a process simulator.

While simpler methods are useful for understanding the key effects involved, rigorous methods are recommended for final designs. This subsection also discusses the range of safety factors that are required if simpler methods are used.

Effects of Operating Variables Conditions that give rise to significant heat effects are (1) an appreciable heat of solution and/or (2) absorption of large amounts of solute in the liquid phase. The second condition is favored when the solute concentration in the inlet gas is large, when the liquid flow rate is relatively low (small L_M/G_M), when the solubility of the solute in the liquid is high, and/or when the operating pressure is high.

If the solute-rich gas entering the bottom of an absorber tower is cold, the liquid phase may be cooled somewhat by transfer of sensible heat to the gas. A much stronger cooling effect can occur when the solute is volatile and the entering gas is not saturated with respect to the solvent. It is possible to experience a condition in which solvent is being evaporated near the bottom of the tower and condensed near the top. Under these conditions a pinch point may develop in which the operating and equilibrium curves approach each other at a point inside the tower.

In the references previously cited, the authors discuss the influence of operating variables upon the performance of towers when large heat effects are involved. Some key observations are as follows:

Operating Pressure Raising the pressure may increase the separation effectiveness considerably. Calculations for the absorption of methanol in water from water-saturated air showed that doubling the pressure doubles the allowable concentration of methanol in the feed gas while still achieving the required concentration specification in the off gas.

Temperature of Lean Solvent The temperature of the entering (lean) solvent has surprisingly little influence upon the temperature profile in an absorber since any temperature changes are usually caused by the heat of solution or the solvent vaporization. In these cases, the temperature profile in the liquid phase is usually dictated solely by the internal-heat effects.

Temperature and Humidity of the Rich Gas Cooling and consequent dehumidification of the feed gas to an absorption tower can be very beneficial. A high humidity (or relative saturation with the solvent) limits the capacity of the gas to take up latent heat and hence is unfavorable to absorption. Thus dehumidification of the inlet gas is worth considering in the design of absorbers with large heat effects.

Liquid-to-Gas Ratio The L/G ratio can have a significant influence on the development of temperature profiles in gas absorbers. High L/G ratios tend to result in less strongly developed temperature profiles due to the increased heat capacity of the liquid phase. As the L/G ratio is increased, the operating line moves away from the equilibrium line and more solute is absorbed per stage or packing segment. However, there is a compensating effect; since more heat is liberated in each stage or packing segment, the temperatures will rise, which causes the equilibrium line to shift up. As the L/G ratio is decreased, the concentration of solute tends to build up in the upper part of the absorber, and the point of highest temperature tends to move upward in the tower until finally the maximum temperature occurs at the top of the tower. Of course, the capacity of the liquid to absorb solute falls progressively as L/G is reduced.

Number of Stages or Packing Height When the heat effects combine to produce an extended zone in the tower where little absorption takes place (i.e., a pinch zone), the addition of trays or packing height will have no useful effect on separation efficiency. In this case, increases in absorption may be obtained by increasing solvent flow, introducing strategically placed coolers, cooling and dehumidifying the inlet gas, and/or raising the tower pressure.

Equipment Considerations When the solute has a large heat of solution and the feed gas contains a high concentration of solute, as in absorption of HCl in water, the effects of heat release during absorption may be so pronounced that the installation of heat-transfer surface to remove the heat of absorption may be as important as providing sufficient interfacial area for the mass-transfer process itself. The added heat-transfer area may consist of internal cooling coils on the trays, or the liquid may be withdrawn from the tower, cooled in an external heat exchanger, and then returned to the tower.

In many cases the rate of heat liberation is largest near the bottom of the tower, where the solute absorption is more rapid, so that cooling surfaces or intercoolers are required only at the lower part of the column. Coggan and Bourne [*Trans. Inst. Chem. Eng.*, **47,** T96, T160 (1969)] found, however, that the optimal position for a single interstage cooler does not necessarily coincide with the position of the maximum temperature of the center of the pinch. They found that in a 12-tray tower, two strategically placed interstage coolers tripled the allowable ammonia feed concentration for a given off-gas specification. For a case involving methanol absorption, it was found that greater separation was possible in a 12-stage column with two intercoolers than in a simple column with 100 stages and no intercoolers.

In the case of HCl absorption, a shell-and-tub heat exchanger often is employed as a cooled wetted-wall vertical-column absorber so that the exothermic heat of reaction can be removed continuously as it is released into a liquid film.

Installation of heat-exchange equipment to precool and dehumidify the feed gas to an absorber also deserves consideration, in order to take advantage of the cooling effects created by vaporization of solvent in the lower sections of the tower.

Classical Isothermal Design Method When the feed gas is sufficiently dilute, the exact design solution may be approximated by the isothermal one over the broad range of L/G ratios, since heat effects are generally less important when washing dilute-gas mixtures. The problem, however, is one of defining the term *sufficiently dilute* for each specific case. For a new absorption duty, the assumption of isothermal operation must be subjected to verification by the use of a rigorous design procedure.

When heat-exchange surface is being provided in the design of an absorber, the isothermal design procedure can be rendered valid by virtue of the exchanger design specification. With ample surface area and a close approach, isothermal operation can be guaranteed.

For preliminary screening and feasibility studies or for rough estimates, one may wish to employ a version of the isothermal design method which assumes that the liquid temperatures in the tower are everywhere equal to the inlet-liquid temperature. In their analysis of packed-tower designs, von Stockar and Wilke [*Ind. Eng. Chem. Fundam.*, **16,** 89 (1977)] showed that the isothermal method tended to underestimate the required height of packing by a factor of as much as

1.5 to 2. Thus, for rough estimates one may wish to employ the assumption that the absorber temperature is equal to the inlet-liquid temperature and then apply a design factor to the result.

Another instance in which the constant-temperature method is used involved the direct application of experimental K_Ga values obtained at the desired conditions of inlet temperatures, operating pressures, flow rates, and feed-stream compositions. The assumption here is that, regardless of any temperature profiles that may exist within the actual tower, the procedure of "working the problem in reverse" will yield a correct result. One should, however, be cautious about extrapolating such data from the original basis and be careful to use compatible equilibrium data.

Classical Adiabatic Design Method The classical adiabatic design method assumes that the heat of solution serves only to heat up the liquid stream and there is no vaporization of the solvent. This assumption makes it feasible to relate increases in the liquid-phase temperature to the solute concentration x by a simple enthalpy balance. The equilibrium curve can then be adjusted to account for the corresponding temperature rise on an xy diagram. The adjusted equilibrium curve will be concave upward as the concentration increases, tending to decrease the driving forces near the bottom of the tower, as illustrated in Fig. 14-10 in Example 6.

Colburn [*Trans. Am. Inst. Chem. Eng.*, **35**, 211 (1939)] has shown that when the equilibrium line is straight near the origin but curved slightly at its upper end, N_{OG} can be computed approximately by assuming that the equilibrium curve is a parabolic arc of slope m_2 near the origin and passing through the point x_1, K_1x_1 at the upper end. The Colburn equation for this case is

$$N_{OG} = \frac{1}{1 - m_2G_M/L_M}$$

$$\times \ln\left[\frac{(1 - m_2G_M/L_M)^2}{1 - K_1G_M/L_M}\left(\frac{y_1 - m_2x_2}{y_2 - m_2x_2}\right) + \frac{m_2G_M}{L_M}\right] \quad (14\text{-}46)$$

Comparison by von Stockar and Wilke [*Ind. Eng. Chem. Fundam.*, **16**, 89 (1977)] between the rigorous and the classical adiabatic design methods for packed towers indicates that the simple adiabatic design methods underestimate packing heights by as much as a factor of 1.25 to 1.5. Thus, when using the classical adiabatic method, one should probably apply a design safety factor.

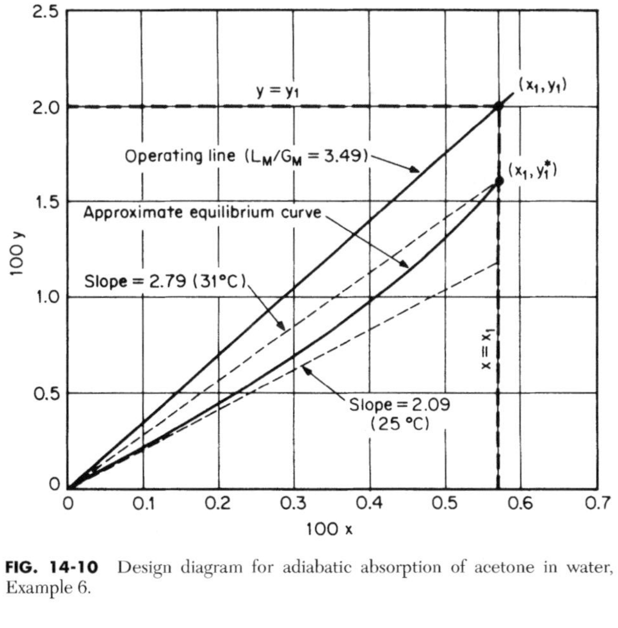

FIG. 14-10 Design diagram for adiabatic absorption of acetone in water, Example 6.

A slight variation of the above method accounts for increases in the solvent content of the gas stream between the inlet and the outlet of the tower and assumes that the evaporation of solvent tends to cool the liquid. This procedure offsets a part of the temperature rise that would have been predicted with no solvent evaporation and leads to the prediction of a shorter tower.

Rigorous Design Methods A detailed discussion of rigorous methods for the design of packed and tray absorbers when large heat effects are involved is beyond the scope of this subsection. In principle, material and energy balances may be executed under the same constraints as for rigorous distillation calculations (see Sec. 13). Further discussion on this subject is given in the subsection "Absorption with Chemical Reaction."

Direct Comparison of Design Methods The following problem, originally presented by Sherwood, Pigford, and Wilke (*Mass Transfer*, McGraw-Hill, New York, 1975, p. 616) was employed by von Stockar and Wilke (op. cit.) as the basis for a direct comparison between the isothermal, adiabatic, semitheoretical shortcut, and rigorous design methods for estimating the height of packed towers.

Example 5: Packed Absorber, Acetone into Water Inlet gas to an absorber consists of a mixture of 6 mole percent acetone in air saturated with water vapor at 15°C and 101.3 kPa (1 atm). The scrubbing liquor is pure water at 15°C, and the inlet gas and liquid rates are given as 0.080 and 0.190 kmol/s respectively. The liquid rate corresponds to 20 percent over the theoretical minimum as calculated by assuming a value of x_1 corresponding to complete equilibrium between the exit liquor and the incoming gas. H_G and H_L are given as 0.42 and 0.30 m respectively, and the acetone equilibrium data at 15°C are $p_A^0 =$ 19.7 kPa (147.4 torr), $\gamma_A = 6.46$, and $m_A = 6.46 \times 19.7/101.3 = 1.26$. The heat of solution of acetone is 7656 cal/gmol (32.05 kJ/gmol), and the heat of vaporization of solvent (water) is 10,755 cal/gmol (45.03 kJ/gmol). The problem calls for determining the height of packing required to achieve a 90 percent recovery of the acetone.

The following table compares the results obtained by von Stockar and Wilke (op. cit.) for the various design methods:

Design method used	N_{OG}	Packed height, m	Design safety factor
Rigorous	5.56	3.63	1.00
Shortcut rigorous	5.56	3.73	0.97
Classical adiabatic	4.01	2.38	1.53
Classical isothermal	3.30	1.96	1.85

It should be clear from this example that there is considerable room for error when approximate design methods are employed in situations involving large heat effects, even for a case in which the solute concentration in the inlet gas is only 6 mole percent.

Example 6: Solvent Rate for Absorption Let us consider the absorption of acetone from air at atmospheric pressure into a stream of pure water fed to the top of a packed absorber at 25°C. The inlet gas at 35°C contains 2 percent by volume of acetone and is 70 percent saturated with water vapor (4 percent H_2O by volume). The mole-fraction acetone in the exit gas is to be reduced to 1/400 of the inlet value, or 50 ppmv. For 100 kmol of feed-gas mixture, how many kilomoles of fresh water should be fed to provide a positive-driving force throughout the packing? How many transfer units will be needed according to the classical adiabatic method? What is the estimated height of packing required if $H_{OG} = 0.70$ m?

The latent heats at 25°C are 7656 kcal/kmol for acetone and 10,490 kcal/kmol for water, and the differential heat of solution of acetone vapor in pure water is given as 2500 kcal/kmol. The specific heat of air is 7.0 kcal/(kmol·K).

Acetone solubilities are defined by the equation

$$K = y^{\circ}/x = \gamma_a p_a/p_T \quad (14\text{-}47)$$

where the vapor pressure of pure acetone in mmHg (torr) is given by (Sherwood et al., *Mass Transfer*, McGraw-Hill, New York, 1975, p. 537):

$$p_A^0 = \exp(18.1594 - 3794.06/T) \quad (14\text{-}48)$$

and the liquid-phase-activity coefficient may be approximated for low concentrations ($x \leq 0.01$) by the equation

$$\gamma_a = 6.5 \exp(2.0803 - 601.2/T) \qquad (14\text{-}49)$$

Typical values of acetone solubility as a function of temperature at a total pressure of 760 mmHg are shown in the following table:

t, °C	25	30	35	40
γ_a	6.92	7.16	7.40	7.63
p_a, mmHg	229	283	346	422
$K = \gamma_a p_a^0/760$	2.09	2.66	3.37	4.23

For dry gas and liquid water at 25°C, the following enthalpies are computed for the inlet- and exit-gas streams (basis, 100 kmol of gas entering):

Entering gas:

Acetone	$2(2500 + 7656) = 20{,}312$ kcal	
Water vapor	$4(10{,}490) = 41{,}960$	
Sensible heat	$(100)(7.0)(35 - 25) = 7{,}000$	
		69,272 kcal

Exit gas (assumed saturated with water at 25°C):

Acetone	$(2/400)(94/100)(2500) =$	12 kcal
Water vapor	$94\left(\dfrac{23.7}{760 - 23.7}\right)(10{,}490) =$	31,600
		31,612 kcal

Enthalpy change of liquid = 69,272 − 31,612 = 37,660 kcal/100 kmol gas. Thus, $\Delta t = t_1 - t_2 = 37{,}660/18L_M$, and the relation between L_M/G_M and the liquid-phase temperature rise is

$$L_M/G_M = (37{,}660)/(18)(100)\,\Delta t = 20.92/\Delta t$$

The following table summarizes the critical values for various assumed temperature rises:

Δt, °C	L_M/G_M	K_1	K_1G_M/L_M	m_2G_M/L_M
0		2.09	0.	0.
2	10.46	2.31	0.221	0.200
3	6.97	2.42	0.347	0.300
4	5.23	2.54	0.486	0.400
5	4.18	2.66	0.636	0.500
6	3.49	2.79	0.799	0.599
7	2.99	2.93	0.980	0.699

Evidently a temperature rise of 7°C would not be a safe design because the equilibrium line nearly touches the operating line near the bottom of the tower, creating a pinch. A temperature rise of 6°C appears to give an operable design, and for this case $L_M = 349$ kmol per 100 kmol of feed gas.

The design diagram for this case is shown in Fig. 14-10, in which the equilibrium curve is drawn so that the slope at the origin m_2 is equal to 2.09 and passes through the point $x_1 = 0.02/3.49 = 0.00573$ at $y_1^\circ = 0.00573 \times 2.79 = 0.0160$.

The number of transfer units can be calculated from the adiabatic design equation, Eq. (14-46):

$$N_{OC} = \frac{1}{1 - 0.599} \ln \left[\frac{(1 - 0.599)^2}{(1 - 0.799)} (400) + 0.599 \right] = 14.4$$

The estimated height of tower packing by assuming $H_{OC} = 0.70$ m and a design safety factor of 1.5 is

$$h_T = (14.4)(0.7)(1.5) = 15.1 \text{ m } (49.6 \text{ ft})$$

For this tower, one should consider the use of two or more shorter packed sections instead of one long section.

Another point to be noted is that this calculation would be done more easily today by using a process simulator. However, the details are presented here to help the reader gain familiarity with the key assumptions and results.

MULTICOMPONENT SYSTEMS

When no chemical reactions are involved in the absorption of more than one soluble component from an insoluble gas, the design conditions (temperature, pressure, liquid-to-gas ratio) are normally determined by the volatility or physical solubility of the least soluble component for which the recovery is specified.

The more volatile (i.e., less soluble) components will only be partially absorbed even for an infinite number of trays or transfer units. This can be seen in Fig. 14-9, in which the asymptotes become vertical for values of mG_M/L_M greater than unity. If the amount of volatile component in the fresh solvent is negligible, then the limiting value of y_1/y_2 for each of the highly volatile components is

$$y_1/y_2 = S/(S - 1) \qquad (14\text{-}50)$$

where $S = mG_M/L_M$ and the subscripts 1 and 2 refer to the bottom and top of the tower, respectively.

When the gas stream is dilute, absorption of each constituent can be considered separately as if the other components were absent. The following example illustrates the use of this principle.

Example 7: Multicomponent Absorption, Dilute Case Air entering a tower contains 1 percent acetaldehyde and 2 percent acetone. The liquid-to-gas ratio for optimum acetone recovery is $L_M/G_M = 3.1$ mol/mol when the fresh-solvent temperature is 31.5°C. The value of y°/x for acetaldehyde has been measured as 50 at the boiling point of a dilute solution, 93.5°C. What will the percentage recovery of acetaldehyde be under conditions of optimal acetone recovery?

Solution. If the heat of solution is neglected, y°/x at 31.5°C is equal to 50(1200/7300) = 8.2, where the factor in parentheses is the ratio of pure-acetaldehyde vapor pressures at 31.5 and 93.5°C respectively. Since L_M/G_M is equal to 3.1, the value of S for the aldehyde is $S = mG_M/L_M = 8.2/3.1 = 2.64$, and $y_1/y_2 = S/(S - 1) = 2.64/1.64 = 1.61$. The acetaldehyde recovery is therefore equal to $100 \times 0.61/1.61 = 38$ percent recovery.

In concentrated systems the change in gas and liquid flow rates within the tower and the heat effects accompanying the absorption of all the components must be considered. A trial-and-error calculation from one theoretical stage to the next usually is required if accurate results are to be obtained, and in such cases calculation procedures similar to those described in Sec. 13 normally are employed. A computer procedure for multicomponent adiabatic absorber design has been described by Feintuch and Treybal [*Ind. Eng. Chem. Process Des. Dev.*, **17**, 505 (1978)]. Also see Holland, *Fundamentals and Modeling of Separation Processes*, Prentice Hall, Englewood Cliffs, N.J., 1975.

In concentrated systems, the changes in the gas and liquid flow rates within the tower and the heat effects accompanying the absorption of all components must be considered. A trial-and-error calculation from one theoretical stage to the next is usually required if accurate and reliable results are to be obtained, and in such cases calculation procedures similar to those described in Sec. 13 need to be employed.

When two or more gases are absorbed in systems involving chemical reactions, the system is much more complex. This topic is discussed later in the subsection "Absorption with Chemical Reaction."

Graphical Design Method for Dilute Systems The following notation for multicomponent absorption systems has been adapted from Sherwood, Pigford, and Wilke (*Mass Transfer*, McGraw-Hill, New York, 1975, p. 415):

L_M^s = moles of solvent per unit time

G_M^0 = moles of rich feed gas to be treated per unit time

X = moles of one solute per mole of solute-free solvent fed to top of tower

Y = moles of one solute in gas phase per mole of rich feed gas

Subscripts 1 and 2 refer to the bottom and the top of the tower, respectively, and the material balance for any one component may be written as

$$L_M^s(X - X_2) = G_M^0(Y - Y_2) \qquad (14\text{-}51)$$

or else as

$$L_M^s(X_1 - X) = G_M^0(Y_1 - Y) \qquad (14\text{-}52)$$

For the special case of absorption from lean gases with relatively large amounts of solvent, the equilibrium lines are defined for each component by the relation

$$Y^0 = K'X \qquad (14\text{-}53)$$

Thus, the equilibrium line for each component passes through the origin with slope K', where

$$K' = K(G_M/G_M^0)/(L_M/L_M^S) \qquad (14\text{-}54)$$

and $K = y^0/x$. When the system is sufficiently dilute, $K' = K$.

The liquid-to-gas ratio is chosen on the basis of the least soluble component in the feed gas that must be absorbed completely. Each component will then have its own operating line with slope equal to L_M^S/G_M^0 (i.e., the operating lines for the various components will be parallel).

A typical diagram for the complete absorption of pentane and heavier components is shown in Fig. 14-11. The oil used as solvent is assumed to be solute-free (i.e., $X_2 = 0$), and the "key component," butane, was identified as that component absorbed in appreciable amounts whose equilibrium line is most nearly parallel to the operating lines (i.e., the K value for butane is approximately equal to L_M^S/G_M^0).

In Fig. 14-11, the composition of the gas with respect to components more volatile than butane will approach equilibrium with the liquid phase at the bottom of the tower. The gas compositions of the components less volatile (heavier) than butane will approach equilibrium with the oil entering the tower, and since $X_2 = 0$, the components heavier than butane will be completely absorbed.

Four theoretical trays have been stepped off for the key component (butane) on Fig. 14-11, and are seen to give a recovery of 75 percent of the butane. The operating lines for the other components have been drawn with the same slope and placed so as to give approximately the same number of theoretical trays. Figure 14-11 shows that equilibrium is easily achieved in fewer than four theoretical trays and that for the heavier components nearly complete recovery is obtained in four theoretical trays. The diagram also shows that absorption of the light components takes place in the upper part of the tower, and the final recovery of the heavier components takes place in the lower section of the tower.

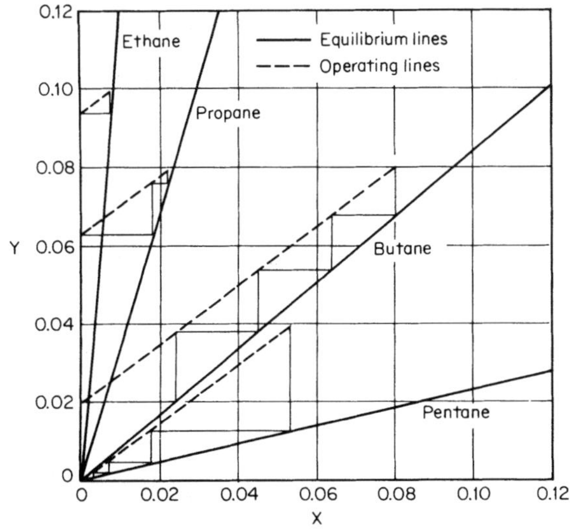

FIG. 14-11 Graphical design method for multicomponent systems; absorption of butane and heavier components in a solute-free lean oil.

Algebraic Design Method for Dilute Systems The design method described above can be performed algebraically by employing the following modified version of the Kremser formula:

$$\frac{Y_1 - Y_2}{Y_1 - mX_2} = \frac{(A^0)^{N+1} - A^0}{(A^0)^{N+1} - 1} \qquad (14\text{-}55)$$

where for dilute gas absorption $A^0 = L_M^S/mG_M^0$ and $m \approx K = y^0/x$.

The left side of Eq. (14-55) represents the efficiency of absorption of any one component of the feed gas mixture. If the solvent is solute-free so that $X_2 = 0$, the left side is equal to the fractional absorption of the component from the rich feed gas. When the number of theoretical trays N and the liquid and gas feed rates L_M^S and G_M^0 have been fixed, the fractional absorption of each component may be computed directly, and the operating lines need not be placed by trial and error as in the graphical method described above.

According to Eq. (14-55), when A^0 is less than unity and N is large,

$$(Y_1 - Y_2)/(Y_1 - mX_2) = A^0 \qquad (14\text{-}56)$$

Equation (14-56) may be used to estimate the fractional absorption of more volatile components when A^0 of the component is greater than A^0 of the key component by a factor of 3 or more.

When A^0 is much larger than unity and N is large, the right side of Eq. (14-55) becomes equal to unity. This signifies that the gas will leave the top of the tower in equilibrium with the incoming oil, and when $X_2 = 0$, it corresponds to complete absorption of the component in question. Thus, the least volatile components may be assumed to be at equilibrium with the lean oil at the top of the tower.

When $A^0 = 1$, the right side of Eq. (14-56) simplifies as follows:

$$(Y_1 - Y_2)/(Y_1 - mX_2) = N/(N + 1) \qquad (14\text{-}57)$$

For systems in which the absorption factor A^0 for each component is not constant throughout the tower, an effective absorption factor for use in the equations just presented can be estimated by the Edmister formula

$$A_e^0 = \sqrt{A_1^0(A_2^0 + 1) + 0.25} - 0.5 \qquad (14\text{-}58)$$

This procedure is a reasonable approximation only when no pinch points exist within the tower and when the absorption factors vary in a regular manner between the bottom and the top of the tower.

Example 8: Multicomponent Absorption, Concentrated Case A hydrocarbon feed gas is to be treated in an existing four-theoretical-tray absorber to remove butane and heavier components. The recovery specification for the key component, butane, is 75 percent. The composition of the exit gas from the absorber and the required liquid-to-gas ratio are to be estimated. The feed-gas composition and the equilibrium K values for each component at the temperature of the (solute-free) lean oil are presented in the following table:

Component	Mole %	*K* value
Methane	68.0	74.137
Ethane	10.0	12.000
Propane	8.0	3.429
Butane	8.0	0.833
Pentane	4.0	0.233
C_6 plus	2.0	0.065

For $N = 4$ and $Y_2/Y_1 = 0.25$, the value of A^0 for butane is found to be equal to 0.89 from Eq. (14-55) by using a trial-and-error method. The values of A^0 for the other components are then proportional to the ratios of their K values to that of butane. For example, $A^0 = 0.89(0.833/12.0) = 0.062$ for ethane. The values of A^0 for each of the other components and the exit-gas composition as computed from Eq. (14-55) are shown in the following table:

Component	A^0	Y_2, mol/mol feed	Exit gas, mole %
Methane	0.010	67.3	79.1
Ethane	0.062	9.4	11.1
Propane	0.216	6.3	7.4
Butane	0.890	2.0	2.4
Pentane	3.182	0.027	0.03
C_6 plus	11.406	0.0012	0.0014

The molar liquid-to-gas ratio required for this separation is computed as $L_M^s/G_M^0 = A^0 \times K = 0.89 \times 0.833 = 0.74$.

We note that this example is the analytical solution to the graphical design problem shown in Fig. 14-11, which therefore is the design diagram for this system.

The simplified design calculations presented in this section are intended to reveal the key features of gas absorption involving multicomponent systems. It is expected that rigorous computations, based upon the methods presented in Sec. 13, will be used in design practice. Nevertheless, it is valuable to study these simplified design methods and examples since they provide insight into the key elements of multicomponent absorption.

ABSORPTION WITH CHEMICAL REACTION

Introduction Many present-day commercial gas absorption processes involve systems in which chemical reactions take place in the liquid phase; an example of the absorption of CO_2 by MEA has been presented earlier in this section. These reactions greatly increase the capacity of the solvent and enhance the rate of absorption when compared to physical absorption systems. In addition, the selectivity of reacting solutes is greatly increased over that of nonreacting solutes. For example, MEA has a strong selectivity for CO_2 compared to chemically inert solutes such as CH_4, CO, or N_2. Note that the design procedures presented here are theoretically and practically related to biofiltration, which is discussed in Sec. 25 (Waste Management).

A necessary prerequisite to understanding the subject of absorption with chemical reaction is the development of a thorough understanding of the principles involved in physical absorption, as discussed earlier in this section and in Sec. 5. Excellent references on the subject of absorption with chemical reactions are the books by Dankwerts (*Gas-Liquid Reactions*, McGraw-Hill, New York, 1970) and Astarita et al. (*Gas Treating with Chemical Solvents*, Wiley, New York, 1983).

Recommended Overall Design Strategy When one is considering the design of a gas absorption system involving chemical reactions, the following procedure is recommended:

1. Consider the possibility that the physical design methods described earlier in this section may be applicable.
2. Determine whether commercial design overall K_Ga values are available for use in conjunction with the traditional design method, being careful to note whether the conditions under which the K_Ga data were obtained are essentially the same as for the new design. Contact the various tower-packing vendors for information as to whether K_Ga data are available for your system and conditions.
3. Consider the possibility of scaling up the design of a new system from experimental data obtained in a laboratory bench-scale or small pilot-plant unit.
4. Consider the possibility of developing for the new system a rigorous, theoretically based design procedure which will be valid over a wide range of design conditions. Note that commercial software is readily available today to develop a rigorous model in a relatively small amount of time.

These topics are further discussed in the subsections that follow.

Dominant Effects in Absorption with Chemical Reaction When the solute is absorbing into a solution containing a reagent that chemically reacts with it, diffusion and reaction effects become closely coupled. It is thus important for the design engineer to understand the key effects. Figure 14-12 shows the concentration profiles that occur when solute A undergoes an irreversible second-order reaction with component B, dissolved in the liquid, to give product C.

$$A + bB \rightarrow cC \tag{14-59}$$

The rate equation is

$$r_A = -k_2 C_A C_B \tag{14-60}$$

Figure 14-12 shows that the fast reaction takes place entirely in the liquid film. In such instances, the dominant mass-transfer mechanism is physical absorption, and physical design methods are applicable but the resistance to mass transfer in the liquid phase is lower due to the reaction. On the other extreme, a slow reaction occurs in the bulk of the liquid, and its rate has little dependence on the resistance to diffusion in either the gas or the liquid films. Here the mass-transfer mechanism is that of chemical reaction, and holdup in the bulk liquid is the determining factor.

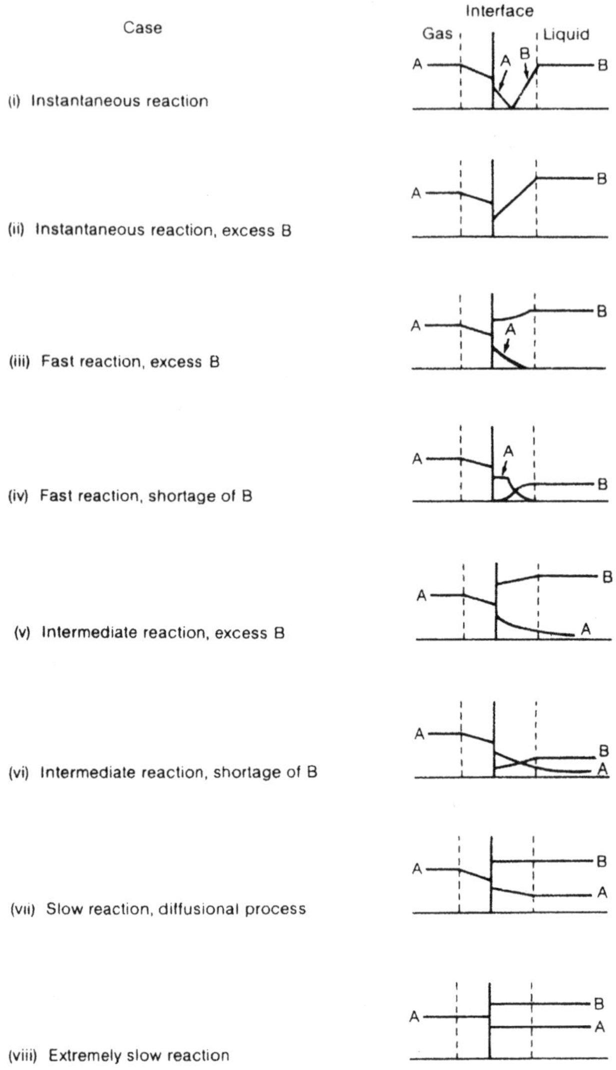

FIG. 14-12 Vapor- and liquid-phase concentration profiles near an interface for absorption with chemical reaction.

The Hatta number is a dimensionless group used to characterize the importance of the speed of reaction relative to the diffusion rate.

$$N_{\text{Ha}} = \frac{\sqrt{D_A k_2 C_{B0}}}{k_L^0} \tag{14-61}$$

As the Hatta number increases, the effective liquid-phase mass-transfer coefficient increases. Figure 14-13, which was first developed by Van Krevelen and Hoftyzer [*Rec. Trav. Chim.*, **67**, 563 (1948)] and later refined by Perry and Pigford and by Brian et al. [*AIChE J.*, **7**, 226 (1961)], shows how the enhancement (defined as the ratio of the effective liquid-phase mass-transfer coefficient to its physical equivalent $\phi = k_L/k_L^0$) increases with N_{Ha} for a second-order, irreversible reaction of the kind defined by Eqs. (14-60) and (14-61). The various curves in Fig. 14-13 were developed based upon penetration theory and

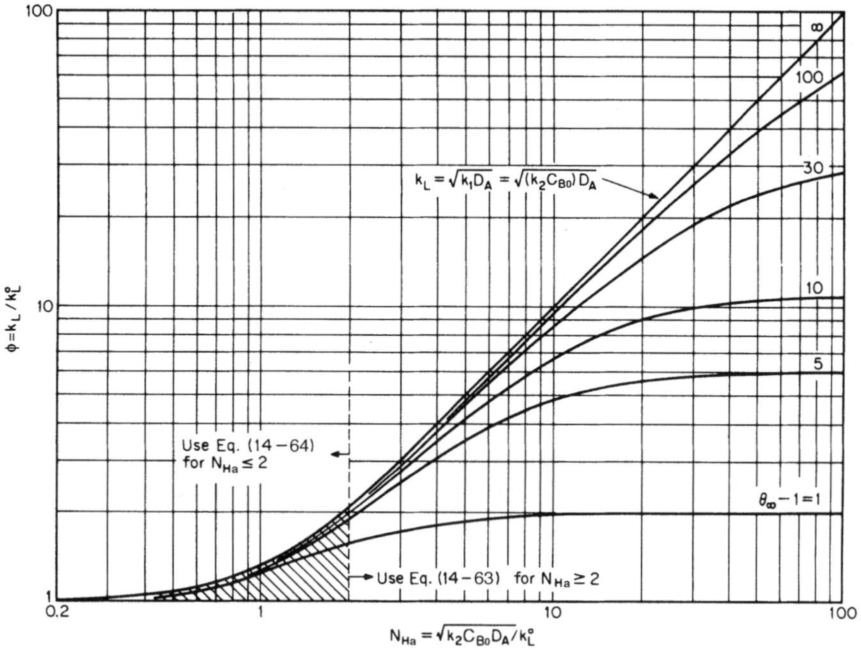

FIG. 14-13 Influence of irreversible chemical reactions on the liquid-phase mass-transfer coefficient k_L. [*Adapted from Van Krevelen and Hoftyzer,* Rec. Trav. Chim., ***67,*** *563 (1948).*]

depend on the parameter $\phi_\infty - 1$, which is related to the diffusion coefficients and reaction coefficients, as shown below.

$$\phi_\infty = \sqrt{\frac{D_A}{D_B}} + \sqrt{\frac{D_A}{D_B}}\left(\frac{C_B}{C_A b}\right) \tag{14-62}$$

For design purposes, the entire set of curves in Fig. 14-13 may be represented by the following two equations:

For, $N_{Ha} \geq 2$:

$$k_L/k_L^0 = 1 + (\phi_\infty - 1)\{1 - \exp[-(N_{Ha} - 1)/(\phi_\infty - 1)]\} \tag{14-63}$$

For, $N_{Ha} \leq 2$:

$$k_L/k_L^0 = 1 + (\phi_\infty - 1)\{1 - \exp[-(\phi_\infty - 1)^{-1}]\}\exp[1 - 2/N_{Ha}] \tag{14-64}$$

Equation (14-64) was originally reported by Porter [*Trans. Inst. Chem. Eng.*, **44,** T25 (1966)], and Eq. (14-64) was derived by Edwards and first reported in the 6th edition of this handbook.

The Van Krevelen-Hoftyzer (Fig. 14-13) relationship was tested by Nijsing et al. [*Chem. Eng. Sci.*, **10,** 88 (1959)] for the second-order system in which CO_2 reacts with either NaOH or KOH solutions. Nijsing's results are shown in Fig. 14-14 and can be seen to be in excellent

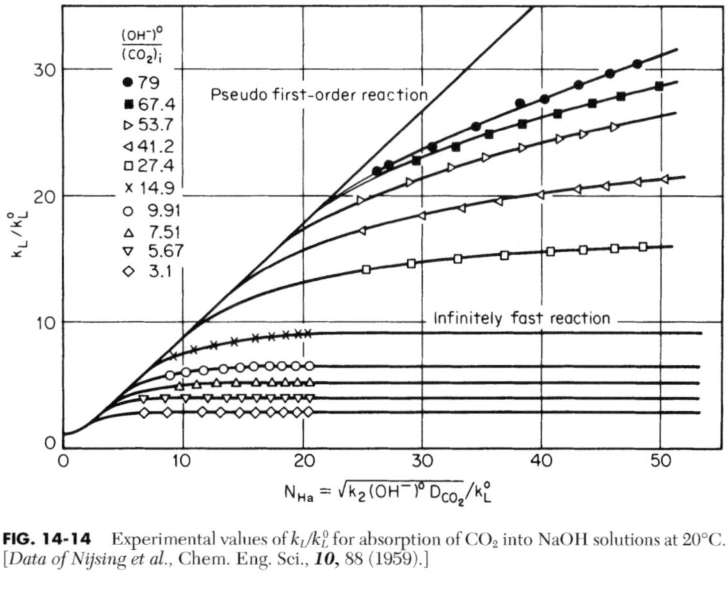

FIG. 14-14 Experimental values of k_L/k_L^0 for absorption of CO_2 into NaOH solutions at 20°C. [*Data of Nijsing et al.,* Chem. Eng. Sci., ***10,*** 88 (1959).]

agreement with the second-order-reaction theory. Indeed, these experimental data are well described by Eqs. (14-62) and (14-63) when values of $b = 2$ and $D_A/D_B = 0.64$ are employed in the equations.

Applicability of Physical Design Methods Physical design models such as the classical isothermal design method or the classical adiabatic design method may be applicable for systems in which chemical reactions are either extremely fast or extremely slow, or when chemical equilibrium is achieved between the gas and liquid phases.

If the chemical reaction is extremely fast and irreversible, the rate of absorption may in some cases be completely governed by gas-phase resistance. For practical design purposes, one may assume, e.g., that this gas-phase mass-transfer-limited condition will exist when the ratio y_i/y is less than 0.05 everywhere in the apparatus.

From the basic mass-transfer flux relationship for species A (Sec. 5)

$$N_A = k_G(y - y_i) = k_L(x_i - x) \tag{14-65}$$

one can readily show that this condition on y_i/y requires that the ratio x/x_i be negligibly small (i.e., a fast reaction) and that the ratio $mk_G/k_L = mk_G/k_L^0\phi$ be less than 0.05 everywhere in the apparatus. The ratio $mk_G/k_L^0\phi$ will be small if the equilibrium backpressure of the solute over the liquid is small (i.e., small m or high reactant solubility), or the reaction enhancement factor $\phi = k_L/k_L^0$ is very large, or both. The reaction enhancement factor ϕ will be large for all extremely fast pseudo-first-order reactions and will be large for extremely fast second-order irreversible reaction systems in which there is sufficiently large excess of liquid reagent.

Figure 14-12, case (ii), illustrates the gas-film and liquid-film concentration profiles one might find in an extremely fast (gas-phase mass-transfer-limited), second-order irreversible reaction system. The solid curve for reagent B represents the case in which there is a large excess of bulk liquid reagent B^0. Figure 14-12, case (iv), represents the case in which the bulk concentration B^0 is not sufficiently large to prevent the depletion of B near the liquid interface.

Whenever these conditions on the ratio y_i/y apply, the design can be based upon the physical rate coefficient k_G or upon the height of one gas-phase mass-transfer unit H_G. The gas-phase mass-transfer-limited condition is approximately valid for the following systems: absorption of NH_3 into water or acidic solutions, absorption of H_2O into concentrated sulfuric acid, absorption of SO_2 into alkali solutions, absorption of H_2S from a gas stream into a strong alkali solution, absorption of HCl into water or alkaline solutions, or absorption of Cl_2 into strong alkali solutions.

When the liquid-phase reactions are extremely slow, the gas-phase resistance can be neglected and one can assume that the rate of reaction has a predominant effect upon the rate of absorption. In this case the differential rate of transfer is given by the equation

$$dn_A = R_A f_H S\, dh = (k_L^0 a/\rho_L)(c_i - c)S\, dh \tag{14-66}$$

where n_A = rate of solute transfer, R_A = volumetric reaction rate (function of c and T), f_H = fractional liquid volume holdup in tower or apparatus, S = tower cross-sectional area, h = vertical distance, k_L^0 = liquid-phase mass-transfer coefficient for pure physical absorption, a = effective interfacial mass-transfer area per unit volume of tower or apparatus, ρ_L = average molar density of liquid phase, c_i = solute concentration in liquid at gas-liquid interface, and c = solute concentration in bulk liquid.

Although the right side of Eq. (14-66) remains valid even when chemical reactions are extremely slow, the mass-transfer driving force may become increasingly small, until finally $c \approx c_i$. For extremely slow first-order irreversible reactions, the following rate expression can be derived from Eq. (14-66):

$$R_A = k_1 c = k_1 c_i/(1 + k_1\rho_L f_H/k_L^0 a) \tag{14-67}$$

where k_1 = first-order reaction rate coefficient.

For dilute systems in countercurrent absorption towers in which the equilibrium curve is a straight line (i.e., $y_i = mx_i$), the differential relation of Eq. (14-66) is formulated as

$$dn_A = -G_M S\, dy = k_1 c f_H S\, dh \tag{14-68}$$

where G_M = molar gas-phase mass velocity and y = gas-phase solute mole fraction.

Substitution of Eq. (14-67) into Eq. (14-68) and integration lead to the following relation for an extremely slow first-order reaction in an absorption tower:

$$y_2 = y_1 \exp\left[-\frac{k_1\rho_L f_H h_T/(mG_m)}{1 + k_1\rho_L f_H/(k_L^0 a)}\right] \tag{14-69}$$

In Eq. (14-69) subscripts 1 and 2 refer to the bottom and top of the tower, respectively.

As discussed above, the Hatta number N_{Ha} usually is employed as the criterion for determining whether a reaction can be considered extremely slow. A reasonable criterion for slow reactions is

$$N_{Ha} = \sqrt{k_1 D_A}/k_L^0 \le 0.3 \tag{14-70}$$

where D_A = liquid-phase diffusion coefficient of the solute in the solvent. Figure 14-12, cases (vii) and (viii), illustrates the concentration profiles in the gas and liquid films for the case of an extremely slow chemical reaction.

Note that when the second term in the denominator of the exponential in Eq. (14-69) is very small, the liquid holdup in the tower can have a significant influence upon the rate of absorption if an extremely slow chemical reaction is involved.

When chemical equilibrium is achieved quickly throughout the liquid phase, the problem becomes one of properly defining the physical and chemical equilibria for the system. It is sometimes possible to design a tray-type absorber by assuming chemical equilibrium relationships in conjunction with a stage efficiency factor, as is done in distillation calculations. Rivas and Prausnitz [*AIChE J.*, **25**, 975 (1979)] have presented an excellent discussion and example of the correct procedures to be followed for systems involving chemical equilibria.

Traditional Design Method The traditional procedure for designing packed-tower gas absorption systems involving chemical reactions makes use of overall mass-transfer coefficients as defined by the equation

$$K_G a = n_A/(h_T S p_T \Delta y_{lm}^\circ) \tag{14-71}$$

where $K_G a$ = overall volumetric mass-transfer coefficient, n_A = rate of solute transfer from the gas to the liquid phase, h_T = total height of tower packing, S = tower cross-sectional area, p_T = total system pressure, and Δy_{lm}° is defined by the equation

$$\Delta y_{lm}^\circ = \frac{(y - y^\circ)_1 - (y - y^\circ)_2}{\ln\left[(y - y^\circ)_1/(y - y^\circ)_2\right]} \tag{14-72}$$

in which subscripts 1 and 2 refer to the bottom and top of the absorption tower respectively, y = mole-fraction solute in the gas phase, and y° = gas-phase solute mole fraction in equilibrium with bulk-liquid-phase solute concentration x. When the equilibrium line is straight, $y^\circ = mx$.

The traditional design method normally makes use of overall $K_G a$ values even when resistance to transfer lies predominantly in the liquid phase. For example, the CO_2-NaOH system which is most commonly used for comparing $K_G a$ values of various tower packings is a liquid-phase-controlled system. When the liquid phase is controlling, extrapolation to different concentration ranges or operating conditions is not recommended since changes in the reaction mechanism can cause k_L to vary unexpectedly and the overall $K_G a$ do not capture such effects.

Overall $K_G a$ data may be obtained from tower-packing vendors for many of the established commercial gas absorption processes. Such data often are based either upon tests in large-diameter test units or upon actual commercial operating data. Since application to untried operating conditions is not recommended, the preferred procedure for applying the traditional design method is equivalent to duplicating a previously successful commercial installation. When this is not possible, a commercial demonstration at the new operating conditions may be required, or else one could consider using some of the more rigorous methods described later.

While the traditional design method is reported here because it has been used extensively in the past, it should be used with extreme

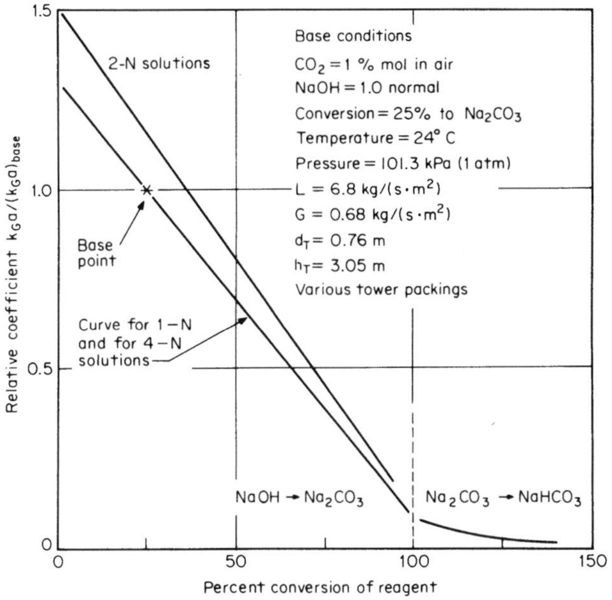

FIG. 14-15 Effects of reagent-concentration and reagent-conversion level upon the relative values of K_Ga in the CO_2-NaOH-H_2O system. [*Adapted from Eckert et al.,* Ind. Eng. Chem., ***59****(2), 41 (1967).*]

caution. In addition to the lack of an explicit liquid-phase resistance term, the method has other limitations. Equation (14-71) assumes that the system is dilute ($y_{BM} \approx 1$) and that the operating and equilibrium lines are straight, which are weak assumptions for reacting systems. Also, Eq. (14-65) is strictly valid only for the temperature and solute partial pressure at which the original test was done even though the total pressure p_T appears in the denominator.

In using Eq. (14-71), therefore, it should be understood that the numerical values of K_Ga will be a complex function of pressure, temperature, the type and size of packing employed, the liquid and gas mass flow rates, and the system composition (e.g., the degree of conversion of the liquid-phase reactant).

Figure 14-15 illustrates the influence of system composition and degree of reactant conversion upon the numerical values of K_Ga for the absorption of CO_2 into sodium hydroxide at constant conditions of temperature, pressure, and type of packing. An excellent experimental study of the influence of operating variables upon overall K_Ga values is that of Field et al. (*Pilot-Plant Studies of the Hot Carbonate Process for Removing Carbon Dioxide and Hydrogen Sulfide*, U.S. Bureau of Mines Bulletin 597, 1962).

Table 14-2 illustrates the observed variations in K_Ga values for different packing types and sizes for the CO_2-NaOH system at a 25 percent reactant conversion for two different liquid flow rates. The lower rate of 2.7 kg/(s·m²) or 2000 lb/(h·ft²) is equivalent to 4 U.S. gal/(min·ft²) and is typical of the liquid rates employed in fume scrubbers. The higher rate of 13.6 kg/(s·m²) or 10,000 lb/(h·ft²) is equivalent to 20 U.S. gal/(min·ft²) and is more typical of absorption towers such as used in CO_2 removal systems, for example. We also note that two gas velocities are represented in the table, corresponding to superficial velocities of 0.59 and 1.05 m/s (1.94 and 3.44 ft/s).

Table 14-3 presents a typical range of K_Ga values for chemically reacting systems. The first two entries in the table represent systems that can be designed by the use of purely physical design methods, because they are completely gas-phase mass-transfer-limited. To ensure a negligible liquid-phase resistance in these two tests, the HCl was absorbed into a solution maintained at less than 8 wt % HCl, and the NH_3 was absorbed into a water solution maintained below pH 7 by the addition of acid. The last two entries in Table 14-3 represent liquid-phase mass-transfer-limited systems.

Scaling Up from Laboratory Data Laboratory experimental techniques offer an efficient and cost-effective route to develop commercial absorption designs. For example, Ouwerkerk (*Hydrocarbon Process.*, April 1978, 89–94) revealed that both laboratory and small-scale pilot plant data were employed as the basis for the design of an 8.5-m (28-ft) diameter commercial Shell Claus off-gas treating (SCOT) tray-type absorber. Ouwerkerk claimed that the cost of developing comprehensive design procedures can be minimized, especially in the development of a new process, by the use of these modern techniques.

In a 1966 paper that is considered a classic, Dankwerts and Gillham [*Trans. Inst. Chem. Eng.*, 44, T42 (1966)] showed that data taken in a small stirred-cell laboratory apparatus could be used in the design of a packed-tower absorber when chemical reactions are involved. They showed that if the packed-tower mass-transfer coefficient in the absence of reaction (k_L^0) can be reproduced in the laboratory unit, then the rate of absorption in the laboratory apparatus will respond to chemical reactions in the same way as in the packed column, even though the means of agitating the liquid in the two systems may be quite different.

According to this method, it is not necessary to investigate the kinetics of the chemical reactions in detail; nor is it necessary to determine the solubilities or diffusivities of the various reactants in their unreacted forms. To use the method for scaling up, it is necessary to independently obtain data on the values of the interfacial area per unit volume a and the physical mass-transfer coefficient k_L^0 for the commercial packed tower. Once these data have been measured and tabulated, they can be used directly for scaling up the experimental laboratory data for any new chemically reacting system.

Dankwerts and Gillham did not investigate the influence of the gas-phase resistance in their study (for some processes, gas-phase resistance

TABLE 14-2 Typical Effects of Packing Type, Size, and Liquid Rate on K_Ga in a Chemically Reacting System, K_Ga, kmol/(h·m³)

	L = 2.7 kg/(s·m²)				L = 13.6 kg/(s·m²)			
Packing size, mm	25	38	50	75–90	25	38	50	75–90
Berl-saddle ceramic	30	24	21		45	38	32	
Raschig-ring ceramic	27	24	21		42	34	30	
Raschig-ring metal	29	24	19		45	35	27	
Pall-ring plastic	29	27	26*	16	45	42	38*	24
Pall-ring metal	37	32	27	21*	56	51	43	27*
Intalox-saddle ceramic	34	27	22	16*	56	43	34	26*
Super-Intalox ceramic	37*		26*		59*		40*	
Intalox-saddle plastic	40*		24*	16*	56*		37*	26*
Intalox-saddle metal	43*	35*	30*	24*	66*	58*	48*	37*
Hy-Pak metal	35	32*	27*	18*	54	50*	42*	27*

Data courtesy of the Norton Company.

Operating conditions: CO_2, 1 percent mole in air; NaOH, 4 percent weight (1 normal); 25 percent conversion to sodium carbonate; temperature, 24°C (75°F); pressure, 98.6 kPa (0.97 atm); gas rate = 0.68 kg/(s·m²) = 0.59 m/s = 500 lb/(h·ft²) = 1.92 ft/s except for values with asterisks, which were run at 1.22 kg/(s·m²) = 1.05 m/s = 900 lb/(h·ft²) = 3.46 ft/s superficial velocity; packed height, 3.05 m (10 ft); tower diameter, 0.76 m (2.5 ft). To convert table values to units of (lb·mol)/(h·ft³), multiply by 0.0624.

TABLE 14-3 Typical K_Ga Values for Various Chemically Reacting Systems, kmol/(h·m³)

Gas-phase reactant	Liquid-phase reactant	K_Ga	Special conditions
HCl	H_2O	353	Gas-phase limited
NH_3	H_2O	337	Gas-phase limited
Cl_2	NaOH	272	8% weight solution
SO_2	Na_2CO_3	224	11% weight solution
HF	H_2O	152	
Br_2	NaOH	131	5% weight solution
HCN	H_2O	114	
HCHO	H_2O	114	Physical absorption
HBr	H_2O	98	
H_2S	NaOH	96	4% weight solution
SO_2	H_2O	59	
CO_2	NaOH	38	4% weight solution
Cl_2	H_2O	8	Liquid-phase limited

Data courtesy of the Norton Company.

Operating conditions (see text): 38-mm ceramic Intalox saddles; solute gases, 0.5–1.0 percent mole; reagent conversions = 33 percent; pressure, 101 kPa (1 atm); temperature, 16–24°C; gas rate = 1.3 kg/(s·m²) = 1.1 m/s; liquid rates = 3.4 to 6.8 kg/(s·m²); packed height, 3.05 m; tower diameter, 0.76 m. Multiply table values by 0.0624 to convert to (lb·mol)/(h·ft³).

may be neglected). However, in 1975 Dankwerts and Alper [*Trans. Inst. Chem. Eng.*, 53, T42 (1975)] showed that by placing a stirrer in the gas space of the stirred-cell laboratory absorber, the gas-phase mass-transfer coefficient k_G in the laboratory unit could be made identical to that in a packed-tower absorber. When this was done, laboratory data for chemically reacting systems having a significant gas-side resistance could successfully be scaled up to predict the performance of a commercial packed-tower absorber.

If it is assumed that the values for k_G, k_L^0, and a have been measured for the commercial tower packing to be employed, the procedure for using the laboratory stirred-cell reactor is as follows:

1. The gas-phase and liquid-phase stirring rates are adjusted so as to produce the same values of k_G and k_L^0 as will exist in the commercial tower.

2. For the reaction system under consideration, experiments are made at a series of bulk-liquid and bulk-gas compositions representing the compositions to be expected at different levels in the commercial absorber (on the basis of material balance).

3. The ratios of $r_A(c_i, B^0)$ are measured at each pair of gas and liquid compositions.

For the dilute-gas systems, one form of the equation to be solved in conjunction with these experiments is

$$h_T = \frac{G_M}{a} \int_{y2}^{y1} \frac{dy}{r_A} \tag{14-73}$$

where h_T = height of commercial tower packing, G_M = molar gas-phase mass velocity, a = effective mass-transfer area per unit volume in the commercial tower, y = mole fraction solute in the gas phase, and r_A = experimentally determined rate of absorption per unit of exposed interfacial area.

By using the series of experimentally measured rates of absorption, Eq. (14-73) can be integrated numerically to determine the height of packing required in the commercial tower.

A number of different types of experimental laboratory units could be used to develop design data for chemically reacting systems. Charpentier [*ACS Symp. Ser.*, **72,** 223–261 (1978)] has summarized the state of the art with respect to methods of scaling up laboratory data and has tabulated typical values of the mass-transfer coefficients, interfacial areas, and contact times to be found in various commercial gas absorbers, as well as in currently available laboratory units.

The laboratory units that have been employed to date for these experiments were designed to operate at a total system pressure of about 101 kPa (1 atm) and at near-ambient temperatures. In practical situations, it may become necessary to design a laboratory absorption unit that can be operated either under vacuum or at elevated pressure and over a range of temperatures in order to apply the Dankwerts method.

It would be desirable to reinterpret existing data for commercial tower packings to extract the individual values of the interfacial area a and the mass-transfer coefficients k_G and $k^0{}_L$ to facilitate a more general usage of methods for scaling up from laboratory experiments. Some progress has already been made, as described later in this section. In the absence of such data, it is necessary to operate a pilot plant or a commercial absorber to obtain k_G, k_L^0, and a as described by Ouwerkerk (op. cit.).

Modern techniques use rigorous modeling computer-based methods to extract fundamental parameters from laboratory-scale measurements and then apply them to the design of commercial absorption towers. These techniques are covered next.

Rigorous Computer-Based Absorber Design While the techniques described earlier in this section are very useful to gain an understanding of the key effects in commercial absorbers, current design methods used in industrial practice for chemically reactive systems are increasingly often based upon computerized rigorous methods, which are commercially available from software vendors. The advantages of these rigorous methods are as follows: (1) Approximations do not have to be made for special cases (e.g., fast chemical reactions or mass-transfer resistance dominated by the gas or liquid phase), and all effects can be simultaneously modeled. (2) Fundamental quantities such as kinetic parameters and mass-transfer coefficients can be extracted from laboratory equipment and applied to commercial absorber towers. (3) Integrated models can be developed for an entire absorption process flowsheet (e.g., the absorber-stripper system with heat integration presented in Fig. 14-3), and consequently the entire system may be optimized.

Computer programs for chemically reacting systems are available from several vendors, notably the following:

Program	Vendor	Reference
AMSIM	Schlumberger Limited	Zhang and Ng, *Proc. Ann. Conv.—Gas Proc. Assoc.*, Denver, Colo.; 1996, p. 22.
ProTreat	Sulphur Experts	Weiland and Dingman, *Proc. Ann. Conv., Gas Proc. Assoc.*, Houston, Tex., 2001, p. 80.
TSWEET	Bryan Research and Engineering	Polasek, Donnelly, and Bullin, Proc. 71st GPA Annual Conv., 1992, p. 58.
RateSep	Aspen Technology	Chen et al., AIChE Annual Meeting, San Francisco, Nov. 12–17, 2006.

The specific approaches used to model the chemically reacting absorption system are slightly different among the different vendors. The general approach used and the benefits obtained are highlighted by considering a specific example: removal of CO_2 from flue gases discharged by a power plant using aqueous monoethanolamine (MEA), as presented by Freguia and Rochelle [*AIChE J.*, **49,** 1676 (2003)].

The development and application of a rigorous model for a chemically reactive system typically involves four steps: (1) development of a thermodynamic model to describe the physical and chemical equilibrium; (2) adoption and use of a modeling framework to describe the mass transfer and chemical reactions; (3) parameterization of the mass-transfer and kinetic models based upon laboratory, pilot-plant, or commercial-plant data; and (4) use of the integrated model to optimize the process and perform equipment design.

Development of Thermodynamic Model for Physical and Chemical Equilibrium The first and perhaps most important step in the development of the thermodynamic model is the *speciation,* or representation of the set of chemical reactions. For CO_2 absorption in aqueous MEA solutions, the set of reactions is

$$CO_2 + MEA + H_2O \leftrightarrow MEACOO^- + H_3O^+ \quad (14\text{-}74a)$$

$$CO_2 + OH^- \leftrightarrow HCO_3^- \quad (14\text{-}74b)$$

$$HCO_3^- + H_2O \leftrightarrow CO_3^{2-} + H_3O^+ \quad (14\text{-}74c)$$

$$MEAH^+ + H_2O \leftrightarrow MEA + H_3O^+ \quad (14\text{-}74d)$$

$$2H_2O \leftrightarrow H_3O^+ + OH^- \quad (14\text{-}74e)$$

In addition, a model is needed that can describe the nonideality of a system containing molecular and ionic species. Freguia and Rochelle adopted the model developed by Chen et al. [*AIChE J.*, **25,** 820 (1979)] and later modified by Mock et al. [*AIChE J.*, **32,** 1655 (1986)] for mixed-electrolyte systems. The combination of the speciation set of reactions [Eqs. (14-74*a*) to (14-74*e*)] and the nonideality model is capable of representing the solubility data, such as presented in Figs. 14-1 and 14-2, to good accuracy. In addition, the model accurately and correctly represents the actual species present in the aqueous phase, which is important for faithful description of the chemical kinetics and species mass transfer across the interface. Finally, the thermodynamic model facilitates accurate modeling of the heat effects, such as those discussed in Example 6.

Rafal et al. (Chapter 7, "Models for Electrolyte Solutions," in *Models for Thermodynamic and Phase Equilibria Calculations*, S. I. Sandler, ed., Marcel Dekker, New York, 1994, p. 686) have provided a comprehensive discussion of speciation and thermodynamic models.

Adoption and Use of Modeling Framework The rate of diffusion and species generation by chemical reaction can be described by film theory, penetration theory, or a combination of the two. The most popular description is in terms of a two-film theory, which is diagrammed in Fig. 14-16 for absorption. Accordingly, there exists a stable interface separating the gas and the liquid. A certain distance from the interface, large fluid motions exist, and these distribute the material rapidly and equally so that no concentration gradients develop. Next to the interface, however, there are regions in which the fluid motion is slow; in these regions, termed *films*, material is transferred by diffusion alone. At the gas-liquid interface, material is transferred instantaneously, so that the gas and liquid are in physical equilibrium at the interface. The rate of diffusion in adsorption is therefore the rate of diffusion in the gas and liquid films adjacent to the interface. The model framework is completed by including terms for species generation (chemical equilibrium *and* chemical kinetics) in the gas and liquid film and bulk regions. Taylor, Krishna, and Kooijman (*Chem. Eng. Progress*, July 2003, p. 28) have provided an excellent discussion of rate-based models; these authors emphasize that the diffusion flux for multicomponent systems must be based upon the Maxwell-Stefan approach. The book by Taylor and Krishna (*Multicomponent Mass Transfer*, Wiley, New York, 1993) provides a detailed discussion of the Maxwell-Stefan approach. More details and discussion have been provided by the program vendors listed above.

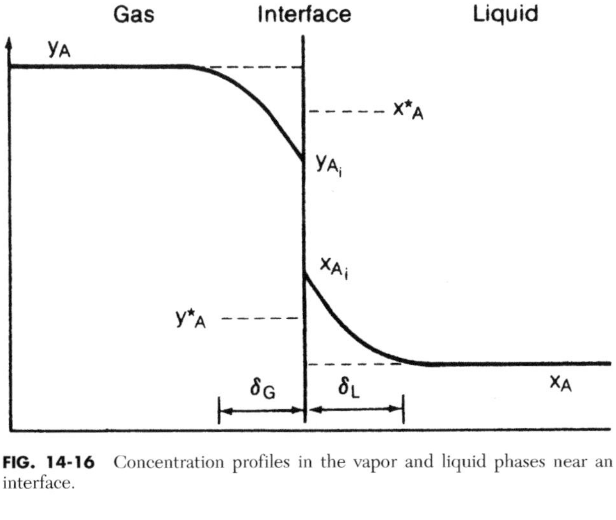

FIG. 14-16 Concentration profiles in the vapor and liquid phases near an interface.

Parameterization of Mass-Transfer and Kinetic Models The mass-transfer and chemical kinetic rates required in the rigorous model are typically obtained from the literature, but must be carefully evaluated; and fine-tuning through pilot-plant and commercial data is highly recommended.

Mass-transfer coefficient models for the vapor and liquid coefficients are of the general form

$$k_{ij}^L = a\rho_L f(D_{ij}^m, \mu_L, \rho_V, a, \text{internal characteristics}) \quad (14\text{-}75a)$$

$$k_{ij}^V = aP\, f(D_{ij}^m, \mu_V, \rho_V, a, \text{internal characteristics}) \quad (14\text{-}75b)$$

where a = effective interfacial area per unit volume, D_{ij}^m are the Stefan-Maxwell diffusion coefficients, P = pressure, ρ = molar density, and μ = viscosity. The functions in Eqs. (14-75*a*) and (14-75*b*) are correlations that depend on the column internals. Popular correlations in the literature are those by Onda at al. [*J. Chem.. Eng. Jap.*, **1,** 56 (1968)] for random packing, Bravo and Fair [*Ind. Eng. Chem. Proc. Des. Dev.*, **21,** 162 (1982)] for structured packing, Chan and Fair [*Ind. Eng. Chem. Proc. Des. Dev.*, **23,** 814 (1984)] for sieve trays, Scheffe and Weiland [*Ind. Eng. Chem. Res.*, **26,** 228 (1987)] for valve trays, and Hughmark [*AIChE J.*, **17,** 1295 (1971)] for bubble-cap trays.

It is highly recommended that the mass-transfer correlations be tested and improved by using laboratory, pilot-plant, or commercial data for the specific application. Commercial software generally provides the capability for correction factors to adjust generalized correlations to the particular application.

Kinetic models are usually developed by replacing a subset of the speciation reactions by kinetically reversible reactions. For example, Freguia and Rochelle replaced equilibrium reactions (14-74*a*) and (14-74*b*) with kinetically reversible reactions and retained the remaining three reactions as very fast and hence effectively at equilibrium. The kinetic constants were tuned using wetted-wall column data from Dang (M.S. thesis, University of Texas, Austin, 2001) and field data from a commercial plant.

Modern commercial software provides powerful capability to deploy literature correlations and to customize models for specific applications.

Deployment of Rigorous Model for Process Optimization and Equipment Design Techniques similar to those described above may be used to develop models for the stripper as well as other pieces of plant equipment, and thus an integrated model for the entire absorption system may be produced. The value of integrated models is that they can be used to understand the combined effects of many variables that determine process performance and to rationally optimize process performance. Freguia and Rochelle have shown that the reboiler duty (the dominant source of process operating costs) may be reduced by 10 percent if the absorber height is increased by 20 percent and by 4 percent if the absorber is intercooled with a duty equal to one-third of the reboiler duty. They also show that the power plant lost work is affected by varying stripper pressure, but not significantly, so any convenient pressure can be chosen to operate the stripper.

In this section, we have used the example of CO_2 removal from flue gases using aqueous MEA to demonstrate the development and application of a rigorous model for a chemically reactive system. Modern software enables rigorous description of complex chemically reactive systems, but it is very important to carefully evaluate the models and to tune them using experimental data.

Use of Literature for Specific Systems A large body of experimental data obtained in bench-scale laboratory units and in small-diameter packed towers has been published since the early 1940s. One might wish to consider using such data for a particular chemically reacting system as the basis for scaling up to a commercial design. Extreme caution is recommended in interpreting such data for the purpose of developing commercial designs, as extrapolation of the data can lead to serious errors. Extrapolation to temperatures, pressures, or liquid-phase reagent conversions different from those that were employed by the original investigators definitely should be regarded with caution.

Bibliographies presented in the General References listed at the beginning of this section are an excellent source of information on specific chemically reacting systems. *Gas-Liquid Reactions* by Dankwerts (McGraw-Hill, New York, 1970) contains a tabulation of references to specific chemically reactive systems. *Gas Treating with Chemical Solvents* by Astarita et al. (Wiley, New York, 1983) deals with the absorption of acid gases and includes an extensive listing of patents. *Gas Purification* by Kohl and Nielsen (Gulf Publishing, Houston, 1997) provides a practical description of techniques and processes in widespread use and typically also sufficient design and operating data for specific applications.

In searching for data on a particular system, a computerized search of *Chemical Abstracts, Engineering Index,* and *National Technical Information Service* (NTIS) databases is recommended. In addition, modern search engines will rapidly uncover much potentially valuable information.

The experimental data for the system CO_2-NaOH-Na_2CO_3 are unusually comprehensive and well known as the result of the work of many experimenters. A serious study of the data and theory for this system therefore is recommended as the basis for developing a good understanding of the kind and quality of experimental information needed for design purposes.

EQUIPMENT FOR DISTILLATION AND GAS ABSORPTION: TRAY COLUMNS

Distillation and gas absorption are the prime and most common gas-liquid mass-transfer operations. Other operations that are often performed in similar equipment include stripping (often considered part of distillation), direct-contact heat transfer, flashing, washing, humidification, and dehumidification.

The most common types of contactors by far used for these are tray and packed towers. These are the focus of this subsection. Other contactors used from time to time and their applications are listed in Table 14-4.

In this subsection, the terms *gas* and *vapor* are used interchangeably. *Vapor* is more precise for distillation, where the gas phase is at equilibrium. Also, the terms *tower* and *column* are used interchangeably.

A crossflow tray (Fig. 14-17) consists of the *bubbling area* and the *downcomer*. Liquid descending the downcomer from the tray above enters the bubbling area. Here, the liquid contacts gas ascending through the tray perforations, forming froth or spray. An *outlet weir* on the downstream side of the bubbling area helps maintain liquid level on the tray. Froth overflowing the weir enters the outlet downcomer. Here, gas disengages from the liquid, and the liquid descends to the tray below. The bubbling area can be fitted with numerous types of tray hardware. The most common types by far are:

Sieve trays (Fig. 14-18*a*) are perforated plates. The velocity of upflowing gas keeps the liquid from descending through the perforations (weeping). At low gas velocities, liquid weeps through the perforations, bypassing part of the tray and reducing tray efficiency. Because of this, sieve trays have relatively poor turndown.

Fixed valve trays (Fig. 14-18*b*) have the perforations covered by a fixed cover, often a section of the tray floor pushed up. Their performance is similar to that of sieve trays.

Moving valve trays (Fig. 14-18*c*) have the perforations covered by movable disks (valves). Each valve rises as the gas velocity increases. The upper limit of the rise is controlled by restricting legs on the bottom of the valve (Fig. 14-18*c*) or by a cage structure around the valve. As the gas velocity falls, some valves close completely, preventing weeping. This gives the valve tray good turndown.

Table 14-5 is a general comparison of the three main tray types, assuming proper design, installation, and operation. Sieve and valve trays are comparable in capacity, efficiency, entrainment, and pressure drop. The turndown of moving valve trays is much better than that of sieve and fixed valve trays. Sieve trays are least expensive; valve trays cost only slightly more. Maintenance, fouling tendency, and effects of corrosion are least troublesome in fixed valve and sieve trays (provided the perforations or fixed valves are large enough) and most troublesome with moving valve trays.

Fixed valve and sieve trays prevail when fouling or corrosion is expected, or if turndown is unimportant. Valve trays prevail when high turndown is required. The energy saved, even during short turndown periods, usually justifies the small additional cost of the moving valve trays.

DEFINITIONS

Tray Area Definitions Some of these are illustrated in Fig. 14-17.

Total tower cross-section area A_T The inside cross-sectional area of the empty tower (without trays or downcomers).

Net area A_N (also called free area) The total tower cross-sectional area A_T minus the area at the top of the downcomer A_{DT}. The

TABLE 14-4 Equipment for Liquid-Gas Systems

Equipment designation	Mode of flow	Gross mechanism	Continuous phase	Primary process applications
Tray column	Cross-flow, countercurrent	Integral	Liquid and/or gas	Distillation, absorption, stripping, DCHT, washing
Packed column	Countercurrent, cocurrent	Differential	Liquid and/or gas	Distillation, absorption, stripping, humidification, dehumidification, DCHT, washing
Wetted-wall (falling-film) column	Countercurrent, cocurrent	Differential	Liquid and/or gas	Distillation, absorption, stripping, evaporation
Spray chamber	Cocurrent, cross-flow, countercurrent	Differential	Gas	Absorption, stripping, humidification, dehumidification
Agitated vessel	Complete mixing	Integral	Liquid	Absorption
Line mixer	Cocurrent	Differential	Liquid or gas	Absorption, stripping

DCHT = direct contact heat transfer.

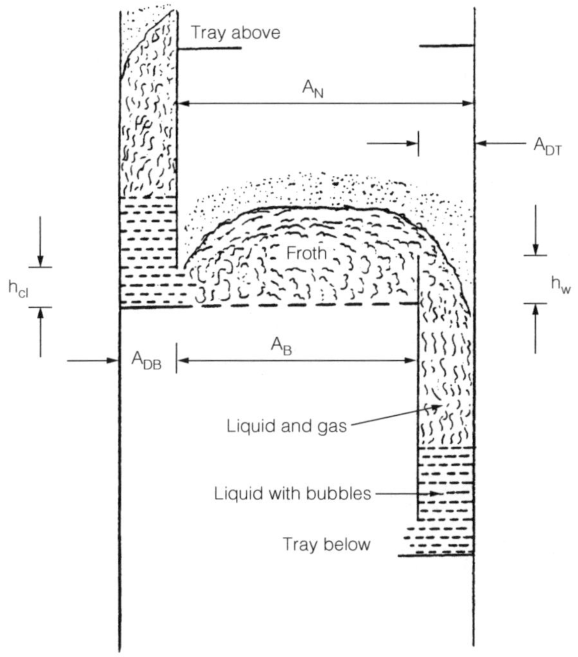

FIG. 14-17 Schematic of a tray operating in the froth regime. (*Based on H. Z. Kister,* Distillation Design, *copyright © 1992 by McGraw-Hill; reprinted by permission.*)

net area represents the smallest area available for vapor flow in the intertray spacing.

Bubbling area $\mathbf{A_B}$ **(also called *active area*)** The total tower cross-sectional area minus the sum of downcomer top area A_{DT}, downcomer seal area A_{DB}, and any other nonperforated areas on the tray. The bubbling area represents the area available for vapor flow just above the tray floor.

Hole area $\mathbf{A_h}$ The total area of the perforations on the tray. The hole area is the smallest area available for vapor passage on a sieve tray.

Slot area $\mathbf{A_S}$ The total (for all open valves) vertical curtain area through which vapor passes in a horizontal direction as it leaves the valves. It is a function of the narrowest opening of each valve and the number of valves that are open. The slot area is normally the smallest area available for vapor flow on a valve tray.

Open slot area $\mathbf{A_{SO}}$ The slot area when all valves are open.

Fractional hole area $\mathbf{A_f}$ The ratio of hole area to bubbling area (sieve trays) or slot area to bubbling area (valve trays).

Vapor and Liquid Load Definitions

F-factor **F** This is the square root of the kinetic energy of the gas, defined by Eq. (14-76). The velocity in Eq. (14-76) is usually (not always) based on the tower cross-sectional area A_T, the net area A_N, or the bubbling area A_B. The user should beware of any data for which the area basis is not clearly specified.

$$F = u\sqrt{\rho_G} \tag{14-76}$$

C-factor **C** The C-factor, defined in Eq. (14-77), is the best gas load term for comparing capacities of systems of different physical properties. It has the same units as velocity (m/s or ft/s) and is directly related to droplet entrainment. As with the F-factor, the user should beware of any data for which the area basis is not clearly specified.

$$C = u\sqrt{\frac{\rho_G}{\rho_L - \rho_G}} \tag{14-77}$$

Weir load For trays (as distinct from downcomers), liquid load is normally defined as

$$Q_L = \frac{\text{volume of liquid}}{\text{length of outlet weir}} = \frac{Q}{L_w} \tag{14-78}$$

This definition describes the flux of liquid horizontally across the tray. Units frequently used are $m^3/(h{\cdot}m)$, $m^3/(s{\cdot}m)$, and gpm/in.

Downcomer liquid load For downcomer design, the liquid load is usually defined as the liquid velocity at the downcomer entrance (m/s or ft/s):

$$Q_D = \frac{\text{volume of liquid}}{\text{downcomer entrance area}} = \frac{Q}{A_{DT}} \tag{14-79}$$

FLOW REGIMES ON TRAYS

Three main flow regimes exist on industrial distillation trays. These regimes may all occur on the same tray under different liquid and gas flow rates (Fig. 14-19). Excellent discussion of the fundamentals and modeling of these flow regimes was presented by Lockett (*Distillation Tray Fundamentals,* Cambridge University Press, Cambridge, 1986). An excellent overview of these as well as of less common flow regimes was given by Prince (*PACE,* June 1975, p. 31; July 1975, p. 18).

Froth regime (or mixed regime; Fig. 14-20a). This is the most common operating regime in distillation practice. Each perforation bubbles vigorously. The bubbles circulate rapidly through the liquid, are of nonuniform sizes and shapes, and travel at varying velocities. The froth surface is mobile and not level, and is generally covered by droplets. Bubbles are formed at the tray perforations and are swept away by the froth.

As gas load increases in the froth regime, jetting begins to replace bubbling in some holes. The fraction of holes that is jetting increases with gas velocity. When jetting becomes the dominant mechanism, the dispersion changes from froth to spray. Prado et al. [*Chemical Engineering Progr.* **83**(3), p. 32, (1987)] showed the transition from froth to spray takes place gradually as jetting replaces bubbling in 45 to 70 percent of the tray holes.

Emulsion regime (Fig. 14-20b). At high liquid loads and relatively low gas loads, the high-velocity liquid bends the swarms of gas bubbles leaving the orifices, and tears them off, so most of the gas becomes emulsified as small bubbles within the liquid. The mixture behaves as a uniform two-phase fluid, which obeys the Francis weir formula [see the subsection "Pressure Drop" and Eq. (14-109) (*Hofhuis and Zuiderweg, IChemE Symp. Ser.* **56,** p. 2.2/1 (1979); Zuiderweg, *Int. Chem. Eng.* **26**(1), 1 (1986)]. In industrial practice, the emulsion regime is the most common in high-pressure and high-liquid-rate operation.

Spray regime (or drop regime, Fig. 14-20c). At high gas velocities and low liquid loads, the liquid pool on the tray floor is shallow and easily atomized by the high-velocity gas. The dispersion becomes a turbulent cloud of liquid droplets of various sizes that reside at high elevations above the tray and follow free trajectories. Some droplets are entrained to the tray above, while others fall back into the liquid pools and become reatomized. In contrast to the liquid-continuous froth and emulsion regimes, the phases are reversed in the spray regime: here the gas is the continuous phase, while the liquid is the dispersed phase.

The spray regime frequently occurs where gas velocities are high and liquid loads are low (e.g., vacuum and rectifying sections at low liquid loads).

Three-layered structure. Van Sinderen, Wijn, and Zanting [*Trans. IChemE,* **81,** Part A, p. 94 (January 2003)] postulate a tray dispersion consisting of a bottom liquid-rich layer where jets/bubbles form; an intermediate liquid-continuous froth layer where bubbles erupt, generating drops; and a top gas-continuous layer of drops. The intermediate layer that dampens the bubbles and

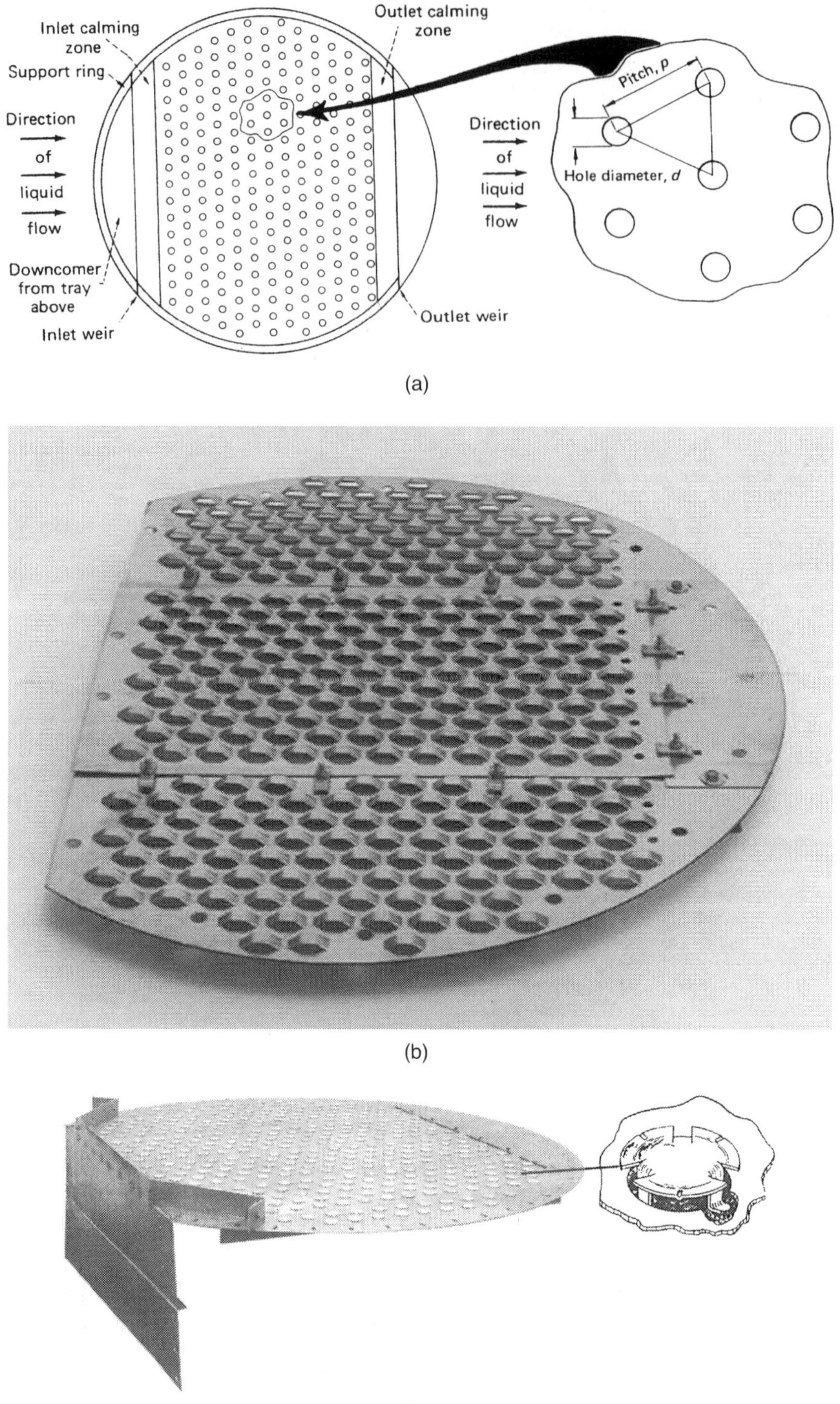

FIG. 14-18 Common tray types. (*a*) Sieve. (*b*) Fixed valve. (*c*) Moving valve with legs. [*Part a, from Henry Z. Kister,* Chem. Eng., *September 8, 1980; reprinted courtesy of* Chemical Engineering. *Part b, Courtesy of Sulzer Chemtech and Fractionation Research Inc. (FRI). Part c, courtesy of Koch-Glitsch LP.*]

TABLE 14-5 Comparison of the Common Tray Types

	Sieve trays	Fixed valve tray	Moving valve tray
Capacity	High	High	High to very high
Efficiency	High	High	High
Turndown	About 2:1. Not generally suitable for operation under variable loads	About 2.5:1. Not generally suitable for operation under variable loads	About 4:1 to 5:1. Some special designs achieve 8:1 or more
Entrainment	Moderate	Moderate	Moderate
Pressure drop	Moderate	Moderate	Slightly higher
Cost	Low	Low	About 20 percent higher
Maintenance	Low	Low	Moderate
Fouling tendency	Low to very low	Low to very low	Moderate
Effects of corrosion	Low	Very low	Moderate
Main applications	(1) Most columns when turndown is not critical (2) High fouling and corrosion potential	(1) Most columns when turndown is not critical (2) High fouling and corrosion potential	(1) Most columns (2) Services where turndown is important

jets disappears at low liquid rates, and the drop layer approaches the tray floor, similar to the classic spray regime.

PRIMARY TRAY CONSIDERATIONS

Number of Passes Tray liquid may be split into two or more flow passes to reduce tray liquid load Q_L (Fig. 14-21). Each pass carries $1/N_p$ fraction of the total liquid load (e.g., $\frac{1}{4}$ in four-pass trays). Liquid in each pass reverses direction on alternate trays. Two-pass trays have perfect symmetry with full remixing in the center downcomers. Four-pass trays are symmetric along the centerline, but the side and central passes are nonsymmetric. Also, the center and off-center downcomers only partially remix the liquid, allowing any maldistribution to propagate. Maldistribution can cause major loss of efficiency and capacity in four-pass trays. Three-pass trays are even more prone to maldistribution due to their complete nonsymmetry. Most designers avoid three-pass trays altogether, jumping from two to four passes. Good practices for liquid and vapor balancing and for avoiding maldistribution in multipass trays were described by Pilling [*Chemical Engineering Progr.*, p. 22 (June 2005)], Bolles [*AIChE J.*, **22**(1), p. 153 (1976)], and Kister (*Distillation Operation*, McGraw-Hill, New York, 1990).

Common design practice is to minimize the number of passes, resorting to a larger number only when the liquid load exceeds 100 to 140 $m^3/(h\cdot m)$ (11 to 15 gpm/in) of outlet weir length [*Davies and Gordon, Petro/Chem Eng.*, p. 228 (December 1961)]. Trays smaller than 1.5-m (5-ft) diameter seldom use more than a single pass; those with 1.5- to 3-m (5- to 10-ft) diameters seldom use more than two passes. Four-pass trays are common in high liquid services with towers larger than 5-m (16-ft) diameter.

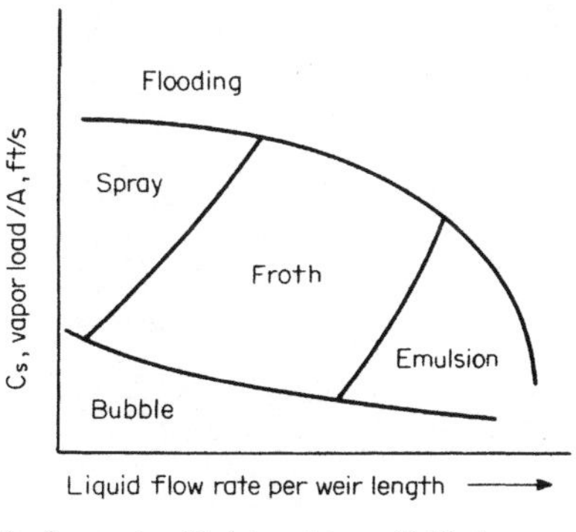

FIG. 14-19 The flow regime likely to exist on a distillation tray as a function of vapor and liquid loads. *(From H. Z. Kister,* Distillation Design, *copyright ©1992 by McGraw-Hill; reprinted by permission.)*

Tray Spacing Taller spacing between successive trays raises capacity, leading to a smaller tower diameter, but also raises tower height. There is an economic tradeoff between tower height and diameter. As long as the tradeoff exists, tray spacing has little effect on tower economies and is set to provide adequate access. In towers with larger than 1.5-m (5-ft) diameter, tray spacing is typically 600 mm (24 in), large enough to permit a worker to crawl between trays. In very large towers (>6-m or 20-ft diameter), tray spacings of 750 mm (30 in) are often used. In chemical towers (as distinct from petrochemical, refinery, and gas plants), 450 mm (18 in) has been a popular tray spacing. With towers smaller than 1.5 m (5 ft), tower walls are reachable from the manways, there is no need to crawl, and it becomes difficult to support thin and tall columns, so smaller tray spacing (typically 380 to 450 mm or 15 to 18 in) is favored. Towers taller than 50 m (160 ft) also favor smaller tray spacings (400 to 450 mm or 16 to 18 in). Finally, cryogenic towers enclosed in cold boxes favor very small spacings, as small as 150 to 200 mm (6 to 8 in), to minimize the size of the cold box.

More detailed considerations for setting tray spacing were discussed by Kister (*Distillation Operation*, McGraw-Hill, New York, 1990) and Mukherjee [*Chem. Eng.* p. 53 (September 2005)].

Outlet Weir The outlet weir should maintain a liquid level on the tray high enough to provide sufficient gas-liquid contact without causing excessive pressure drop, downcomer backup, or a capacity limitation. Weir heights are usually set at 40 to 80 mm (1.5 to 3 in). In this range, weir heights have little effect on distillation efficiency [Van Winkle, *Distillation*, McGraw-Hill, New York, 1967; Kreis and Raab, *IChemE Symp. Ser.* **56,** p. 3.2/63 (1979)]. In operations where long residence times are necessary (e.g., chemical reaction, absorption, stripping) taller weirs do improve efficiency, and weirs 80 to 100 mm (3 to 4 in) are more common (Lockett, *Distillation Tray Fundamentals,* Cambridge University Press, Cambridge, England, 1986).

Adjustable weirs (Fig. 14-22*a*) are used to provide additional flexibility. They are uncommon with conventional trays, but are used with some proprietary trays. Swept-back weirs (Fig. 14-22*b*) are used to extend the effective length of side weirs, either to help balance liquid flows to nonsymmetric tray passes or/and to reduce the tray liquid loads. Picket fence weirs (Fig. 14-22*c*) are used to shorten the effective length of a weir, either to help balance multipass trays' liquid flows (they are used in center and off-center weirs) or to raise tray liquid load and prevent drying in low-liquid-load services. To be effective, the pickets need to be tall, typically around 300 to 400 mm (12 to 16 in) above the top of the weir. An excellent discussion of weir picketing practices was provided by Summers and Sloley (*Hydroc. Proc.*, p. 67, January 2007).

Downcomers A downcomer is the drainpipe of the tray. It conducts liquid from one tray to the tray below. The fluid entering the downcomer is far from pure liquid; it is essentially the froth on the tray, typically 20 to 30 percent liquid by volume, with the balance being gas. Due to the density difference, most of this gas disengages in the downcomer and vents back to the tray from the downcomer entrance. Some gas bubbles usually remain in the liquid even at the bottom of the downcomer, ending on the tray below [Lockett and Gharani, *IChemE Symp. Ser.* **56,** p. 2.3/43 (1979)].

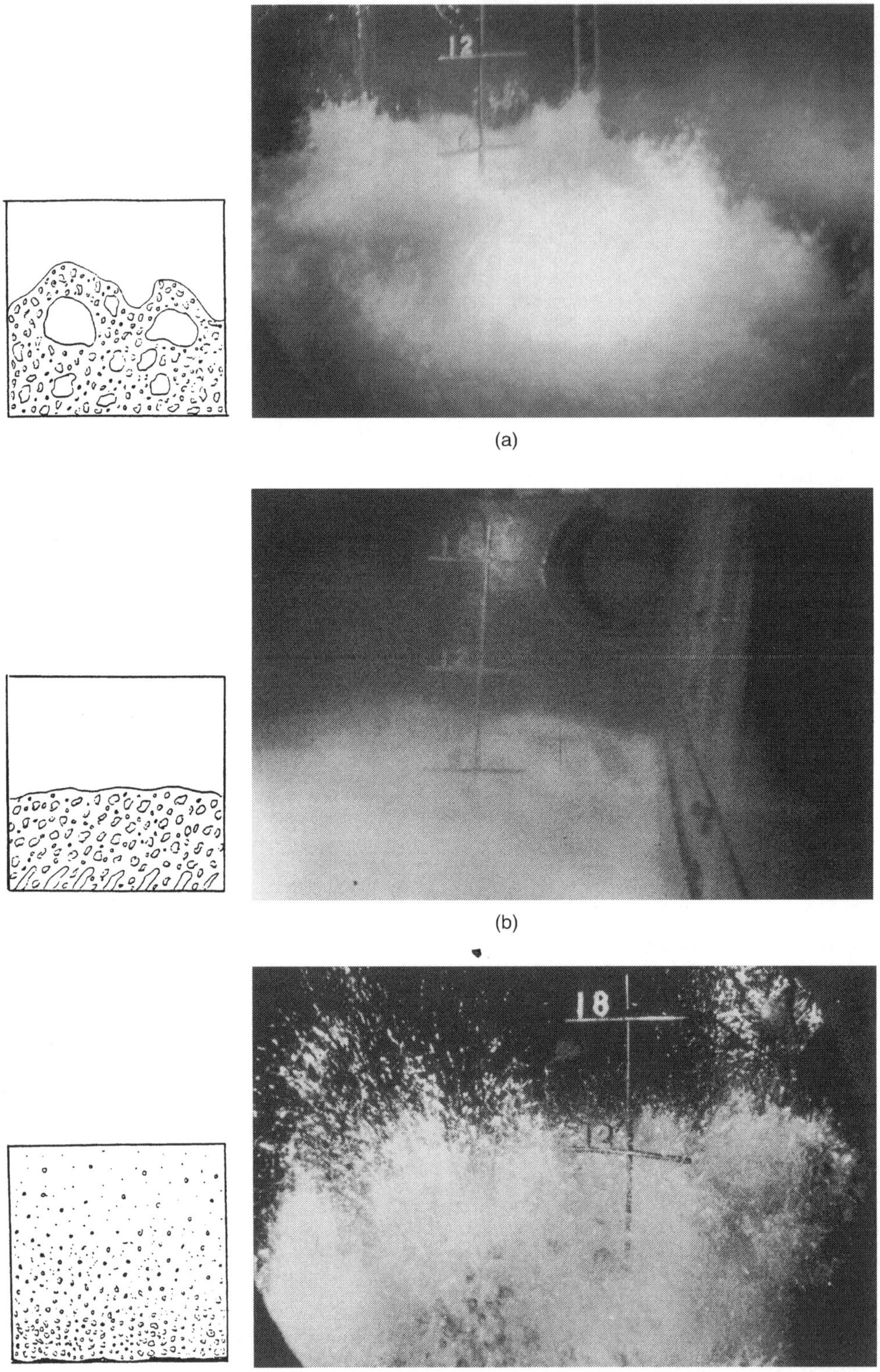

FIG. 14-20 Distillation flow regimes: schematics and photos. (*a*) Froth. (*b*) Emulsion. (*c*) Spray. [*Schematics from H. Z. Kister,* Distillation Design, *copyright © 1992 by McGraw-Hill, Inc.; reprinted by permission. Photographs courtesy of Fractionation Research Inc. (FRI).*]

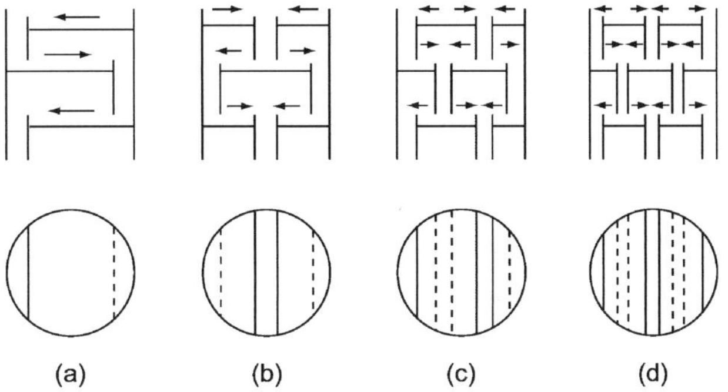

FIG. 14-21 Flow passes on trays. (*a*) Single-pass. (*b*) Two-pass. (*c*) Three-pass. (*d*) Four-pass.

The straight, segmental vertical downcomer (Fig. 14-23*a*) is the most common downcomer geometry. It is simple and inexpensive and gives good utilization of tower area for downflow. Circular downcomers (downpipes) (Fig. 14-23*b*), are cheaper, but poorly utilize tower area and are only suitable for very low liquid loads. Sloped downcomers (Fig. 14-23*c, d*) improve tower area utilization for downflow. They provide sufficient area and volume for gas-liquid disengagement at the top of the downcomer, gradually narrowing as the gas disengages, minimizing the loss of bubbling area at the foot of the downcomer. Sloped downcomers are invaluable when large downcomers are required such as at high liquid loads, high pressures, and foaming systems. Typical ratios of downcomer top to bottom areas are 1.5 to 2.

Antijump baffles (Fig. 14-24) are sometimes installed just above center and off-center downcomers of multipass trays to prevent liquid from one pass skipping across the downcomer onto the next pass. Such liquid jump adds to the liquid load on each pass, leading to premature flooding. These baffles are essential with proprietary trays that induce forward push (see below).

Clearance under the Downcomer Restricting the downcomer bottom opening prevents gas from the tray from rising up the downcomer and interfering with its liquid descent (*downcomer unsealing*). A common design practice makes the downcomer clearance 13 mm (0.5 in) lower than the outlet weir height (Fig. 14-25) to ensure submergence at all times [Davies and Gordon, *Petro/Chem Eng.*, p. 250 (November 1961)]. This practice is sound in the froth and emulsion regimes, where tray dispersions are liquid-continuous, but is ineffective in the spray regime where tray dispersions are gas-continuous and there is no submergence. Also, this practice can be unnecessarily restrictive at high liquid loads where high crests over the weirs sufficiently protect the downcomers from gas rise. Generally, downcomer clearances in the spray regime need to be smaller, while those in the emulsion regime can be larger, than those set by the above practice. Seal pans and inlet weirs are devices sometimes used to help with downcomer sealing while keeping downcomer clearances large. Details are in Kister's book (*Distillation Operation,* McGraw-Hill, New York, 1990).

Hole Sizes Small holes slightly enhance tray capacity when limited by entrainment flood. Reducing sieve hole diameters from 13 to 5 mm ($\frac{1}{2}$ to $\frac{3}{16}$ in) at a fixed hole area typically enhances capacity by 3 to 8 percent, more at low liquid loads. Small holes are effective for reducing entrainment and enhancing capacity in the spray regime (Q_L < 20 m³/hm of weir). Hole diameter has only a small effect on pressure drop, tray efficiency, and turndown.

On the debit side, the plugging tendency increases exponentially as hole diameters diminish. Smaller holes are also more prone to corrosion. While 5-mm ($\frac{3}{16}$-in) holes easily plug even by scale and rust, 13-mm ($\frac{1}{2}$-in) holes are quite robust and are therefore very common. The small holes are only used in clean, noncorrosive services. Holes smaller than 5 mm are usually avoided because they require drilling (larger holes are punched), which is much more expensive. For highly fouling services, 19- to 25-mm ($\frac{3}{4}$- to 1-in) holes are preferred.

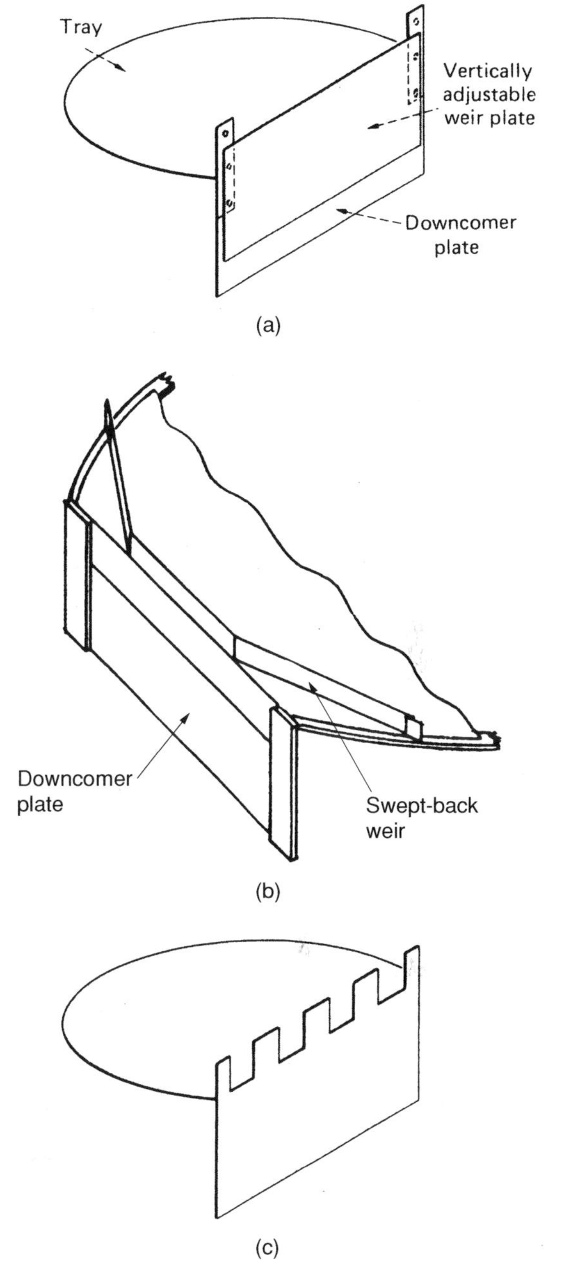

FIG. 14-22 Unique outlet weir types. (*a*) Adjustable. (*b*) Swept back. (*c*) Picket fence. (*Parts a, c, from H. Z. Kister,* Distillation Operation, *copyright © 1990 by McGraw-Hill; reprinted by permission. Part b, courtesy of Koch-Glitsch LP.*)

Similar considerations apply to fixed valves. Small fixed valves have a slight capacity advantage, but are far more prone to plugging than larger fixed valves.

For round moving valves, common orifice size is 39 mm (1$^{17}\!/_{32}$ in). The float opening is usually of the order of 8 to 10 mm (0.3 to 0.4 in).

In recent years there has been a trend toward minivalves, both fixed and moving. These are smaller and therefore give a slight capacity advantage while being more prone to plugging.

Fractional Hole Area Typical sieve and fixed valve tray hole areas are 8 to 12 percent of the bubbling areas. Smaller fractional hole

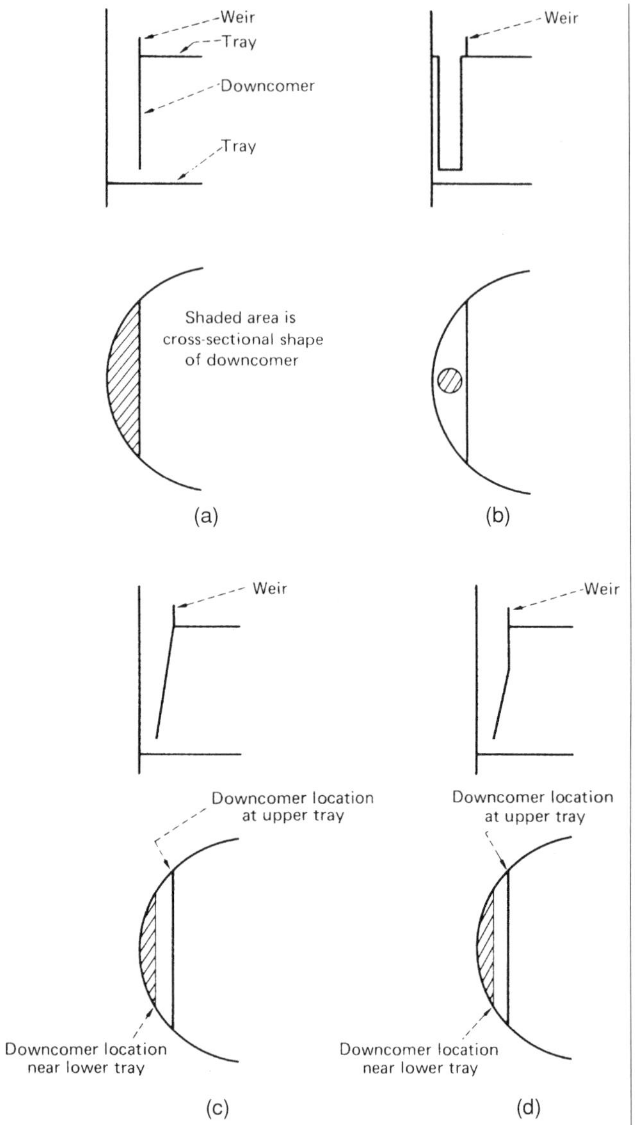

FIG. 14-23 Common downcomer types. (*a*) Segmental. (*b*) Circular. (*c, d*) Sloped. (*From Henry Z. Kister,* Chem. Eng., *December 29, 1980; reprinted courtesy of* Chemical Engineering.)

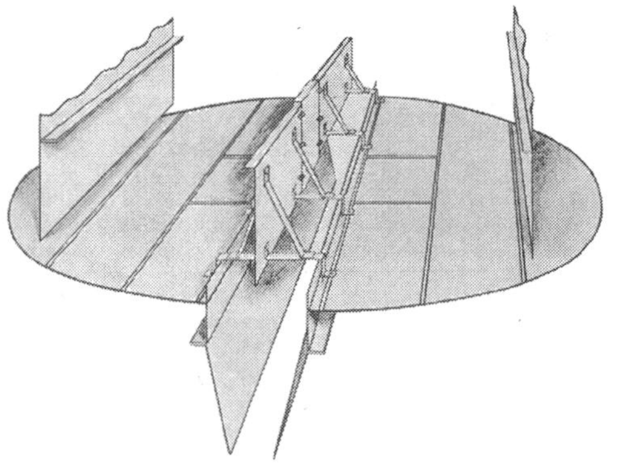

FIG. 14-24 Antijump baffle. (*Reprinted courtesy of Koch-Glitsch LP.*)

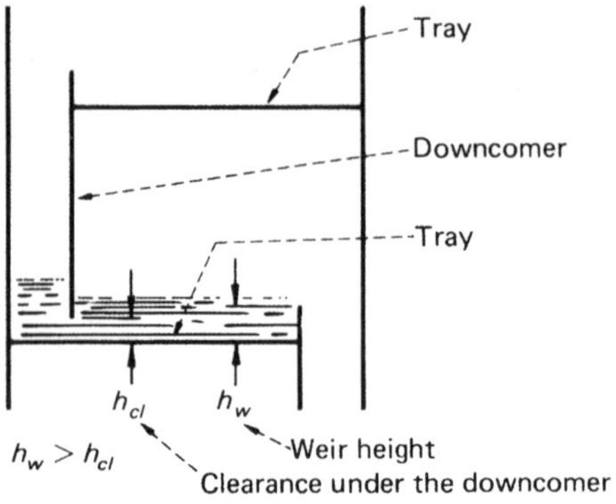

FIG. 14-25 A common design practice of ensuring a positive downcomer seal. (*From Henry Z. Kister,* Chem. Eng., *December 29, 1980; reprinted courtesy of* Chemical Engineering.)

areas bring about a capacity reduction when limited by entrainment or downcomer backup flood or by excessive pressure drop. At above 12 percent of the bubbling areas, the capacity gains from higher hole areas become marginal while weeping and, at high liquid loads also channeling, escalate.

Typical open-slot areas for moving valve trays are 14 to 15 percent of the bubbling area. Here the higher hole areas can be afforded due to the high turndown of the valves.

Moving valves can have a sharp or a smooth ("venturi") orifice. The venturi valves have one-half the dry pressure drop of the sharp-orifice valves, but are far more prone to weeping and channeling than the sharp-orifice valves. Sharp orifices are almost always preferred.

Multipass Balancing There are two balancing philosophies: equal bubbling areas and equal flow path lengths. Equal bubbling areas means that all active area panels on Fig. 14-21*d* are of the same area, and each panel has the same hole (or open-slot) area. In a four-pass tray, one-quarter of the gas flows through each panel. To equalize the L/G ratio on each panel, the liquid needs to be split equally to each panel. Since the center weirs are longer than the side weirs, more liquid tends to flow toward the center weir. To equalize, side weirs are often swept back (Fig. 14-22*b*) while center weirs often contain picket fences (Fig. 14-22*c*).

The alternative philosophy (equal flow path lengths) provides more bubbling and perforation areas in the central panels of Fig. 14-21*d* and less in the side panels. To equalize the L/G ratio, less liquid needs to flow toward the sides, which is readily achieved, as the center weirs are naturally longer than the side weirs. Usually there is no need for swept-back weirs, and only minimal picket-fencing is required at the center weir.

Equal flow path panels are easier to fabricate and are cheaper, while equal bubbling areas have a robustness and reliability advantage due to the ease of equally splitting the fluids. The author had good experience with both when well-designed. Pass balancing is discussed in detail by Pilling [*Chem. Eng. Prog.*, p. 22 (June 2005)] and by Jaguste and Kelkar [*Hydroc. Proc.*, p. 85 (March 2006)].

TRAY CAPACITY ENHANCEMENT

High-capacity trays evolved from conventional trays by including one or more capacity enhancement features such as those discussed below. These features enhance not only the capacity but usually also the complexity and cost. These features have varying impact on the efficiency, turndown, plugging resistance, pressure drop, and reliability of the trays.

Truncated Downcomers/Forward Push Trays Truncated downcomers/forward push trays include the Nye™ Tray, Maxfrac™ (Fig. 14-26*a*), Triton™, and MVGT™. In all these, the downcomer from the tray above terminates about 100 to 150 mm (4 to 6 in) above the tray floor. Liquid from the downcomer issues via holes or slots,

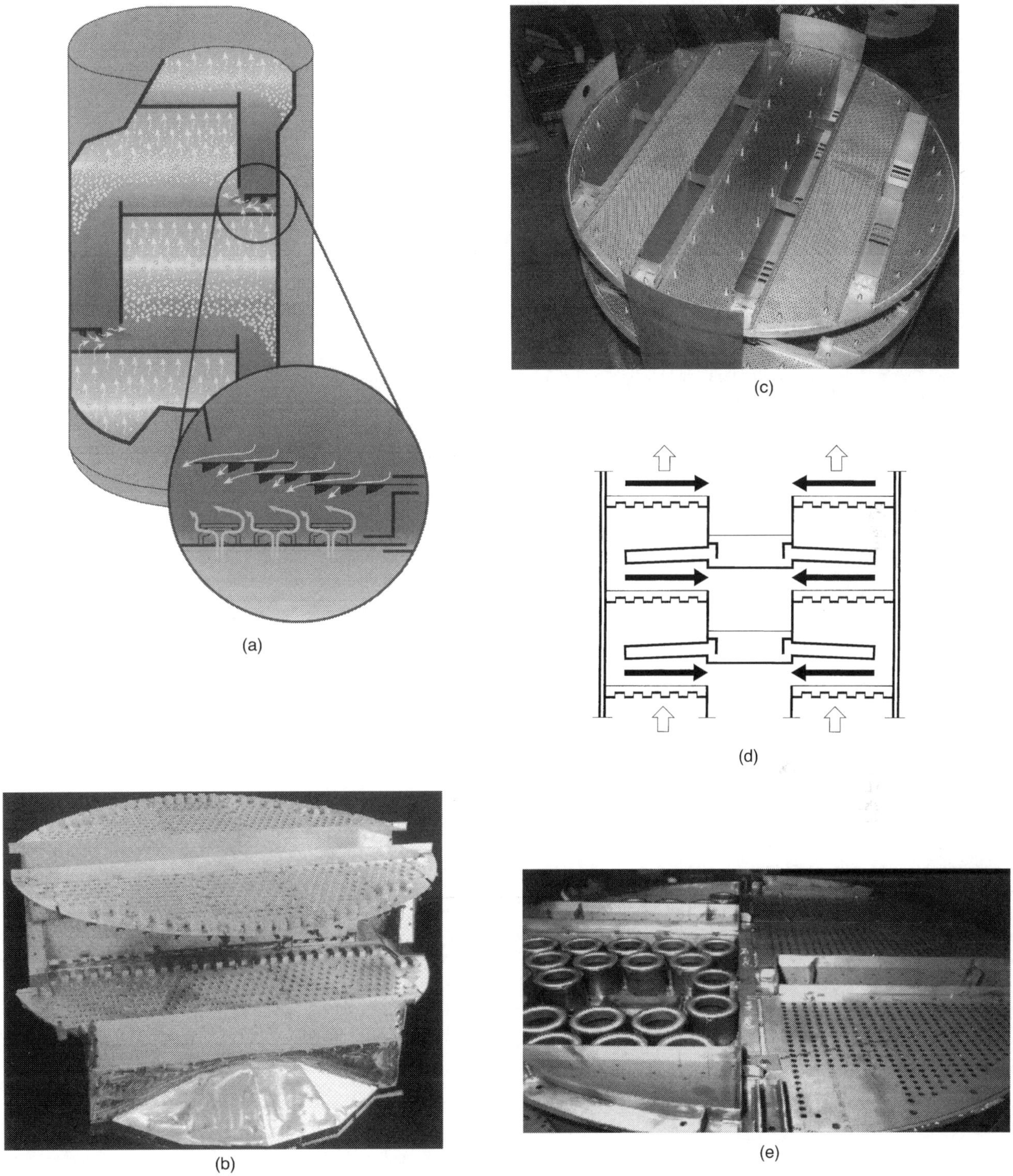

FIG. 14-26 Tray capacity enhancement. (*a*) Truncated downcomer/forward-push principle illustrated with a schematic of the Maxfrac™ tray. (*b*) High top-to-bottom area ratio illustrated with a two-pass Superfrac™ tray. Note the baffle in the front side downcomer that changes the side downcomer shape from segmental to multichordal. Also note the bubble promoters on the side of the upper tray and in the center of the lower tray, which give forward push to the tray liquid. (*c*) Top view of an MD™ tray with four downcomers. The decks are perforated. The holes in the downcomer lead the liquid to the active area of the tray below, which is rotated 90°. (*d*) Schematic of the Slit™ tray, type A, showing distribution pipes. Heavy arrows depict liquid movement; open arrows, gas movement. (*e*) The ConSep™ tray. The right-hand side shows sieve panels. On the left-hand side, these sieve panels were removed to permit viewing the contact cyclones that catch the liquid from the tray below. (*Parts a, b, courtesy of Koch-Glitsch LP; part c, courtesy of UOP LLC; part d, courtesy of Kühni AG; part e, courtesy of Sulzer Chemtech Ltd. and Shell Global Solutions International BV.*)

directed downward or in the direction of liquid flow. The tray floor under each downcomer is equipped with fixed valves or side perforations. Gas issuing in this region, typically 10 to 20 percent of the total tray gas, is deflected horizontally in the direction of liquid flow by the downcomer floor. This horizontal gas flow pushes liquid droplets toward the tower wall directly above the outlet downcomer. The tower wall catches this liquid, and directs it downward into the downcomer. This deentrains the gas space. In multipass trays, antijump baffles (Fig. 14-24), typically 300 mm or taller, are installed above center and off-center downcomers to catch the liquid and prevent its jumping from pass to pass. The rest of the tray features are similar to those of conventional trays. The tray floor may contain fixed valves, moving valves, or sieve holes.

Trays from this family are proprietary, and have been extensively used in the last two to three decades with great success. Compared to equivalent conventional trays, the truncated downcomer/forward push trays give about 8 to 12 percent more gas-handling capacity at much the same efficiency.

High Top-to-Bottom Downcomer Area and Forward Push Sloping downcomers from top to bottom raises the available tray bubbling area and, therefore, the gas-handling capacity (see "Downcomers"). As long as the ratio of top to bottom areas is not excessive, sloping does not lower downcomer capacity. Downcomer choke flood restricts the downcomer entrance, not exit, because there is much less gas at the downcomer bottom. However, a high top-to-bottom area ratio makes the downcomer bottom a very short chord, which makes distribution of liquid to the tray below difficult. To permit high top-to-bottom area ratios, some trays use a special structure (Fig. 14-26*b*) to change the downcomer shape from segmental to semiarc or multichordal. This high ratio of top to bottom areas, combined with forward push (above) imparted by bubblers and directional fixed or moving valves, and sometimes directional baffles, is used in trays including Superfrac™ III (Fig. 14-26*b*) and IV and V-Grid Plus™. When the downcomer inlet areas are large, these trays typically gain 15 to 20 percent capacity compared to equivalent conventional trays at much the same efficiency. Trays from this family are proprietary, and have been used successfully for about a decade.

Large Number of Truncated Downcomers These include the MD™ (Fig. 14-26*c*) and Hi-Fi™ trays. The large number of downcomers raises the total weir length, moving tray operation toward the peak capacity point of 20 to 30 m^3/hm (2 to 3 gpm/in) of outlet weir (see Fig. 14-29). The truncated downcomers extend about halfway to the tray below, discharging their liquid via holes or slots at the downcomer floor. The area directly under the downcomers is perforated or valved, and there is enough open height between the tray floor and the bottom of the downcomer for this perforated or valved area to be effective in enhancing the tray bubbling area.

Trays from this family are proprietary and have been successfully used for almost four decades. Their strength is in high-liquid-load services where reducing weir loads provides major capacity gains. Compared to conventional trays, they can gain as much as 20 to 30 percent capacity but at an efficiency loss. The efficiency loss is of the order of 10 to 20 percent due to the large reduction in flow path length (see "Efficiency"). When using these trays, the separation is maintained by either using more trays (typically at shorter spacing) or raising reflux and boilup. This lowers the net capacity gains to 10 to 20 percent above conventional trays. In some variations, forward push slots and antijump baffles are incorporated to enhance the capacity by another 10 percent.

Radial Trays These include the Slit™ tray and feature radial flow of liquid. In the efficiency-maximizing A variation (Fig. 14-26*d*), a multipipe distributor conducts liquid from each center downcomer to the periphery of the tray below, so liquid flow is from periphery to center on each tray. The capacity-maximizing B variation has central and peripheral (ring) downcomers on alternate trays, with liquid flow alternating from center-to-periphery to periphery-to-center on successive trays. The trays are arranged at small spacing (typically, 200 to 250 mm, or 8 to 10 in) and contain small fixed valves. Slit trays are used in chemical and pharmaceutical low-liquid-rate applications (<40 m^3/hm or 4 gpm/in of outlet weir), typically at pressures ranging from moderate vacuum to slight superatmospheric.

Centrifugal Force Deentrainment These trays use a contact step similar to that in conventional trays, followed by a separation step that disentrains the tray dispersion by using centrifugal force. Separation of entrained liquid before the next tray allows very high gas velocities, as high as 25 percent above the system limit (see "System Limit"), to be achieved. The capacity of these trays can be 40 percent above that of conventional trays. The efficiency of these trays can be 10 to 20 percent less than that of conventional trays due to their typical short flow paths (see "Efficiency").

These trays include the Ultrfrac™, the ConSep™ (Fig. 14-26*e*), and the Swirl Tube™ trays. This technology has been sporadically used in eastern Europe for quite some time. It is just beginning to make inroads into distillation in the rest of the world, and looks very promising.

OTHER TRAY TYPES

Bubble-Cap Trays (Fig. 14-27*a*) These are flat perforated plates with risers (chimneylike pipes) around the holes, and caps in the form of inverted cups over the risers. The caps are usually (but not always) equipped with slots through which some of the gas comes out, and may be round or rectangular. Liquid and froth are trapped on the tray to a depth at least equal to the riser or weir height, giving the bubble-cap tray a unique ability to operate at very low gas and liquid rates.

The bubble-cap tray was the workhorse of distillation before the 1960s. It was superseded by the much cheaper (as much as 10 times) sieve and valve trays. Compared to the bubble-cap trays, sieve and valve trays also offer slightly higher capacity and efficiency and lower entrainment and pressure drop, and are less prone to corrosion and fouling. Today, bubble-cap trays are only used in special applications where liquid or gas rates are very low. A large amount of information on bubble-cap trays is documented in several texts (e.g., Bolles in B. D. Smith, *Design of Equilibrium Stage Processes,* McGraw-Hill, 1963; Bolles, *Pet. Proc.,* February 1956, p. 65; March 1956, p. 82; April 1956, p. 72; May 1956, p. 109; Ludwig, *Applied Process Design for Chemical and Petrochemical Plants,* 2d ed., vol. 2, Gulf Publishing, Houston, 1979).

Dual-Flow Trays These are sieve trays with no downcomers (Fig. 14-27*b*). Liquid continuously weeps through the holes, hence their low efficiency. At peak loads they are typically 5 to 10 percent less efficient than sieve or valve trays, but as the gas rate is reduced, the efficiency gap rapidly widens, giving poor turndown. The absence of downcomers gives dual-flow trays more area, and therefore greater capacity, less entrainment, and less pressure drop, than conventional trays. Their pressure drop is further reduced by their large fractional hole area (typically 18 to 30 percent of the tower area). However, this low pressure drop also renders dual-flow trays prone to gas and liquid maldistribution.

In general, gas and liquid flows pulsate, with a particular perforation passing both gas and liquid intermittently, but seldom simultaneously. In large-diameter (>2.5-m, or 8-ft) dual-flow trays, the pulsations sometimes develop into sloshing, instability, and vibrations. The Ripple Tray™ is a proprietary variation in which the tray floor is corrugated to minimize this instability.

With large holes (16 to 25 mm), these trays are some of the most fouling-resistant and corrosion-resistant devices in the industry. This defines their main application: highly fouling services, slurries, and corrosive services. Dual-flow trays are also the least expensive and easiest to install and maintain.

A wealth of information for the design and rating of dual-flow trays, much of it originating from FRI data, was published by Garcia and Fair [*Ind. Eng. Chem. Res.* **41**:1632 (2002)].

Baffle Trays Baffle trays ("shed decks," "shower decks") (Fig. 14-28*a*) are solid half-circle plates, sloped slightly in the direction of outlet flow, with weirs at the end. Gas contacts the liquid as it showers from the plate. This contact is inefficient, typically giving 30 to 40 percent of the efficiency of conventional trays. This limits their application mainly to heat-transfer and scrubbing services. The capacity is high and pressure drop is low due to the high open area (typically 50 percent of the tower cross-sectional area). Since there is not much

FIG. 14-27 Other trays. (*a*) Bubble-cap tray. (*b*) Dual-flow tray. [*Part a, courtesy of Koch-Glitsch LP; part b, courtesy of Fractionation Research Inc. (FRI).*]

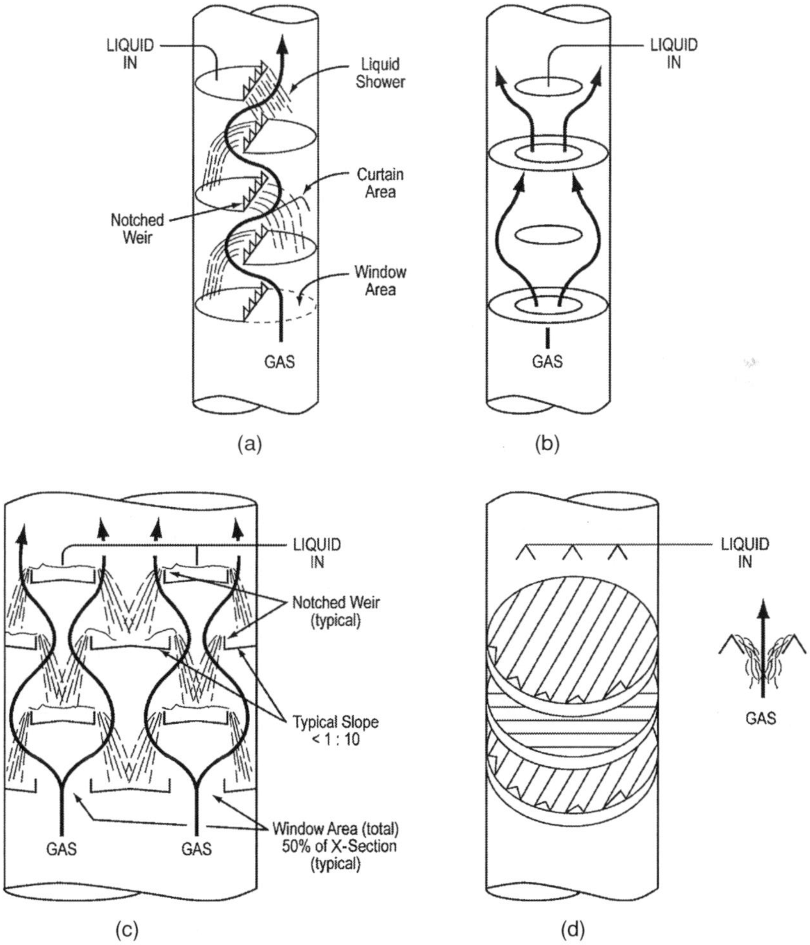

FIG. 14-28 Baffle tray variations. (*a*) Segmental. (*b*) Disk and doughnut. (*c*) Multipass. (*d*) Angle irons.

that can plug up, the baffle trays are perhaps the most fouling-resistant device in the industry, and their main application is in extremely fouling services. To be effective in these services, their liquid rate needs to exceed 20 m^3/hm (2 gpm/in) of outlet weir and dead spots formed due to poor support design (Kister, *Distillation Troubleshooting*, Wiley, 2006) eliminated.

There are several geometric variations. The disk and doughnut trays (Fig. 14-28*b*) replace the half-circle segmental plates by alternate plates shaped as disks and doughnuts, each occupying about 50 percent of the tower cross-sectional area. In large towers, multipass baffle trays (Fig. 14-28*c*) are common. Another variation uses angle irons, with one layer oriented at 90° to the one below (Fig. 14-28*d*). Multipass baffle trays, as well as angle irons, require good liquid (and to a lesser extent, also good gas) distribution, as has been demonstrated from field heat-transfer measurements [Kister and Schwartz, *Oil & Gas J.*, p. 50 (May 20, 2002)]. Excellent overviews of the fundamentals and design of baffle trays were given by Fair and Lemieux [Fair, *Hydro. Proc.*, p. 75 (May 1993); Lemieux, *Hydroc. Proc.*, p. 106 (September 1983)]. Mass-transfer efficiency data with baffle trays by Fractionation Research Inc. (FRI) have been released and presented together with their correlation (Fair, Paper presented at the AIChE Annual Meeting, San Francisco, November 2003).

FLOODING

Flooding is by far the most common upper capacity limit of a distillation tray. Column diameter is set to ensure the column can achieve the required throughput without flooding. Towers are usually designed to operate at 80 to 90 percent of the flood limit.

Flooding is an excessive accumulation of liquid inside a column. Flood symptoms include a rapid rise in pressure drop (the accumulating liquid increases the liquid head on the trays), liquid carryover from the column top, reduction in bottom flow rate (the accumulating liquid does not reach the tower bottom), and instability (accumulation is non-steady-state). This liquid accumulation is generally induced by one of the following mechanisms.

Entrainment (Jet) Flooding Froth or spray height rises with gas velocity. As the froth or spray approaches the tray above, some of the liquid is aspirated into the tray above as entrainment. Upon a further increase in gas flow rate, massive entrainment of the froth or spray begins, causing liquid accumulation and flood on the tray above.

Entrainment flooding can be subclassified into *spray entrainment flooding* (common) and *froth entrainment flooding* (uncommon). Froth entrainment flooding occurs when the froth envelope approaches the tray above, and is therefore only encountered with small tray spacings (<450 mm or 18 in) in the froth regime. At larger (and often even lower) tray spacing, the froth breaks into spray well before the froth envelope approaches the tray above.

The entrainment flooding prediction methods described here are based primarily on spray entrainment flooding. Considerations unique to froth entrainment flooding can be found elsewhere (Kister, *Distillation Design*, McGraw-Hill, New York, 1992).

Spray Entrainment Flooding Prediction Most entrainment flooding prediction methods derive from the original work of Souders and Brown [*Ind. Eng. Chem.* **26**(1), 98 (1934)]. Souders and Brown theoretically analyzed entrainment flooding in terms of droplet settling velocity. Flooding occurs when the upward vapor velocity is high enough to suspend a liquid droplet, giving

$$C_{SB} = u_{S,\text{flood}}\sqrt{\frac{\rho_G}{\rho_L - \rho_G}} \qquad (14\text{-}80)$$

The Souders and Brown constant C_{SB} is the C-factor [Eq. (14-77)] at the entrainment flood point. Most modern entrainment flooding correlations retain the Souders and Brown equation (14-80) as the basis, but depart from the notion that C_{SB} is a constant. Instead, they express C_{SB} as a weak function of several variables, which differ from one correlation to another. Depending on the correlation, C_{SB} and $u_{S,\text{flood}}$ are based on either the net area A_N or on the bubbling area A_B.

The constant C_{SB} is roughly proportional to the tray spacing to a power of 0.5 to 0.6 (Kister, *Distillation Design*, McGraw-Hill, New York, 1992). Figure 14-29 demonstrates the effect of liquid rate and fractional hole area on C_{SB}. As liquid load increases, C_{SB} first increases, then peaks, and finally declines. Some interpret the peak as the transition from the froth to spray regime [Porter and Jenkins, *I. Chem. E. Symp. Ser.* **56,** Summary Paper, London (1979)]. C_{SB} increases slightly with fractional hole area at lower liquid rates, but there is little effect of fractional hole area on C_{SB} at high liquid rates. C_{SB}, slightly increases as hole diameter is reduced.

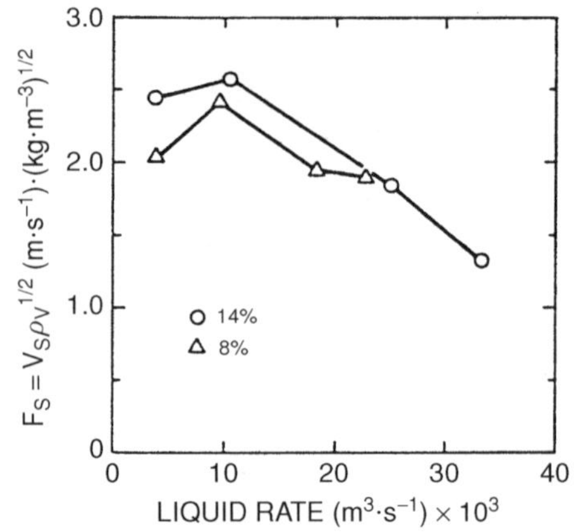

FIG. 14-29 Effect of liquid rate and fractional hole area on flood capacity. FRI sieve tray test data, cyclohexane/*n*-heptane, 165 kPa (24 psia), D_T = 1.2 m (4 ft), S = 610 mm (24 in), h_w = 51 mm (2 in), d_H = 12.7 mm (0.5 in), straight downcomers, A_D/A_T = 0.13. *(From T. Yanagi and M. Sakata,* Ind. Eng. Chem. Proc. Des. Dev. ***21,*** 712; copyright © 1982, *American Chemical Society, reprinted by permission.)*

For sieve trays, the entrainment flood point can be predicted by using the method by Kister and Haas [*Chem. Eng. Progr.*, **86**(9), 63 (1990)]. The method is said to reproduce a large database of measured flood points to within ± 15 percent. $C_{SB,\text{flood}}$ is based on the net area. The equation is

$$C_{SB,\text{flood}} = 0.0277(d_h^2\sigma/\rho_L)^{0.125}(\rho_G/\rho_L)^{0.1}(TS/h_{ct})^{0.5} \qquad (14\text{-}81)$$

where d_h = hole diameter, mm
σ = surface tension, mN/m (dyn/cm)
ρ_G, ρ_L = vapor and liquid densities, kg/m^3
TS = tray spacing, mm
h_{ct} = clear liquid height at the froth-to-spray transition, mm; obtained from:

$$h_{ct} = h_{ct,H_2O}\,(996/\rho_L)^{0.5(1-n)} \qquad (14\text{-}82)$$

$$h_{ct,H_2O} = \frac{0.497\,A_f^{-0.791}d_h^{\,0.833}}{1 + 0.013\,Q_L^{-0.59}A_f^{-1.79}} \qquad (14\text{-}83)$$

$$n = 0.00091 d_h/A_f \qquad (14\text{-}84)$$

In Eq. (14-83), Q_L = m^3 liquid downflow/(h·m weir length) and A_f = fractional hole area based on active ("bubbling") area; for instance, $A_f = A_h/A_a$.

The Kister and Haas method can also be applied to valve trays, but the additional approximations reduce its data prediction accuracy for valve trays to within ±20 percent. For valve trays, adaptations of Eqs. (14-81) to (14-84) are required:

$$d_h = \frac{4 \times (\text{area of opening of one fully open valve})}{\text{wetted perimeter of opening of one fully open valve}} \qquad (14\text{-}85)$$

$$A_f = \frac{\text{no. valves} \times (\text{area of opening of one fully open valve})}{\text{active (bubbling) area}} \qquad (14\text{-}86)$$

A correlation for valve tray entrainment flooding that has gained respect and popularity throughout the industry is the Glitsch "Equation 13" (Glitsch, Inc., *Ballast Tray Design Manual,* 6th ed., 1993;

available from Koch-Glitsch, Wichita, Kans.). This equation has been applied successfully for valve trays from different manufacturers, as well as for sieve trays with large fractional hole areas (12 to 15 percent). With tray spacings of 600 mm and higher, its flood prediction accuracy for valve trays has generally been within ±10 percent in the author's experience. The Glitsch correlation is

$$\frac{\%\ \text{flood}}{100} = \frac{C_B}{\text{CAF}} + 1.359\frac{Q\ \text{FPL}}{A_B\ \text{CAF}} \tag{14-87}$$

where

$$\text{CAF} = 0.3048\ \text{CAF}_0\ \text{SF} \tag{14-88}$$

C_B is the operating C-factor based on the bubbling area, m/s; Q is the liquid flow rate, m³/s; A_B is the bubbling area, m²; FPL is the flow path length, m, i.e., the horizontal distance between the inlet downcomer and the outlet weir. The flow path length becomes shorter as the number of passes increases. CAF_0 and CAF are the flood C-factors. CAF_0 is obtained from Fig. 14-30 in English units (ft/s). Equation (14-88) converts CAF_0 to the metric CAF (m/s), and corrects it by using a system factor SF. Values of SF are given in Table 14-9.

The Fair correlation [*Pet/Chem Eng.* **33**(10), 45 (September 1961)] for decades has been the standard of the industry for entrainment flood prediction. It uses a plot (Fig. 14-31) of surface-tension-corrected Souders and Brown flood factor C_{SB} against the dimensionless flow parameter shown in Fig. 14-31. The flow parameter represents a ratio of liquid to vapor kinetic energies:

$$F_{LG} = \frac{L}{G}\left(\frac{\rho_G}{\rho_L}\right)^{0.5} \tag{14-89}$$

Low values of F_{LG} indicate vacuum operation, high values indicate operation at higher pressures or at high liquid/vapor loadings. The liquid/gas ratio L/G is based on mass flow rates. For multipass trays, the ratio needs to be divided by the number of passes. The strength of the correlation is at the lower flow parameters. At higher flow parameters (high L/G ratios, high pressures, emulsion flow), Fig. 14-31 gives excessively conservative predictions, with the low values of C_{sbf} to the right likely to result from downcomer flow restrictions rather than excessive entrainment. The curves may be expressed in equation form as [Lygeros and Magoulas, *Hydrocarbon Proc.* **65**(12), 43 (1986)]:

$$C_{sbf} = 0.0105 + 8.127(10^{-4})(\text{TS}^{0.755})\exp[-1.463\ F_{LG}^{0.842}] \tag{14-90}$$

where TS = plate spacing, mm.

Figure 14-31 or Eq. (14-90) may be used for sieve, valve, or bubble-cap trays. The value of the capacity parameter (ordinate term in Fig. 14-31) may be used to calculate the maximum allowable vapor velocity through the net area of the plate:

$$U_{nf} = C_{sbf}\left(\frac{\sigma}{20}\right)^{0.2}\left(\frac{\rho_L - \rho_g}{\rho_g}\right)^{0.5} \tag{14-91}$$

where U_{nf} = gas velocity through net area at flood, m/s
C_{sbf} = capacity parameter corrected for surface tension, m/s

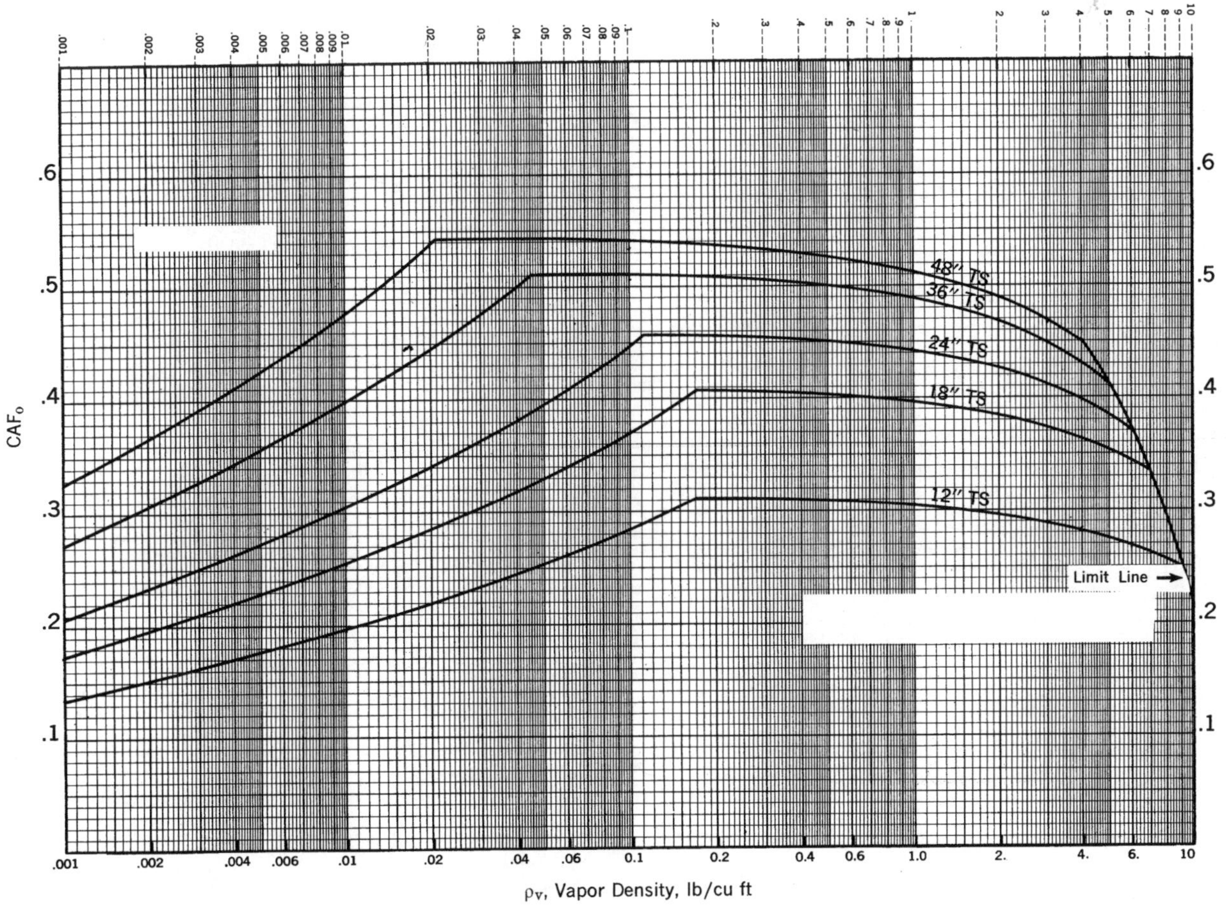

FIG. 14-30 Flood capacity of moving valve trays. (*Courtesy of Koch-Glitch LP.*)

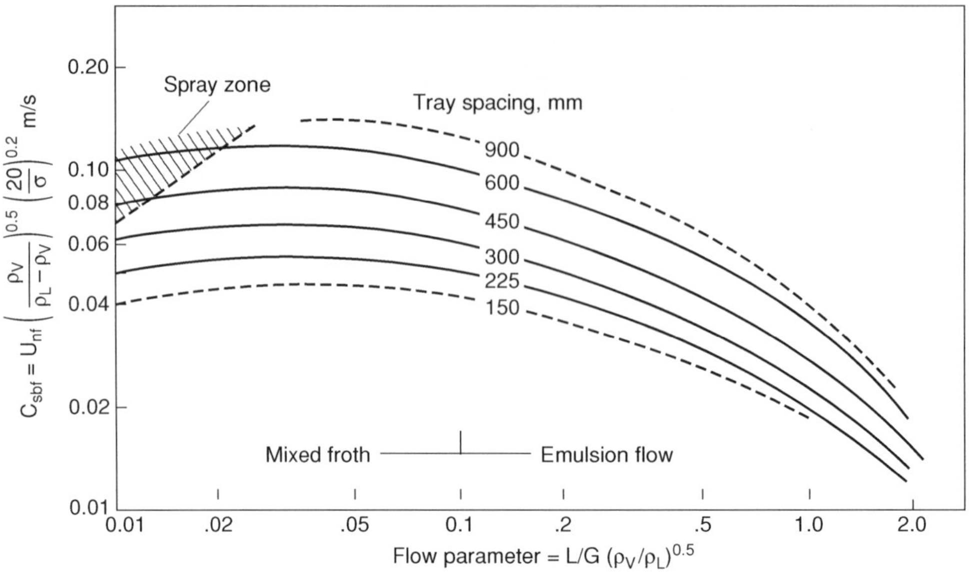

FIG. 14-31 Fair's entrainment flooding correlation for columns with crossflow trays (sieve, valve, bubble-cap). [*Fair, Pet/Chem Eng 33(10), 45 (September 1961).*]

σ = liquid surface tension, mN/m (dyn/cm)
ρ_ℓ = liquid density, kg/m^3
ρ_G = gas density, kg/m^3

The application of the correlation is subject to the following restrictions:

1. System is low or nonfoaming.
2. Weir height is less than 15 percent of tray spacing.
3. Sieve-tray perforations are 13 mm (½ in) or less in diameter.
4. Ratio of slot (bubble cap), perforation (sieve), or full valve opening (valve plate) area A_h to active area A_a is 0.1 or greater. Otherwise the value of U_{nf} obtained from Fig. 14-31 should be corrected:

A_h/A_a	$U_{nf}/U_{nf,\text{Fig. 14-31}}$
0.10	1.00
0.08	0.90
0.06	0.80

where A_h = total slot, perforated, or open-valve area on tray.

Example 9: Flooding of a Distillation Tray An available sieve tray column of 2.5-m diameter is being considered for an ethylbenzene/styrene separation. An evaluation of loading at the top tray will be made. Key dimensions of the single-pass tray are:

Column cross section, m^2	4.91
Downcomer area, m^2	0.25
Net area, m^2	4.66
Active area, m^2	4.41
Hole area, m^2	0.617
Hole diameter, mm	4.76
Weir length, m	1.50
Weir height, mm	38
Tray spacing, mm	500

Conditions and properties at the top tray are:

Temperature, °C	78
Pressure, torr	100
Vapor flow, kg/h	25,500
Vapor density, kg/m^3	0.481
Liquid flow, kg/h	22,000
Liquid density, kg/m^3	841
Surface tension, mN/m	25

Solution. The method of Kister and Haas gives:

$$Q_L = \frac{22{,}000}{841 \times 1.50} = 17.44 \text{ m}^3\text{/h·m weir}$$

$$A_f = \frac{0.617}{4.41} = 0.14$$

By Eq. (14-83), $h_{ct,H_2O} = 7.98$ mm

$$\text{Eq. (14-84): } n = 0.0309$$

$$\text{Eq. (14-82): } h_{ct} = 8.66 \text{ mm}$$

Finally, by Eq. (14-81),

$$C_{SB,\text{flood}} = 0.0277[(4.76^2)(25/841)]^{0.125} \times (0.481/841)^{0.1}(500/8.66)^{0.5}$$
$$= 0.0947 \text{ m/s}$$

Alternatively, applying the Fair correlation:

The flow parameter $F_{LG} = 0.021$ [Eq. (14-89)]. From Fig. 14-31, $C_{sbf} = 0.095$ m/s. Then, based on the net area,

$$C_{SB} = 0.095(25/20)^{0.2} = 0.0993 \text{ m/s}$$

about 5 percent higher than the answer obtained from Kister and Haas.

For the design condition, the C-factor based on the net area is

$$C = \frac{25{,}500}{3600(0.481)(4.66)}\sqrt{\frac{0.481}{841 - 0.481}} = 0.0756 \text{ m/s}$$

or about 80 percent of flood. The proposed column is entirely adequate for the service required.

System Limit (Ultimate Capacity) This limit is discussed later under "System Limit."

Downcomer Backup Flooding Aerated liquid backs up in the downcomer because of tray pressure drop, liquid height on the tray, and frictional losses in the downcomer apron (Fig. 14-32). All these increase with increasing liquid rate. Tray pressure drop also increases as the gas rate rises. When the backup of aerated liquid exceeds the tray spacing, liquid accumulates on the tray above, causing downcomer backup flooding.

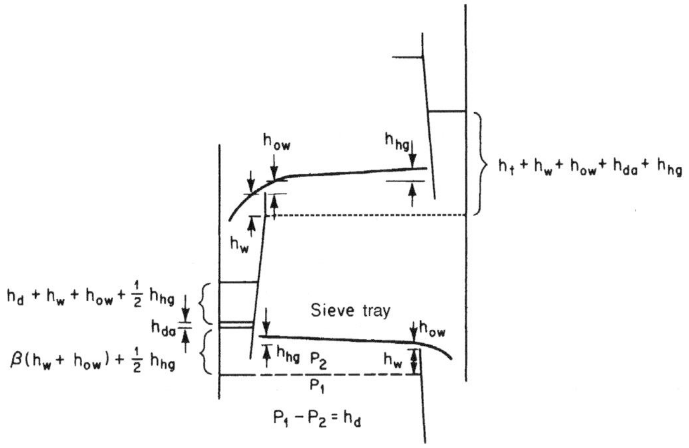

FIG. 14-32 Pressure-drop contributions for trays. h_d = pressure drop through cap or sieve, equivalent height of tray liquid; h_w = height of weir; h_{ow} = weir crest; h_{hg} = hydraulic gradient; h_{da} = loss under downcomer.

Downcomer backup is calculated from the pressure balance

$$h_{dc} = h_t + h_w + h_{ow} + h_{da} + h_{hg} \qquad (14\text{-}92)$$

where h_{dc} = clear liquid height in downcomer, mm liquid
h_t = total pressure drop across the tray, mm liquid
h_w = height of weir at tray outlet, mm liquid
h_{ow} = height of crest over weir, mm liquid
h_{da} = head loss due to liquid flow under downcomer apron, mm liquid
h_{hg} = liquid gradient across tray, mm liquid

The heights of head losses in Eq. (14-92) should be in consistent units, e.g., millimeters or inches of liquid under operating conditions on the tray.

As noted, h_{dc} is calculated in terms of equivalent clear liquid. Actually, the liquid in the downcomer is aerated and actual backup is

$$h'_{dc} = \frac{h_{dc}}{\phi_{dc}} \qquad (14\text{-}93)$$

where ϕ_{dc} is an average *relative* froth density (ratio of froth density to liquid density) in the downcomer. Design must not permit h'_{dc} to exceed the value of tray spacing plus weir height; otherwise, flooding can be precipitated.

The value of ϕ_{dc} depends upon the tendency for gas and liquid to disengage (froth to collapse) in the downcomer. For cases favoring rapid bubble rise (low gas density, low liquid viscosity, low system foamability) collapse is rapid, and fairly clear liquid fills the bottom of the downcomer (Fig. 14-17). For such cases, it is usual practice to employ a higher value of ϕ_{dc}. For cases favoring slow bubble rise (high gas density, high liquid viscosity, high system foamability), lower values of ϕ_{dc} should be used. As the critical point is approached in high-pressure distillations and absorptions, special precautions with downcomer sizing are mandatory. Table 14-6 lists values of ϕ_{dc} commonly used by the industry.

Downcomer Choke Flooding This is also called downcomer entrance flood or downcomer velocity flood. A downcomer must be sufficiently large to transport all the liquid downflow. Excessive friction losses in the downcomer entrance, and/or excessive flow rate of gas venting from the downcomer in counterflow, will impede liquid downflow, initiating liquid accumulation (termed downcomer choke flooding) on the tray above. The prime design parameter is the downcomer top area. Further down the downcomer, gas disengages from the liquid and the volumes of aerated liquid downflow and vented gas

TABLE 14-6 Criteria for Downcomer Aeration Factors

Foaming tendency	Bolles' criterion*		Glitsch's criterion†		Fair et al.'s criterion‡	
	Examples	ϕ_{dc}	Examples	ϕ_{dc}	Examples	ϕ_{dc}
Low	Low-molecular-weight hydrocarbons§ and alcohols	0.6	$\rho_G < 1.0$ lb/ft³	0.6	Rapid bubble rise systems, such as low gas density, low liquid viscosity	0.5
Moderate	Distillation of medium-molecular-weight hydrocarbons	0.5	$1.0 < \rho_G < 3.0$ lb/ft³	0.5		
High	Mineral oil absorbers	0.4	$\rho_G > 3.0$ lb/ft³	0.4		
Very high	Amines, glycols	0.3			Slow bubble rise systems, such as high gas density, high liquid viscosity, foaming systems	0.2–0.3

*"Distillation Theory and Practice—an Intensive Course," University of New South Wales/University of Sydney, August 9–11, 1977.

† Glitsch, Inc., *Ballast Tray Design Manual*, 6th ed., 1993; available from Koch-Glitsch LP, Wichita, Kans.

‡ R. H. Perry and D. W. Green (eds.), *Perry's Chemical Engineers' Handbook*, 7th ed., McGraw-Hill, 1997.

§ The author believes that low-molecular-weight hydrocarbons refers to light hydrocarbons at near atmospheric pressure or under vacuum. The foam stability of light-hydrocarbon distillation at medium and high pressure is best inferred from the Glitsch criterion.

To convert from lb/ft³, to kg/m³, multiply by 16.0.

SOURCE: From H. Z. Kister, *Distillation Design*, copyright © 1992 by McGraw-Hill, Inc.; reprinted by permission.

upflow are greatly reduced. With sloped downcomers, the downcomer bottom area is normally set at 50 to 60 percent of the top downcomer area. This taper is small enough to keep the downcomer top area the prime choke variable.

There is no satisfactory published correlation for downcomer choke. The best that can be done in the absence of data or correlation is to apply the criteria for maximum velocity of clear liquid at the downcomer entrance. Kister *(Distillation Operation,* McGraw-Hill, New York, 1990) surveyed the multitude of published criteria for maximum downcomer velocity and incorporated them into a single set of guidelines (Table 14-7). The values for 30-in spacing were revised to reflect the author's recent experiences. The values given in Table 14-7 are not conservative. For a conservative design, multiply the values from Table 14-7 by a safety factor of 0.75. For very highly foaming systems, where antifoam application is undesirable, there are benefits for reducing downcomer design velocities down to 0.1 to 0.15 ft/s.

Another criterion sometimes used is to provide sufficient residence time in the downcomer to allow adequate disengagement of gas from the descending liquid, so that the liquid is relatively gas-free by the time it enters the tray below. Inadequate removal of gas from the liquid may choke the downcomer. Kister (loc. cit.) reviewed various published criteria for downcomer residence times and recommended those by Bolles (private communication, 1977) and Erbar and Maddox (Maddox, *Process Engineer's Absorption Pocket Handbook,* Gulf Publishing, Houston, 1985). Both sets of guidelines are similar and are summarized in Table 14-8. The residence times in Table 14-8 are *apparent residence times*, defined as the ratio of the total downcomer volume to the clear liquid flow in the downcomer.

As a segmental downcomer becomes smaller, its width decreases faster than its length, turning the downcomer into a long, narrow slot. This geometry increases the resistance to liquid downflow and to the upflow of disengaging gas. Small downcomers are also extremely sensitive to foaming, fouling, construction tolerances, and the introduction of debris. Generally, segmental downcomers smaller than 5 percent of the column cross-sectional area should be avoided. Additional discussion of small downcomers is available (Kister, *Distillation Operation,* McGraw-Hill, New York, 1990).

Derating ("System") Factors With certain systems, traditional flooding equations consistently give optimistic predictions. To allow for this discrepancy, an empirical derating or system factor (SF < 1.0) is applied. To obtain the actual or derated flood load, the flood gas load (entrainment flooding) or flood liquid load (downcomer choke) obtained from the traditional equations is multiplied by the derating factor. In the case of downcomer backup flood, the froth height from the traditional flood equation is divided by the derating factor.

Derating factors are vaguely related to the foaming tendency, but are also applied to nonfoaming systems where standard flooding equations consistently predict too high. Sometimes, derating factors are used solely as overdesign factors. Brierley *(Chem. Eng. Prog.,* July 1994, p. 68) states that some derating factors actually evolved from plant misoperation or from misinterpretation of plant data. Kister (loc. cit.) compiled the derating factors found in the literature into Table 14-9.

The application of derating factors is fraught with inconsistent practices and confusion. Caution is required. The following need to be carefully specified:

1. The flooding mechanism to which the derating factor applies (entrainment, downcomer backup, downcomer choke, or all these) must be specified.
2. Avoiding double derating. For instance, the values in Table 14-9 may apply with Eq. (14-81) because Eq. (14-81) does not take foaminess into account. However, they will double-derate a flood calculation that is made with a correlation or criteria that already take foaminess into account, such as the criteria for downcomer choke in Tables 14-7 and 14-8. Similarly, two different factors from Table 14-9 may apply to a single system; only one should be used.
3. Derating factors vary from source to source, and may depend on the correlation used as well as the system. For instance, some caustic wash applications have a track record of foaming more severely than other caustic wash applications (see note in Table 14-9). The derating factors in Table 14-9 are a useful guide, but are far from absolute.

ENTRAINMENT

Entrainment (Fig. 14-33) is liquid transported by the gas to the tray above. As the lower tray liquid is richer with the less-volatile components, entrainment counteracts the mass-transfer process, reducing tray efficiency. At times entrainment may transport nonvolatile impurities upward to contaminate the tower overhead product, or damage rotating machinery located in the path of the overhead gas.

Effect of Gas Velocity Entrainment increases with gas velocity to a high power. Generally, smaller powers, indicative of a relatively gradual change, are typical of low-pressure systems. Higher powers, which indicate a steep change, are typical of high-pressure systems.

Due to the steep change of entrainment with gas velocity at high pressure, the gas velocity at which entrainment becomes significant tends to coincide with the flood point. At low pressure, the rate of change of entrainment with gas velocity is much slower, and entrainment can be significant even if the tray is operating well below the flood point. For this reason, excessive entrainment is a common problem in low-pressure and vacuum systems, but is seldom troublesome with high-pressure systems. If encountered at high pressure, entrainment usually indicates flooding or abnormality.

Effect of Liquid Rate As the liquid rate is raised at constant gas rate, entrainment first diminishes, then passes through a minimum, and finally increases [Sakata and Yanagi, *I. Chem. E. Symp. Ser.* **56,** 3.2/21 (1979); Porter and Jenkins, *I. Chem. E. Symp. Ser.* **56,** Summary Paper, 1979; Friend, Lemieux, and Schreiner, *Chem. Eng.,* October 31, 1960, p. 101]. The entrainment minima coincide with the maxima in plots of entrainment flood *F*-factor against liquid load (Fig. 14-29). At the low liquid loads (spray regime), an increase in liquid load suppresses atomization, drop formation, and consequently entrainment. At higher liquid loads, an increase in liquid load reduces the effective tray spacing, thereby increasing entrainment. The entrainment minima have been interpreted by many workers as the tray dispersion change from predominantly spray to the froth regime [Porter and Jenkins, loc. cit.; Kister and Haas, *I. Chem. E. Symp. Ser.* **104,** p. A483 (1987)].

Effect of Other Variables Entrainment diminishes with higher tray spacing and increases with hole diameter [Kister and Haas, *I.*

TABLE 14-7 Maximum Downcomer Velocities

Foaming tendency	Example	Clear liquid velocity in downcomer, ft/s: 18-in spacing	24-in spacing	30-in spacing
Low	Low-pressure (<100 psia) light hydrocarbons, stabilizers, air-water simulators	0.4–0.5	0.5–0.6	0.5–0.6*
Medium	Oil systems, crude oil distillation, absorbers, midpressure (100–300 psia) hydrocarbons	0.3–0.4	0.4–0.5	0.4–0.5*
High	Amines, glycerine, glycols, high-pressure (>300 psia) light hydrocarbons	0.2–0.25	0.2–0.25	0.2–0.3

*Revised from previous versions.

To convert from ft/s to m/s, multiply by 0.3048; from in to mm, multiply by 25.4; from psia to bar, multiply by 0.0689.

SOURCE: From H. Z. Kister, *Distillation Operation,* copyright 1990 by McGraw-Hill, Inc.; reprinted by permission.

TABLE 14-8 Recommended Minimum Residence Time in the Downcomer

Foaming tendency	Example	Residence time, s
Low	Low-molecular-weight hydrocarbons,* alcohols	3
Medium	Medium-molecular-weight hydrocarbons	4
High	Mineral oil absorbers	5
Very high	Amines and glycols	7

*The author believes that low-molecular-weight hydrocarbons refers to light hydrocarbons at atmospheric conditions or under vacuum. The foaming tendency of light-hydrocarbon distillation at medium pressure [>7 bar (100 psia)] is medium; at high pressure [>21 bar (300 psia)], it is high.

SOURCE: W. L. Bolles (Monsanto Company), private communication, 1977.

Chem. E. Symp. Ser. **104,** p. A483 (1987); *Ind. Eng. Chem. Res.* **27,** p. 2331 (1988); Lemieux and Scotti, *Chem. Eng. Prog.* **65**(3), 52 (1969)]. The hole diameter effect is large in the spray regime but small in the froth regime. In the spray regime, entrainment also increases as the fractional hole area is lowered, but this variable has little effect in the froth regime [Yanagi and Sakata, *Ind. Eng. Chem. Proc. Des. Dev.* **21,** 712 (1982); and Kister and Haas, loc. cit.].

Entrainment Prediction For spray regime entrainment, the Kister and Haas correlation was shown to give good predictions to a wide commercial and pilot-scale data bank [*I. Chem. E. Symp. Ser.* **104,** A483 (1987)]. The correlation is

$$E_S = 4.742^{(10/\sqrt{\sigma})^{1.64}} \chi^{(10/\sqrt{\sigma})} \tag{14-94}$$

where
$$\chi = 872\left(\frac{u_B h_{Lt}}{\sqrt{d_H S}}\right)^4 \left(\frac{\rho_G}{Q_L \rho_L}\right)\left(\frac{\rho_L - \rho_G}{\sigma}\right)^{0.25} \tag{14-95}$$

and
$$h_{Lt} = \frac{h_{ct}}{1 + 0.00262 h_w} \tag{14-96}$$

The terms in Eqs. (14-94) through (14-96) are in the metric units described in the Nomenclature table at the beginning of this section.

The recommended range of application of the correlation is given in Table 14-10. The clear liquid height at the froth-to-spray transition h_{ct} is calculated using the corrected Jeronimo and Sawistowski [*Trans. Inst. Chem. Engrs.* **51,** 265 (1973)] correlation as per Eqs. (14-82) to (14-84).

For decades, the Fair correlation [*Pet/Chem. Eng.*, **33**(10), 45 (September 1961)] has been used for entrainment prediction. In the spray regime the Kister and Haas correlation was shown to be more accurate [Koziol and Mackowiak, *Chem. Eng. Process.*, **27,** p. 145 (1990)]. In the froth regime, the Kister and Haas correlation does not apply, and Fair's correlation remains the standard of the industry. Fair's correlation (Fig. 14–34) predicts entrainment in terms of the flow parameter [Eq. (14-89)] and the ratio of gas velocity to entrainment flooding gas velocity. The ordinate values Ψ are fractions of gross liquid downflow, defined as follows:

$$\psi = \frac{e}{L_m + e} \tag{14-97}$$

where e = absolute entrainment of liquid, mol/time
L_m = liquid downflow rate without entrainment, mol/time

Figure 14-34 also accepts the validity of the Colburn equation [*Ind. Eng. Chem.*, **28,** 526 (1936)] for the effect of entrainment on efficiency:

$$\frac{E_a}{E_{mv}} = \frac{1}{1 + E_{mv}[\psi/(1-\psi)]} \tag{14-98}$$

where E_{mv} = Murphree vapor efficiency [see Eq. (14-134)]
E_a = Murphree vapor efficiency, corrected for recycle of liquid entrainment

The Colburn equation is based on complete mixing on the tray. For incomplete mixing, e.g., liquid approaching plug flow on the tray, Rahman and Lockett [*I. Chem. E. Symp. Ser. No. 61,* 111 (1981)] and Lockett et al. [*Chem. Eng. Sci.*, **38,** 661 (1983)] have provided corrections.

TABLE 14-9 Derating ("System") Factors

System	Factor	Reference	Notes
Nonfoaming regular systems	1.0	1–4	
High pressure (ρ_G >1.8 lb/ft^3)	2.94/ $\rho_G^{0.32}$	2	Do not double-derate.
Low-foaming			
Depropanizers	0.9	4	
H_2S strippers	0.9	3, 4	
	0.85	2	
Fluorine systems (freon, BF_3)	0.9	1, 4	
Hot carbonate regenerators	0.9	2, 4	
Moderate-foaming			
Deethanizers			
Absorbing type, top section	0.85	1–4	
Absorbing type, bottom section	1.0	3	
	0.85	1, 2, 4	
Refrigerated type, top section	0.85	4	
	0.8	3	
Refrigerated type, bottom section	1.0	1, 3	
	0.85	4	
Demethanizers			
Absorbing type, top section	0.85	1–3	
Absorbing type, bottom section	1.0	3	
	0.85	1, 2	
Refrigerated type, top section	0.8	3	
Refrigerated type, bottom section	1.0	3	
Oil absorbers			
Above 0°F	0.85	1–4	Ref. 2 proposes these for "absorbers" rather than "oil absorbers."
Below 0°F	0.95	3	
	0.85	1, 4	
	0.8	2	
Crude towers	1.0	3	
	0.85	4	
Crude vacuum towers	1.0	3	
	0.85	2	
Furfural refining towers	0.85	2	
	0.8	4	
Sulfolane systems	1.0	3	
	0.85	4	
Amine regenerators	0.85	1–4	
Glycol regenerators	0.85	1, 4	
	0.8	3	
	0.65	2	
Hot carbonate absorbers	0.85	1, 4	
Caustic wash	0.65	2	The author suspects that this low factor refers only to some caustic wash applications but not to others.
Heavy-foaming			
Amine absorbers	0.8	2	
	0.75	3, 4	
	0.73	1	
Glycol contactors	0.73	1	Ref. 2 recommends 0.65 for glycol contactors in glycol synthesis gas service, 0.5 for others.
	0.65	3, 4	
	0.50	2	
Sour water strippers	0.5–0.7	3	
	0.6	2	
Oil reclaimer	0.7	2	
MEK units	0.6	1, 4	
Stable foam			
Caustic regenerators	0.6	2	
	0.3	1, 4	
Alcohol synthesis absorbers	0.35	2, 4	

References:
1. Glitsch, Inc., *Ballast Tray Design Manual,* Bulletin 4900, 6th ed., 1993. Available from Koch-Glitsch, Wichita, Kans.
2. Koch Engineering Co., Inc., *Design Manual—Flexitray,* Bulletin 960-1, Wichita, Kans., 1982.
3. Nutter Engineering, *Float Valve Design Manual,* 1976. Available from Sulzer ChemTech, Tulsa, Okla.
4. M. J. Lockett, *Distillation Tray Fundamentals,* Cambridge University Press, Cambridge, England, 1986.

To convert lb/ft^3 to kg/m^3, multiply by 16.0.

SOURCE: H. Z. Kister, *Distillation Design*, copyright © 1992 by McGraw-Hill, Inc. Reprinted by permission.

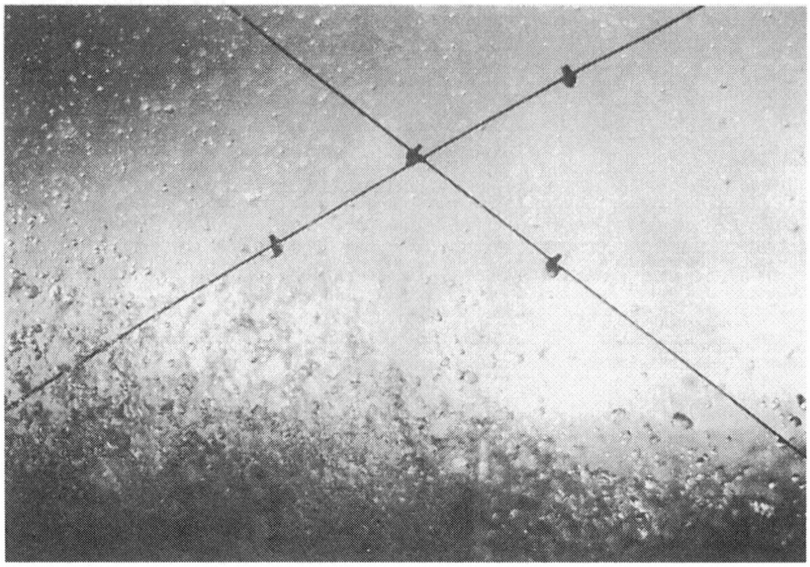

FIG. 14-33 Entrainment. [*Reprinted courtesy of Fractionation Research Inc. (FRI).*]

Fair (paper presented at the AIChE Annual Meeting, Chicago, Ill., November 1996) correlated data for efficiency reduction due to the rise of entrainment near entrainment flood, getting

$$\ln \psi = A + B\Phi + C\Phi^2 \qquad (14\text{-}99)$$

where Φ is the fractional approach to entrainment flood and A, B, and C are constants given by

	A	B	C
Highest likely efficiency loss	−3.1898	−4.7413	7.5312
Median (most likely) efficiency loss	−3.2108	−8.9049	11.6291
Lowest likely efficiency loss	4.0992	−29.9141	25.3343

Either the Kister and Haas or the Fair method can be used to evaluate Φ. The correlation has been tested with sieve trays in the flow parameter range of 0.024 to 0.087.

Example 10: Entrainment Effect on Tray Efficiency For the column in Example 9, estimate the efficiency loss should the operation be pushed from the design 80 percent of flood to 90 percent of flood. The midrange dry Murphree tray efficiency is 70 percent.

Solution The vapor and liquid densities and L/V ratio remain unchanged from Example 9, and so is the flow parameter (calculated 0.021 in Example 9). At 80 and 90 percent of flood, respectively, Fig. 14-34 gives $\psi = 0.15$ and 0.24. The respective efficiency reductions are calculated from Eq. (14-98),

$$\frac{E_a}{E_{mv}} = \frac{1}{1 + 0.70[0.15/(1 - 0.15)]} = 0.89$$

$$\frac{E_a}{E_{mv}} = \frac{1}{1 + 0.70[0.24/(1 - 0.24)]} = 0.82$$

signifying an efficiency loss from 62 to 57 percent.

Alternatively, at 80 and 90 percent of entrainment flood, the median value of ψ from Eq. (14-99) is

$$\ln \psi = -3.2108 - 8.9049(0.80) + 11.6291(0.80^2)$$

giving $\psi = 0.056$ and from Eq. (14-98) $E_a/E_{mv} = 0.96$

$$\ln \psi = -3.2108 - 8.9049(0.90) + 11.6291(0.90^2)$$

giving $\psi = 0.164$ and from Eq. (14-98) $E_a/E_{mv} = 0.88$

signifying an efficiency reduction from 67 to 62 percent.

TABLE 14-10 Recommended Range of Application for the Kister and Haas Spray Regime Entrainment Correlation

Flow regime	Spray only
Pressure	20–1200 kPa (3–180 psia)
Gas velocity	0.4–5 m/s (1.3–15 ft/s)
Liquid flow rate	3–40 m³/(m·h) (0.5–4.5 gpm/in)
Gas density	0.5–30 kg/m³ (0.03–2 lb/ft³)
Liquid density	450–1500 kg/m³ (30–90 lb/ft³)
Surface tension	5–80 mN/m
Liquid viscosity	0.05–2 cP
Tray spacing	400–900 mm (15–36 in)
Hole diameter	3–15 mm (0.125–0.75 in)
Fractional hole area	0.07–0.16
Weir height	10–80 mm (0.5–3 in)

PRESSURE DROP

In vacuum distillation, excessive pressure drop causes excessive bottom temperatures which, in turn, increase degradation, polymerization, coking, and fouling, and also loads up the column, vacuum system, and reboiler. In the suction of a compressor, excessive pressure drop increases the compressor size and energy usage. Such services attempt to minimize tray pressure drop. Methods for estimating pressure drops are similar for most conventional trays. The total pressure drop across a tray is given by

$$h_t = h_d + h'_L \qquad (14\text{-}100)$$

where h_t = total pressure drop, mm liquid

h_d = pressure drop across the dispersion unit (dry hole for sieve trays; dry valve for valve trays), mm liquid

h'_L = pressure drop through aerated mass over and around the disperser, mm liquid

It is convenient and consistent to relate all of these pressure-drop terms to height of equivalent clear liquid (deaerated basis) on the tray, in either millimeters or inches of liquid.

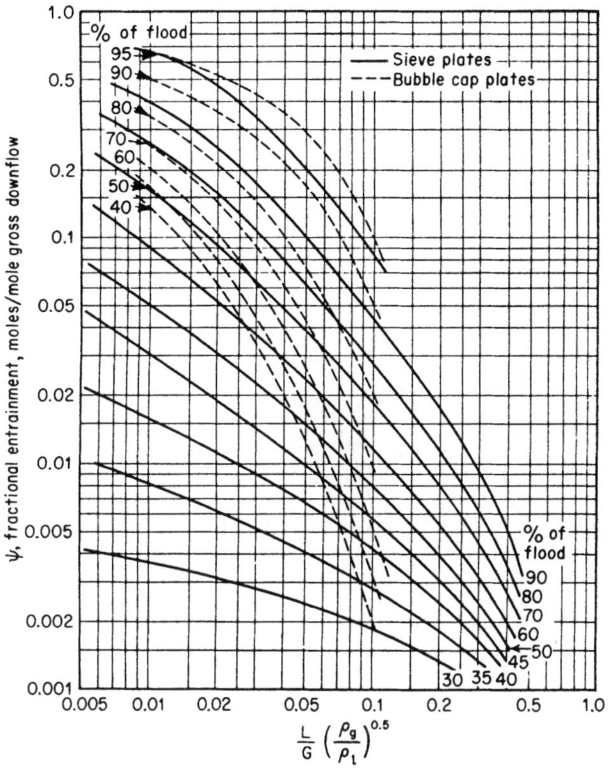

FIG. 14-34 Entrainment correlation. L/G = liquid-gas mass ratio; and ρ_l and ρ_g = liquid and gas densities. [*Fair,* Pet./Chem. Eng., **33**(10), 45 (*September 1961*).]

Pressure drop across the disperser is calculated by variations of the standard orifice equation:

$$h_d = K\left(\frac{\rho_G}{\rho_L}\right)U_h^2 \tag{14-101}$$

where U_h = linear gas velocity through slots (valve trays) or perforations (sieve tray), m/s.

For sieve trays, $K = 50.8/C_v^2$. Values of C_v are taken from Fig. 14-35. Values from Fig. 14-35 may be calculated from

$$C_v = 0.74(A_h/A_a) + \exp[0.29(t_t/d_h) - 0.56] \tag{14-102}$$

For Sulze's fixed valve trays, Summers and van Sinderen (*Distillation 2001: Topical Conference Proceedings,* AIChE Spring National Meeting, p. 444, Houston, April 22–26, 2001) provided the following equation for K:

$$K = 58 + 386A_f \quad \text{for MVG fixed valves} \tag{14-103a}$$

$$K = 58 + 461A_f \quad \text{for SVG and LVG fixed valves} \tag{14-103b}$$

Figure 14-36 illustrates the pressure drop of a typical moving valve tray as a function of gas velocity. At low velocities, all valves are closed. Gas rises through the crevices between the valves and the tray deck, with increasing pressure drop as the gas velocity rises. Once point A, the closed balance point (CBP), is reached, some valves begin to open. Upon further increase in gas velocity, more valves open until point B, the open balance point (OBP), is reached. Between points A and B, gas flow area increases with gas velocity, keeping pressure drop constant. Further increases in gas velocity increase pressure drop similar to that in a sieve tray.

The term K in Eq. (14-101) depends on valve slot area, orifice geometry, deck thickness, and the type, shape, and weight of the valves. These are best obtained from the manufacturer's literature,

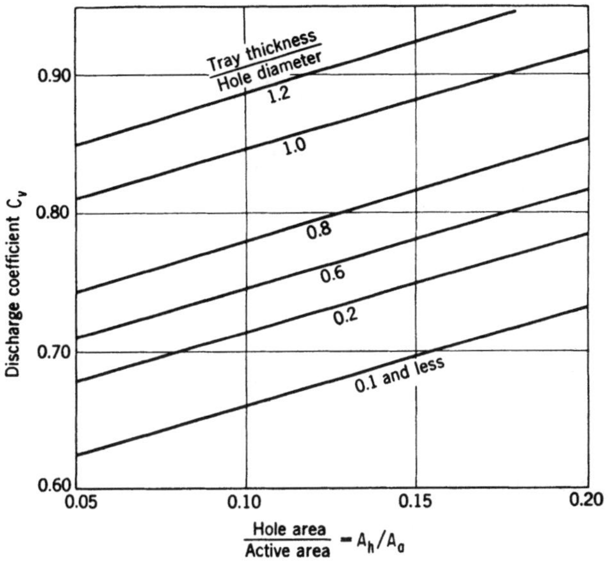

FIG. 14-35 Discharge coefficients for gas flow, sieve trays. [*Liebson, Kelley, and Bullington,* Pet. Refiner, **36**(3), 288 (1957).]

but can also be calculated from Bolles' [*Chem. Eng. Prog.* **72**(9), 43 (1976)], Lockett's (*Distillation Tray Fundamentals,* Cambridge University Press, Cambridge, England, 1986), and Klein's (*Chem. Eng.,* May 3, 1982, p. 81) methods.

For valve trays, Klein gives the following values for K (in $s^2 \cdot mm/m^2$) in Eq. (14-101), when based on the total hole area (not slot area):

	Sharp orifice	Venturi valve
All valves open (K_O)	$254.5(2.64/t_t)^{0.5}$	122
All valves closed (K_C)	1683	841

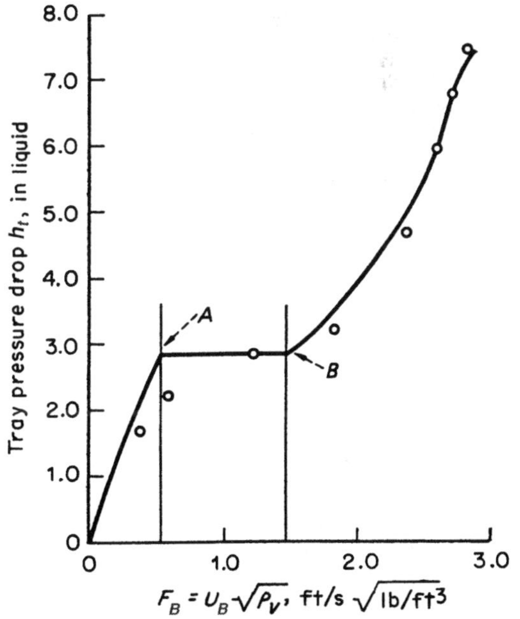

FIG. 14-36 Typical moving valve tray pressure-drop profile. (*From G. F. Klein,* Chem. Eng., *May 3, 1982, p. 81; reprinted courtesy of* Chemical Engineering.)

The velocity at which the valves start to open (point A) is given by

$$U_{h,\text{closed}} = 1.14[t_v(R_{vw}/K_C)(\rho_M/\rho_G)]^{0.5} \tag{14-104}$$

where $U_{h,\text{ closed}}$ = hole area at point A, m/s; t_v = valve thickness, mm; R_{vw} = ratio of valve weight with legs to valve weight without legs, given in Table 14-11; K_C = orifice coefficient with all valves closed (see above), $s^2 \cdot mm/m^2$; ρ_M = valve metal density, kg/m^3 (about 8000 kg/m^3 for steel); ρ_G = gas density, kg/m^3.

The velocity at which all the valves are open $U_{h,\text{open}}$ can be calculated from

$$U_{h,\text{open}} = U_{h,\text{closed}}(K_C/K_O)^{0.5} \tag{14-105}$$

Pressure drop through the aerated liquid [h'_L, in Eq. (14-100)] is calculated by

$$h'_L = \beta h_{ds} \tag{14-106}$$

where β = aeration factor, dimensionless
h_{ds} = calculated height of clear liquid, mm (dynamic seal)

The aeration factor β has been determined from Fig. 14-37 for valve and sieve trays. For sieve trays, values of β in the figure may be calculated from

$$\beta = 0.0825 \ln \frac{Q}{L_w} - 0.269 \ln F_h + 1.679 \tag{14-107}$$

where L_w = weir length, m
F_h = F-factor for flow through holes, $F_h = U_h\rho_G^{0.5}$, m/s (kg/m^3)$^{0.5}$

For sieve and valve trays,

$$h_{ds} = h_w + h_{ow} + 0.5h_{hg} \tag{14-108}$$

where h_w = weir height, mm
h_{ow} = height of crest over weir, mm clear liquid
h_{hg} = hydraulic gradient across tray, mm clear liquid

The value of weir crest h_{ow} may be calculated from the Francis weir equation and its modifications for various weir types. For a segmental weir and for height in millimeters of clear liquid,

$$h_{ow} = 664\left(\frac{Q}{L_w}\right)^{2/3} \tag{14-109}$$

where Q = liquid flow, m^3/s
L_w = weir length, m

For serrated weirs,

$$h_{ow} = 851\left(\frac{Q'}{\tan \theta/2}\right)^{0.4} \tag{14-110}$$

where Q' = liquid flow, m^3/s per serration
θ = angle of serration, deg

For circular weirs,

$$h_{ow} = 44{,}300\left(\frac{Q}{d_w}\right)^{0.704} \tag{14-111}$$

where q = liquid flow, m^3/s
d_w = weir diameter, mm

For most sieve and valve trays, the hydraulic gradient is small and can be dropped from Eq. (14-108). Some calculation methods are available and are detailed in previous editions of this handbook. A rule of thumb by the author is 17 mm/m (0.2 in/ft) of flow path length. This rule only applies in the liquid-loaded froth and emulsion regimes (Q_L >50 m^3/hm or >5.5 gpm/in of outlet weir length). At lower liquid loads, the hydraulic gradient is less.

TABLE 14-11 R_{VW} Values for Eq. (14-104)

Valve type	Sharp	Venturi
Three-leg	1.23	1.29
Four-leg	1.34	1.45
Cages (no legs)	1.00	1.00

As noted, the weir crest h_{ow} is calculated on an equivalent clear-liquid basis. A more realistic approach is to recognize that in general a froth or spray flows over the outlet weir (settling can occur upstream of the weir if a large "calming zone" with no dispersers is used). Bennett et al. [*AIChE J.*, **29,** 434 (1983)] allowed for froth overflow in a comprehensive study of pressure drop across sieve trays; their correlation for residual pressure drop h'_L in Eq. (14-100) is presented in detail in the previous (seventh) edition of this handbook, including a worked example. Although more difficult to use, the method of Bennett et al. was recommended when determination of pressure drop is of critical importance.

Example 11: Pressure Drop, Sieve Tray For the conditions of Example 9, estimate the pressure drop for flow across one tray. The thickness of the tray metal is 2 mm. The superficial F-factor is 2.08 m/s(kg/m^3)$^{1/2}$.

Solution Equations (14-100), (14-106), and (14-107), where $h_t = h_d + \beta(h_w + h_{ow})$, are used. For F_S = 2.08, F_B = 2.32 and F_H = 16.55. From Example 9, L_w = 1.50 m and h_w = 38 mm. For a liquid rate of 22,000 kg/hr, $Q = 7.27(10^{-3})$ m^3/s, and $Q/L_w = 4.8(10^{-3})$. By Eq. (14-107) or Fig. 14-37, β = 0.48. From Eq. (14-102) or Fig. 14-35, C_v = 0.75. Then, by Eq. (14-101), h_d = 29.0 mm liquid. Using Eq. (14-109), h_{ow} = 18.9 mm. Finally, $h_t = h_d + \beta(h_w + h_{ow}) = 29.0 + 0.48(38 + 18.9) = 56.4$ mm liquid.

When straight or serrated segmental weirs are used in a column of circular cross section, a correction may be needed for the distorted pattern of flow at the ends of the weirs, depending on liquid flow rate. The correction factor F_w from Fig. 14-38 is used directly in Eq. (14-109). Even when circular downcomers are utilized, they are often fed by the overflow from a segmental weir.

Loss under Downcomer The head loss under the downcomer apron, as millimeters of liquid, may be estimated from

$$h_{da} = 165.2\left(\frac{Q}{A_{da}}\right)^2 \tag{14-112}$$

where Q = volumetric flow of liquid, m^3/s and A_{da} = most restrictive (minimum) area of flow under the downcomer apron, m^2. Equation (14-112) was derived from the orifice equation with an orifice coefficient of 0.6. Although the loss under the downcomer is small, the clearance is significant from the aspect of tray stability and liquid distribution.

The term A_{da} should be taken as the most restrictive area for liquid flow in the downcomer outlet. Usually, this is the area under the downcomer apron (i.e., the downcomer clearance times the length of the segmental downcomer), but not always. For instance, if an inlet weir is used and the area between the segmental downcomer and the inlet weir is smaller than the area under the downcomer apron, the smaller area should be used.

OTHER HYDRAULIC LIMITS

Weeping Weeping is liquid descending through the tray perforations, short-circuiting the contact zone, which lowers tray efficiency. At the tray floor, liquid static head that acts to push liquid down the perforations is counteracted by the gas pressure drop that acts to hold the liquid on the tray. When the static head overcomes the gas pressure drop, weeping occurs.

Some weeping usually takes place under all conditions due to sloshing and oscillation of the tray liquid. Generally, this weeping is too small to appreciably affect tray efficiency. The *weep point* is the gas velocity at which weeping first becomes noticeable. At this point, little efficiency is lost. As gas velocity is reduced below the weep point, the weep rate increases. When the weep rate becomes large enough to significantly reduce tray efficiency, the lower tray operating limit is reached.

The main factor that affects weeping is the fractional hole area. The larger it is, the smaller the gas pressure drop and the greater the weeping tendency. Larger liquid rates and taller outlet weirs increase

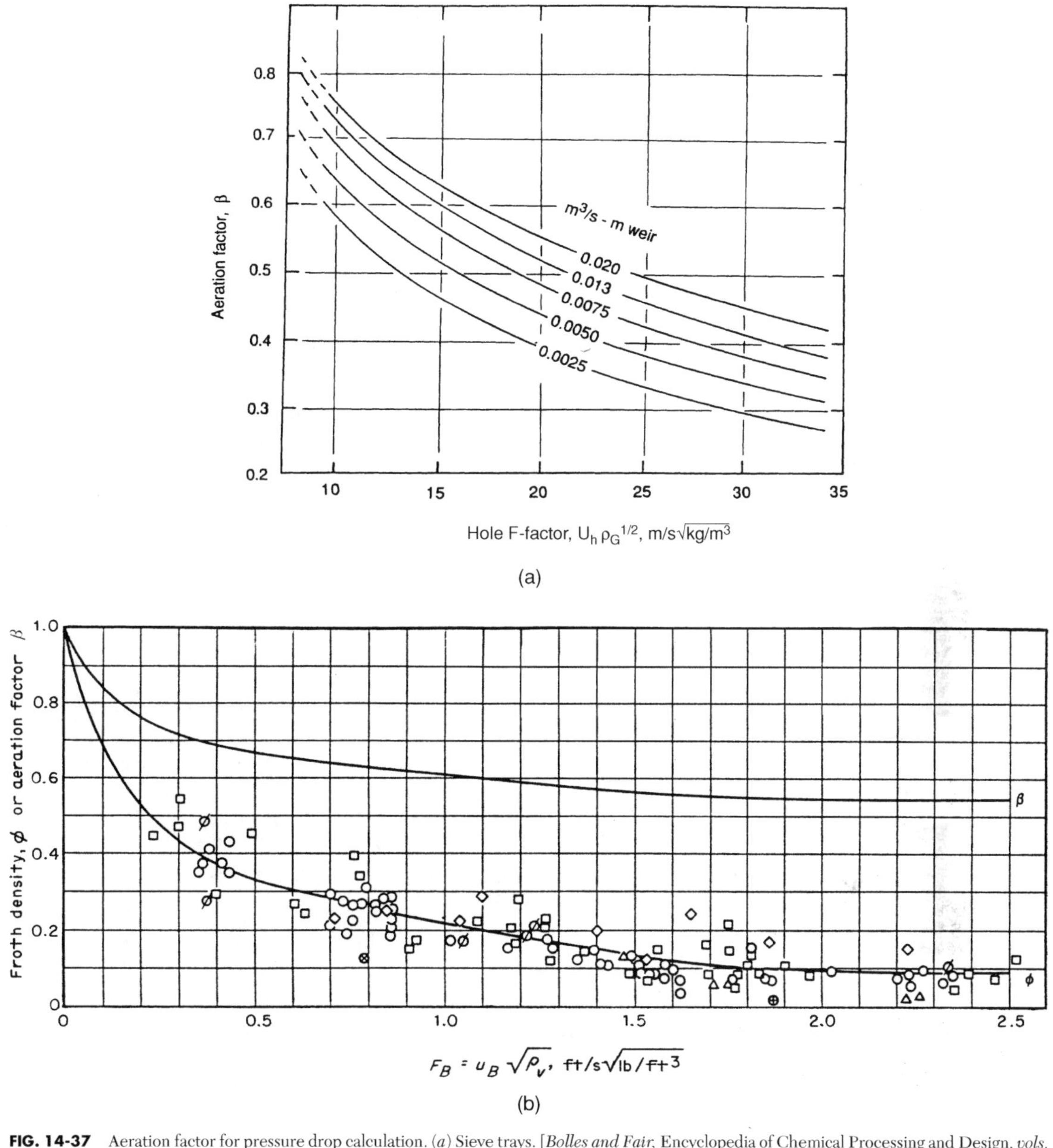

FIG. 14-37 Aeration factor for pressure drop calculation. (*a*) Sieve trays. [*Bolles and Fair,* Encyclopedia of Chemical Processing and Design, *vols. 16, 86. J. M. McKetta (ed.), Marcel Dekker, New York, 1982.*] (*b*) Valve trays. (*From G. F. Klein,* Chem. Eng., *May 3, 1982, p. 81; reprinted courtesy of* Chemical Engineering.)

liquid heads and therefore weeping. Hole diameter has a complex effect on weeping, detailed by Lockett and Banik [*Ind. Eng. Chem. Proc. Des. Dev.* **25,** 561 (1986)].

Tests by Lockett and Banik (loc. cit.) show that weeping is often nonuniform, with some hydraulic conditions favoring weeping from the tray inlet and others from the tray outlet. Weeping from the tray inlet is particularly detrimental to tray efficiency because the weeping liquid bypasses two trays.

Weep Rate Prediction Lockett and Banik (loc. cit.) and Hsieh and McNulty *(Chem. Eng. Progr.,* July 1993, p. 71*)* proposed correlations for predicting weep rates from sieve trays. Colwell and O'Bara (Paper presented at the AIChE Meeting, Houston, April 1989) recommended the Lockett and Banik correlation for low pressures (<1100 kPa or 165 psia), and the Hsieh and McNulty correlation for high pressures (>1100 kPa or 165 psia). They also corrected the Lockett and Banik correlation to improve its accuracy near the weep point.

The *Lockett and Banik correlation* (as corrected by Colwell and O'Bara) is

$$\frac{W}{A_h} = \frac{29.45}{\sqrt{Fr_h}} - 44.18 \qquad Fr_h < 0.2 \tag{14-113a}$$

$$\frac{W}{A_h} = \frac{1.841}{Fr_h^{1.533}} \qquad Fr_h > 0.2 \tag{14-113b}$$

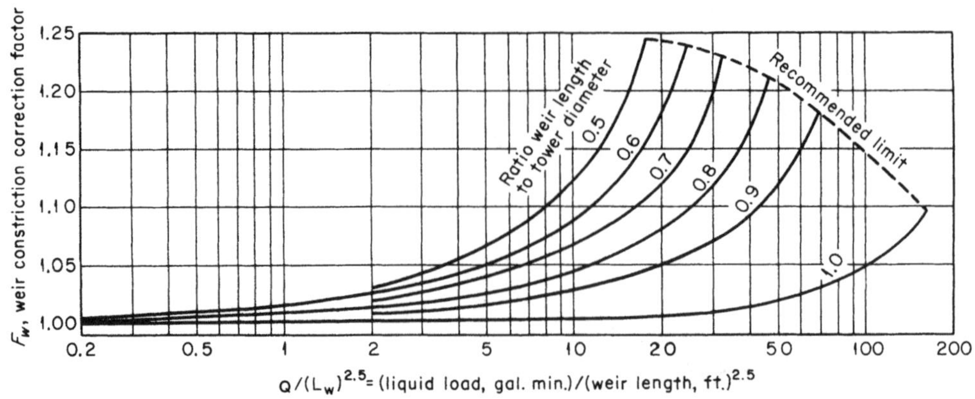

FIG. 14-38 Correction for effective weir length. To convert gallons per minute to cubic meters per second, multiply by 6.309×10^{-5}; to convert feet to meters, multiply by 0.3048. [*Bolles*, Pet. Refiner, **25,** *613 (1946).*]

where
$$\text{Fr}_h = 0.373 \frac{u_h^2}{h_c} \frac{\rho_V}{\rho_L - \rho_V} \quad (14\text{-}114)$$

Equations (14-113) and (14-114) use English units: W is the weep rate, gpm; A_h is the hole area, ft^2; u_h is the hole velocity, ft/s; and h_c is the clear liquid height, in. Colwell's [*Ind. Eng. Chem. Proc. Des. Dev.* **20**(2), 298 (1981)] method below has been recommended for obtaining the clear liquid height h_c in Eq. (14-114).

$$h_c = \phi_f \left\{ 0.527 \left(\frac{Q_L(1-f_w)}{C_d \phi_f} \right)^{2/3} + h_w \right\} \quad (14\text{-}115)$$

where ϕ_f is given by Eq. (14-119) and C_d is given by Eq. (14-116)

$$\left. \begin{aligned} C_d &= 0.61 + 0.08 \frac{h_{fow}}{h_w} & \frac{h_{fow}}{h_w} &< 8.135 \\ C_d &= 1.06\left(1 + \frac{h_w}{h_{fow}}\right)^{1.5} & \frac{h_{fow}}{h_w} &> 8.135 \end{aligned} \right\} \quad (14\text{-}116)$$

$$h_{fow} = h_f - h_w \quad (14\text{-}117)$$

where h_f is given by Eq. (14-122). The froth density ϕ_f is calculated from

$$\eta = 12.6 \text{Fr}^{0.4} \left(\frac{A_B}{A_h} \right)^{0.25} \quad (14\text{-}118)$$

$$\phi_f = \frac{1}{\eta + 1} \quad (14\text{-}119)$$

$$\text{Fr} = 0.37 \frac{\rho_V u_B^2}{h_c(\rho_L - \rho_V)} \quad (14\text{-}120)$$

The term f_w in Eq. (14-115) is the ratio of weep rate from the tray to the total liquid flow entering the tray, calculated as follows:

$$f_w = W/\text{GPM} \quad (14\text{-}121)$$

Some trial and error is required in this calculation because the clear liquid height h_c and the froth density ϕ_f depend on each other, and the weep fraction f_w depends on the clear liquid height h_c. Clear liquid height is related to froth height and froth density by

$$h_c = \phi_f h_f \quad (14\text{-}122)$$

The terms in Eqs. (14-115) to (14-122) are in the English units and are explained in the Nomenclature.

With large-diameter trays and low liquid loads, a small ratio of W/A_h corresponds to a large fractional weep. Under these conditions, the Lockett and Banik correlation is inaccurate. The correlation is unsuitable for trays with very small (<3-mm or $\frac{1}{8}$-in) holes. The correlation appears to fit most data points to an accuracy of ±15 to ±30 percent. The *Hsieh and McNulty correlation* (loc. cit.) is

$$\sqrt{J_G^*} + m\sqrt{J_L^*} = C_w \quad (14\text{-}123)$$

where
$$J_G^* = u_h \left[\frac{\rho_G}{gZ(\rho_L - \rho_G)} \right]^{0.5} \quad (14\text{-}124)$$

and
$$J_L^* = \frac{W}{448.83 A_h} \left[\frac{\rho_L}{gZ(\rho_L - \rho_G)} \right]^{0.5} \quad (14\text{-}125)$$

$$Z = h_c^{1.5}/(12 d_H^{0.5}) \quad (14\text{-}126)$$

The terms in Eqs. (14-123) to (14-126) are in English units and are explained in the Nomenclature. For sieve trays, $m = 1.94$ and $C_w = 0.79$. Note that the constants are a slight revision of those presented in the original paper (C. L. Hsieh, private communication, 1991). Clear liquid height is calculated from Colwell's correlation [Eqs. (14-115) to (14-122)]. The Hsieh and McNulty correlation applies to trays with 9 percent and larger fractional hole area. For trays with smaller hole area, Hsieh and McNulty expect the weeping rate to be smaller than predicted.

Weeping from Valve Trays An analysis of weeping from valve trays [Bolles, *Chem. Eng. Progs.* **72**(9), 43 (1976)] showed that in a well-designed valve tray, the weep point is below the gas load at which the valves open; and throughout the valve opening process, the operating point remains above the weep point. In contrast, if the tray contains too many valves, or the valves are too light, excessive valve opening occurs before the gas pressure drop is high enough to counter weeping. In this case, weeping could be troublesome.

Weep point correlations for valve trays were presented by Bolles (loc. cit.) and by Klein (*Chem. Eng.*, Sept. 17, 1984, p. 128). Hsieh and McNulty (loc. cit.) gave a complex extension of their weep rate correlation to valve trays.

Dumping As gas velocity is lowered below the weep point, the fraction of liquid weeping increases until all the liquid fed to the tray weeps through the holes and none reaches the downcomer. This is the *dump point*, or the *seal point*. The dump point is well below the range of acceptable operation of distillation trays. Below the dump point, tray efficiency is slashed, and mass transfer is extremely poor. Operation below the dump point can be accompanied by severe hydraulic instability due to unsealing of downcomers.

Extensive studies on dumping were reported by Prince and Chan [*Trans. Inst. Chem. Engr.* **43,** T49 (1965)]. The Chan and Prince dump-point correlation was recommended and is presented in detail elsewhere (Kister, *Distillation Design,* McGraw-Hill, 1992). Alternatively, the dump point can be predicted by setting the weep rate equal to 100 percent of the liquid entering the tray in the appropriate weep correlation.

Turndown The turndown ratio is the ratio of the normal operating (or design) gas throughput to the minimum allowable gas throughout. The minimum allowable throughput is usually set by excessive weeping, while normal operating throughput is a safe margin away from the relevant flooding limit.

Sieve and fixed valve trays have a poor turndown ratio (about 2:1). Their turndown can be improved by blanking some rows of tray holes, which reduces the tendency to weep, but will also reduce the tray's maximum capacity. Turndown of *moving valve trays* is normally between about 4:1 to 5:1. Special valve designs can achieve even better turndown ratios, between 6:1 and 10:1, and even more. Turndown can also be enhanced by blanking strips (which require valve removal) or valve leg crimping. Sloley and Fleming (*Chem. Eng. Progr.,* March 1994, p. 39) stress that correct implementation of turndown enhancement is central to achieving a desired turndown. When poorly implemented, turndown may be restricted by poor vapor-liquid contact rather than by weeping.

Vapor Channeling All the correlations in this section assume an evenly distributed tray vapor. When the vapor preferentially channels through a tray region, premature entrainment flood and excessive entrainment take place due to a high vapor velocity in that region. At the same time, other regions become vapor-deficient and tend to weep, which lowers tray efficiency.

Work by Davies [*Pet. Ref.* **29**(8), p. 93, and **29**(9), p. 121 (1950)] based on bubble-cap tray studies suggests that the vapor pressure drop of the tray (the *dry pressure drop*) counteracts channeling. The higher the dry tray pressure drop, the greater the tendency for vapor to spread uniformly over the bubbling area. If the dry tray pressure drop is too small compared with the channeling potential, channeling prevails.

Perhaps the most common vapor channeling mechanism is *vapor crossflow channeling* (*VCFC,* Fig. 14-39). The hydraulic gradient on the tray induces preferential vapor rise at the outlet and middle of the tray, and a vapor-deficient region near the tray inlet. The resulting high vapor velocities near the tray outlet step up entrainment, while the low vapor velocities near the tray inlet induce weeping. Interaction between adjacent trays (Fig. 14-39) accelerates both the outlet entrainment and the inlet weeping. The net result is excessive entrainment and premature flooding at the tray middle and outlet, simultaneous with weeping from the tray inlet, accompanied by a loss of efficiency and turndown.

VCFC takes place when the following four conditions exist simultaneously [Kister, Larson, and Madsen, *Chem. Eng. Progr.,* p. 86 (Nov. 1992); Kister, *The Chemical Engineer,* **544,** p. 18 (June 10, 1993)]:

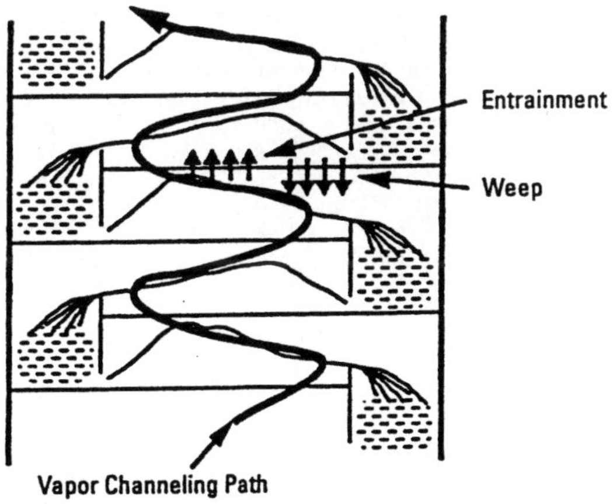

FIG. 14-39 Vapor crossflow channeling. Note entrainment near the tray middle and outlet, and weep near the tray inlet. (*Kister, H. Z., K. F. Larson, and P. Madsen,* Chem. Eng. Prog., *Nov. 1992, p. 86; reproduced with permission.*)

1. Absolute pressure < 500 kPa (70 psia).
2. High liquid rates [>50 m^3/(m·h) or 6 gpm/in of outlet weir].
3. High ratio (>2:1) of flow path length to tray spacing.
4. Low dry tray pressure drop. On sieve and fixed valve trays, this means high (>11 percent) fractional hole area. On moving valve trays, this means venturi valves (smooth orifices) or long-legged valves (>15 percent slot area). On all trays, the channeling tendency and severity escalate rapidly as the dry pressure drop diminishes (e.g., as fractional hole area increases).

Hartman (*Distillation 2001: Topical Conference Proceedings,* AIChE Spring National Meeting, p. 108, Houston, Tex. (April 22–26, 2001)] reports VCFC even with conventional valve trays (14 percent slot area) at very high ratio (3.6:1) of flow path length to tray spacing and tray truss obstruction.

VCFC is usually avoided by limiting fractional hole areas, avoiding venturi valves, and using forward-push devices. Resitarits and Pappademos [Paper presented at the AIChE Annual Meeting, Reno, Nev. (November 2001)] cited tray inlet inactivity as a contributor to VCFC, and advocate inlet forward-push devices to counter it.

TRANSITION BETWEEN FLOW REGIMES

Froth-Spray Froth-spray transition has been investigated for sieve trays using a variety of techniques. The gradual nature of this transition bred a multitude of criteria for defining it, and made its correlation difficult. Excellent overviews were given by Lockett (*Distillation Tray Fundamentals,* Cambridge University Press, Cambridge, England, 1986) and Prado, Johnson, and Fair [*Chem. Eng. Progr.* **83**(3), p. 32 (1987)]. Porter and Jenkins [*I. Chem. E. Symp. Ser.* **56,** Summary Paper (1979)] presented a simple correlation for the froth-to-spray transition.

$$F_{LG}\frac{N_pA_B}{L_w} = 0.0191 \qquad (14\text{-}127)$$

The terms of this equation are in English units and are explained in the Nomenclature. This correlation is based on the premise that froth-to-spray transition occurs when the entrainment vs. liquid load relationship passes through a minimum (see "Entrainment"). Alternatively, it was argued that the minimum represents a transition from the froth regime to a partially developed spray region (Kister, Pinczewski, and Fell, Paper presented in the 90th National AIChE Meeting, Houston, April 1981). If this alternative argument is valid, then when the correlation predicts froth, it is highly unlikely that the column operates in the spray regime; but when it predicts spray, the column may still be operating in the froth regime. Recent entrainment studies by Ohe [*Distillation 2005: Topical Conference Proceedings,* AIChE Spring National Meeting, p. 283, Atlanta (April 10–13, 2005)] argue that the entrainment minima represent minimum liquid residence times on the tray, and are unrelated to the froth-spray transition.

A second correlation is by Pinczeweski and Fell [*Ind. Eng. Chem. Proc. Des. Dev.* **21,** p. 774 (1982)]

$$u_B\sqrt{\rho_G} = 2.25\left(\frac{Q_L\sqrt{\rho_L}}{100}\right)^n \qquad (14\text{-}128)$$

The terms of Eq. (14-128) are in English units and are explained in the Nomenclature. The exponent n is calculated from Eq. (14-84). Equation (14-128) is based on transition data obtained from orifice jetting measurements for the air-water system and on entrainment minimum data for some hydrocarbon systems.

A third recent correlation by Johnson and Fair (loc. cit.) is

$$U_a^\circ = C_1\,\rho_G^{-0.50}\,\rho_L^{0.692}\,\sigma^{0.06}A_f^{0.25}\left(\frac{q}{L_w}\right)^{0.05} d_h^{-0.1} \qquad (14\text{-}129)$$

where U_a° = gas velocity through active area at inversion, m/s
ρ_G = gas density, kg/m^3
ρ_L = liquid density, kg/m^3
σ = surface tension, mN/m
A_f = hole/active area ratio

q/L_w = liquid flow, $m^3/(s{\cdot}m)$ weir
d_h = hole diameter, mm
C_1 = 0.0583 for 25.4-mm overflow weirs
= 0.0568 for 50.4-mm overflow weirs
= 0.0635 for 101.6-mm overflow weirs

Froth-Emulsion Froth-emulsion transition occurs [Hofhuis and Zuiderweg, *I. Chem. E. Symp. Ser.* **56,** p. 2, 2/1 (1979)] when the aerated mass begins to obey the Francis weir formula. Using this criterion, the latest version of this transition correlation is

$$F_{LG} \frac{N_p A_B}{L_w h_c} = 0.0208 \qquad (14\text{-}130)$$

The terms of this equation are in English units and are explained in the Nomenclature; h_c is calculated from the Hofhuis and Zuiderweg (loc. cit.) equation.

$$h_c = 2.08 \left(F_{LG} \frac{N_p A_B}{L_w} p \right)^{0.25} h_w^{0.5} \qquad (14\text{-}131)$$

An inspection of the experimental data correlated shows that this, too, is a gradual transition, which occurs over a range of values rather than at a sharp point.

Valve Trays The amount of work reported thus far on valve tray regime transition is small and entirely based on air-water tests. Correlations proposed to date require the knowledge of liquid holdup at transition, which is generally not available, and are therefore of limited application for commercial columns.

TRAY EFFICIENCY

Definitions

Overall Column Efficiency This is the ratio of the number of theoretical stages to the number of actual stages

$$E_{OC} = N_t/N_a \qquad (14\text{-}132)$$

Since tray efficiencies vary from one section to another, it is best to apply Eq. (14-132) separately for the rectifying and stripping sections. In practice, efficiency data and prediction methods are often too crude to give a good breakdown between the efficiencies of different sections, and so Eq. (14-132) is applied over the entire column.

Point Efficiency This is defined by Eq. (14-133) (Fig. 14-40*a*):

$$E_{OG} = \left(\frac{y_n - y_{n-1}}{y_n^\circ - y_{n-1}} \right)_{\text{point}} \qquad (14\text{-}133)$$

where y_n° is the composition of vapor in equilibrium with the liquid at point n. The term y_n is actual vapor composition at that point. The point efficiency is the ratio of the change of composition at a point to the change that would occur on a theoretical stage. As the vapor composition at a given point cannot exceed the equilibrium composition, fractional point efficiencies are always lower than 1. If there is a composition gradient on the tray, point efficiency will vary between points on the tray.

Murphree Tray Efficiency [*Ind. Eng. Chem.* **17,** 747 (1925)] This is the same as point efficiency, except that it applies to the entire tray instead of to a single point (Fig. 14-40*b*):

$$E_{MV} = \left(\frac{y_n - y_{n-1}}{y_n^\circ - y_{n-1}} \right)_{\text{tray}} \qquad (14\text{-}134)$$

If both liquid and vapor are perfectly mixed, liquid and vapor compositions on the tray are uniform, and the Murphree tray efficiency will coincide with the point efficiency at any point on the tray. In practice, a concentration gradient exists in the liquid, and x_n at the tray outlet is lower than x_n' on the tray (see Fig. 14-40*b*). This frequently lowers y_n° relative to y_n, thus enhancing tray efficiency [Eq. (14-134)] compared with point efficiency. The value of y_n° may even drop below y_n. In this case, E_{MV} exceeds 100 percent [Eq. (14-134)].

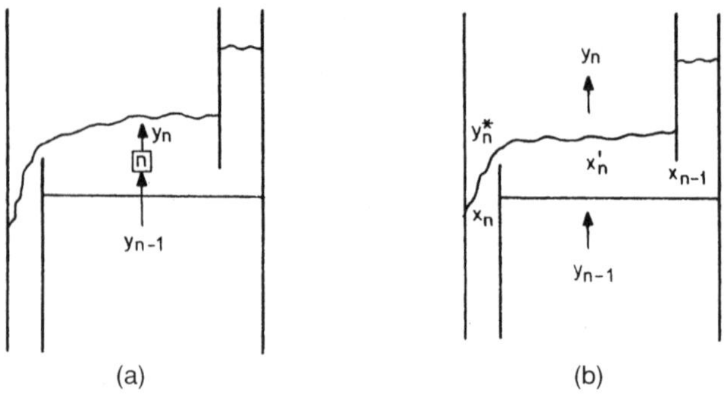

FIG. 14-40 Point and Murphree efficiencies. (*a*) Point. (*b*) Murphree. (*From H. Z. Kister,* Distillation Design, *copyright © 1992 by McGraw-Hill; reprinted by permission.*)

Overall column efficiency can be calculated from the Murphree tray efficiency by using the relationship developed by Lewis [*Ind. Eng. Chem.* **28,** 399 (1936)].

$$E_{OC} = \frac{\ln[1 + E_{MV}(\lambda - 1)]}{\ln \lambda} \qquad (14\text{-}135)$$

where

$$\lambda = m \frac{G_M}{L_M} \qquad (14\text{-}136)$$

Equation (14-135) is based on the assumption of constant molar overflow and a constant value of E_{MV} from tray to tray. It needs to be applied separately to each section of the column (rectifying and stripping) because G_M/L_M, and therefore λ, varies from section to section. Where molar overflow or Murphree efficiencies vary throughout a section of column, the section needs to be divided into subsections small enough to render the variations negligible.

The point and Murphree efficiency definitions above are expressed in terms of vapor concentrations. Analogous definitions can be made in terms of liquid concentrations. Further discussion is elsewhere (Lockett, *Distillation Tray Fundamentals,* Cambridge University, Press, Cambridge, England, 1986).

Fundamentals Figure 14-41 shows the sequence of steps for converting phase resistances to a tray efficiency. Gas and liquid film resistances are added to give the point efficiency. Had both vapor and liquid on the tray been perfectly mixed, the Murphree tray efficiency would have equaled the point efficiency. Since the phases are not perfectly mixed, a model of the vapor and liquid mixing patterns is needed for converting point efficiency to tray efficiency. Liquid mixture patterns are plug flow, backmixing, and stagnant zones, while vapor-mixing patterns are perfect mixing and plug flow.

Lewis (loc. cit.) was the first to derive quantitative relationships between the Murphree and the point efficiency. He derived three mixing cases, assuming plug flow of liquid in all. The Lewis cases give the maximum achievable tray efficiency. In practice, efficiency is lower due to liquid and vapor nonuniformities and liquid mixing.

Most tray efficiency models are based on Lewis case 1 with vapor perfectly mixed between trays. For case 1, Lewis derived the following relationship:

$$E_{MV,\text{dry}} = \frac{\exp(\lambda E_{OG}) - 1}{\lambda} \qquad (14\text{-}137)$$

The "dry" Murphree efficiency calculated thus far takes into account the vapor and liquid resistances and the vapor-liquid contact patterns, but is uncorrected for the effects of entrainment and weeping. This correction converts the dry efficiency to a "wet" or actual Murphree tray efficiency. Colburn [Eq. (14-98), under "Entrainment"] incorporated the effect of entrainment on efficiency, assuming perfect mixing of liquid on the tray.

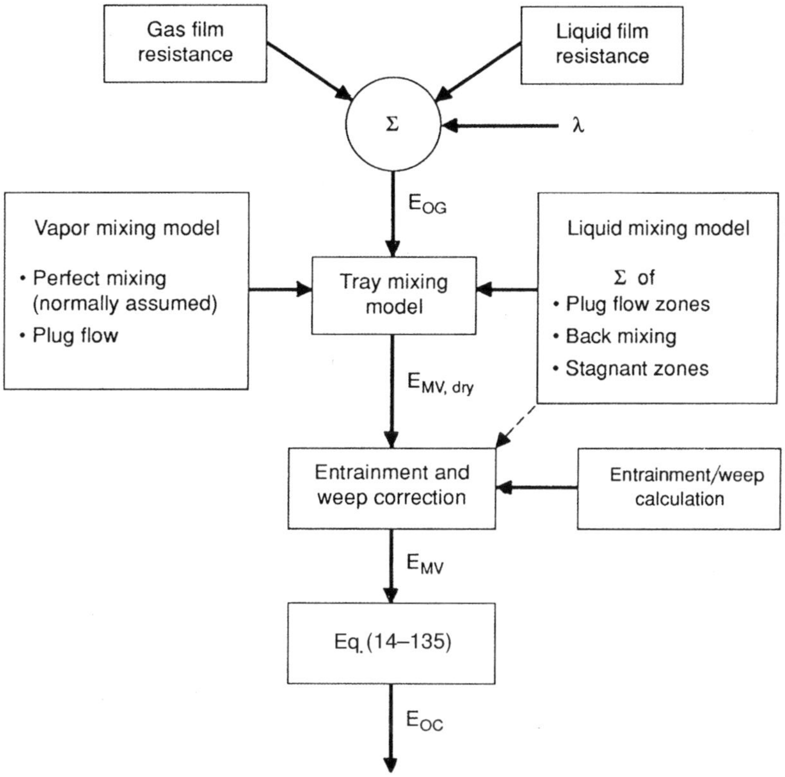

FIG. 14-41 Sequence of steps for theoretical prediction of tray efficiency. (*From H. Z. Kister,* Distillation Design, *copyright © 1992 by McGraw-Hill; reprinted by permission.*)

Factors Affecting Tray Efficiency Below is a summary based on the industry's experience. A detailed discussion of the fundamentals is found in Lockett's book (*Distillation Tray Fundamentals,* Cambridge University Press, Cambridge, England, 1986). A detailed discussion of the reported experience, and the basis of statements made in this section are in Kister's book (*Distillation Design,* McGraw-Hill, New York, 1992).

Errors in Vapor-Liquid Equilibrium (VLE) Errors in relative volatility are the most underrated factor affecting tray efficiency. Figure 14-42 shows the direct effect of the errors [Deibele and Brandt, *Chem. Ing. Tech.* **57**(5), p. 439 (1985); Roy P. and G. K. Hobson, *I. Chem. E. Symp. Ser.* **104,** p. A273 (1987)]. At very low relative volatilities ($\alpha < 1.2$), small errors in VLE have a huge impact on tray efficiency. For instance, at $\alpha = 1.1$, a −3 percent error gives a tray efficiency 40 to 50 percent higher than its true value (Fig. 14-42). Since VLE errors are seldom lower than ±2 to 3 percent, tray efficiencies of low-volatility systems become meaningless unless accompanied by VLE data. Likewise, comparing efficiencies derived for a low-volatility system by different sources is misleading unless one is using identical VLE.

Figure 14-42 shows that errors in relative volatility are a problem only at low relative volatilities; for $\alpha > 1.5$ to 2.0, VLE errors have negligible direct impact on tray efficiency.

Most efficiency data reported in the literature are obtained at total reflux, and there are no indirect VLE effects. For measurements at finite reflux ratios, the indirect effects below compound the direct effect of Fig. 14-42. Consider a case where $\alpha_{apparent} < \alpha_{true}$ and test data at a finite reflux are analyzed to calculate tray efficiency. Due to the volatility difference $R_{min,apparent} > R_{min,true}$. Since the test was conducted at a fixed reflux flow rate, $(R/R_{min})_{apparent} < (R/R_{min})_{true}$. A calculation based on the apparent R/R_{min} will give more theoretical stages than a calculation based on the true R/R_{min}. This means a higher apparent efficiency than the true value.

The indirect effects add to those of Fig. 14-42, widening the gap between true and apparent efficiency. The indirect effects exponentially escalate as minimum reflux is approached. Small errors in VLE or reflux ratio measurement (this includes column material balance as well as reflux rate) alter R/R_{min}. Near minimum reflux, even small R/R_{min} errors induce huge errors in the number of stages, and therefore in tray efficiency. Efficiency data obtained near minimum reflux are therefore meaningless and potentially misleading.

Liquid Flow Patterns on Large Trays The most popular theoretical models (below) postulate that liquid crosses the tray in plug flow with superimposed backmixing, and that the vapor is perfectly mixed. Increasing tray diameter promotes liquid plug flow and suppresses backmixing.

The presence of stagnant zones on large-diameter distillation trays is well established, but the associated efficiency loss is poorly understood; in some cases, significant efficiency losses, presumably due to stagnant zones, were reported [Weiler, Kirkpatrick, and Lockett, *Chem. Eng. Progr.* **77**(1), 63 (1981)], while in other cases, no efficiency difference was observed [Yanagi and Scott, *Chem. Eng. Progr.,* **69**(10), 75 (1973)]. Several techniques are available for eliminating stagnant regions (see Kister, *Distillation Design,* McGraw-Hill, New York, 1992, for some), but their effectiveness for improving tray efficiency is uncertain.

Weir Height Taller weirs raise the liquid level on the tray in the froth and emulsion regimes. This increases interfacial area and vapor contact time, which should theoretically enhance efficiency. In the spray regime, weir height affects neither liquid level nor efficiency. In distillation systems, the improvement of tray efficiency due to taller weirs is small, often marginal.

Length of Liquid Flow Path Longer liquid flow paths enhance the liquid-vapor contact time, the significance of liquid plug flow, and therefore raise efficiency. Typically, doubling the flow path length

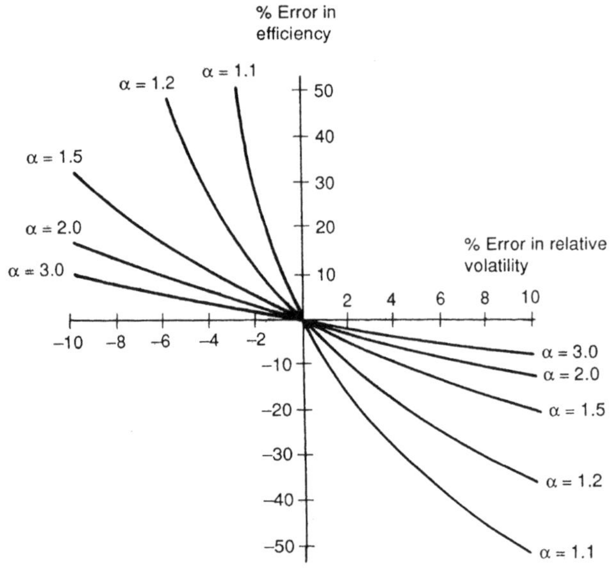

FIG. 14-42 Direct effect of errors in relative volatility on error in tray efficiency. *(From H. Z. Kister,* Distillation Design, *copyright © 1992 by McGraw-Hill; reprinted by permission.)*

(such as going from two-pass to one-pass trays at a constant tower diameter) raises tray efficiency by 5 to 15 percent.

Fractional Hole Area Efficiency increases with a reduction in fractional hole area. Yanagi and Sakata [*Ind. Eng. Chem. Proc. Des. Dev.* **21,** 712 (1982)] tests in commercial-scale towers show a 5 to 15 percent increase in tray efficiency when fractional hole area was lowered from 14 to 8 percent (Fig. 14-43).

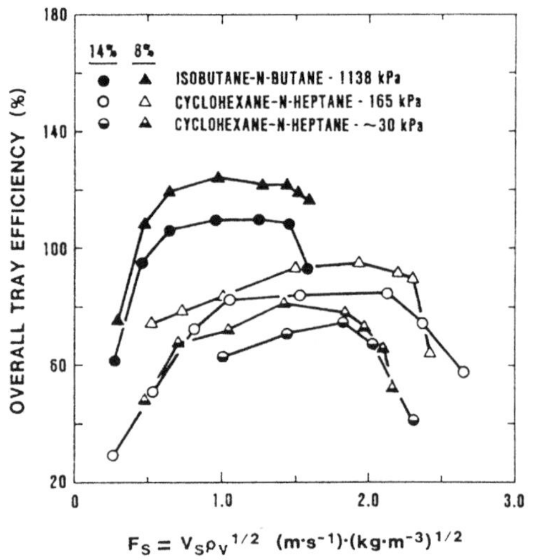

FIG. 14-43 Efficiency reduction when fractional hole area is increased, also showing little effect of vapor and liquid loads on efficiency in the normal operating range (between excessive weeping and excessive entrainment). Also shown is the small increase in efficiency with pressure. FRI data, total reflux, D_T = 1.2 m, S = 610 mm, h_w = 50.8 mm, d_H = 12.7 mm. *(Reprinted with permission from T. Yanagi and M. Sakata,* Ind. Eng. Chem. Proc. Des. Dev. ***21,*** *712; copyright © 1982, American Chemical Society.)*

Hole Diameter The jury is out on the effect of hole diameter on tray efficiency. There is, however, a consensus that the effect of hole diameter on efficiency is small, often negligible.

Vapor-Liquid Loads and Reflux Ratio Vapor and liquid loads, as well as the reflux ratio, have a small effect on tray efficiency (Fig. 14-43) as long as no capacity or hydraulic limits (flood, weep, channeling, etc.) are violated.

Viscosity, Relative Volatility Efficiency increases as liquid viscosity and relative volatility diminish. These effects are reflected in the O'Connell correlation (below).

Surface Tension There is uncertainty regarding the effect of surface tension on tray efficiency. Often, it is difficult to divorce the surface tension effects from those of other physical properties.

Pressure Tray efficiency slightly increases with pressure (Fig. 14-43), reflecting the rise of efficiency with a reduction in liquid viscosity and in relative volatility, which generally accompany a distillation pressure increase.

At pressures exceeding 10 to 20 bar (150 to 300 psia), and especially at high liquid rates, vapor entrainment into the downcomer liquid becomes important, and tray efficiency decreases with further increases in pressure [Zuiderweg, *Int. Chem. Eng.* **26**(1), 1 (1986)].

Maldistribution Maldistribution can cause major efficiency reduction in multipass trays (>two passes). Further discussion is given under "Number of Passes."

OBTAINING TRAY EFFICIENCY

Efficiency prediction methods are listed here in decreasing order of reliability.

Rigorous Testing Rigorous testing of a plant column is generally the most reliable method of obtaining tray efficiency. Test procedures can be found elsewhere (AIChE Equipment Testing Procedures Committee, *AIChE Equipment Testing Procedure—Tray Distillation Columns,* 2d ed., 1987; Kister, *Distillation Operation,* McGraw-Hill, New York, 1990).

Scale-up from an Existing Commercial Column As long as data are for the same system under similar process conditions, loadings, and operating regime, data obtained in one column directly extend to another. Fractional hole area and the number of tray passes will have a small but significant effect on efficiency, and any changes in these parameters need to be allowed for during scale-up. The empirical information in the section "Factors Affecting Tray Efficiency" can be used to estimate the magnitude of the changes on efficiency.

Scale-up from Existing Commercial Column to Different Process Conditions During scale-up, test data are analyzed by computer simulation. The number of theoretical stages is varied until the simulated product compositions and temperature profile match the test data. Tray efficiency is determined by the ratio of theoretical stages to actual trays. In this procedure, errors in VLE are offset by compensating errors in tray efficiency. For instance, if the relative volatility calculated by the simulation is too high, fewer stages will be needed to match the measured data, i.e., "apparent" tray efficiency will be lower. Scale-up will be good as long as the VLE and efficiency errors continue to offset each other equally. This requires that process conditions (feed composition, feed temperature, reflux ratio, etc.) remain unchanged during scale-up.

When process conditions change, the VLE and efficiency errors no longer offset each other equally. If the true relative volatility is higher than simulated, then the scale-up will be conservative. If the true relative volatility is lower than simulated, scale-up will be optimistic. A detailed discussion is found in Kister, *Distillation Design,* McGraw-Hill, New York, 1992.

Experience Factors These are tabulations of efficiencies previously measured for various systems. Tray efficiency is insensitive to tray geometry (above), so in the absence of hydraulic anomalies and issues with VLE data, efficiencies measured in one tower are extensible to others distilling the same system. A small allowance to variations in tray geometry as discussed above is in order. Caution is required with mixed aqueous-organic systems, where concentration may have a marked effect on physical properties, relative volatility, and efficiency. Table 14-12 shows typical tray efficiencies reported in the literature.

TABLE 14-12 Representative Tray Efficiencies

Tray	System	Column diameter, ft	Tray spacing, in	Pressure, psia	Efficiency, %	% hole (slot) area	Ref.
Sieve	Methanol-water	3.2	15.7	14.7	70–90	10.8	2
	Ethanol-water	2.5	14	14.7	75–85	10.4	1
	Methanol-water	3.2	15.7	14.7	90–100	4.8	2
	Ethylbenzene-styrene	2.6	19.7	1.9	70	12.3	5
	Benzene-toluene	1.5	15.7	14.7	60–75	8	10
	Methanol-*n* propanol-sec butanol	6.0	18	18	64*		6
	Mixed xylene + C_8-C_{10} paraffins and naphthenes	13.0	21	25	86*		4
	Cyclohexane-*n*-heptane	4.0	24	5	60–70	14	9
				24	80	14	9
		4.0	24	5	70–80	8	8
				24	90	8	8
	Isobutane-*n*-butane	4.0	24	165	110	14	9
		4.0	24	165	120	8	8
		4.0	24	300	110	8	8
		4.0	24	400	100	8	8
	n-Heptane-toluene	1.5	15.7	14.7	60–75	8	10
	Methanol-water	2.0	13.6	14.7	68–72	10	11
	Isopropanol-water	2.0	13.6	14.7	59–63		11
	Toluene-methylcyclohexane	2.0	13.6	14.7	70–82		11
	Toluene stripping from water	4	24	14.7	31–42	8	13
Valve	Methanol-water	3.2	15.7	14.7	70–80	14.7	2
	Ethanol-water	2.5	14	14.7	75–85		1
	Ethyl benzene-styrene	2.6	19.7	1.9	75–85		3
	Cyclohexane-*n*-heptane	4.0	24	24	70–96*		7
	Isobutane-*n*-butane	4.0	24	165	108–121*		7
	Cyclohexane-*n*-heptane	4.0	24	24	77–93†	14.7	12
				5	57–86†	14.7	12
	Isobutane-*n*-butane	4.0	24	165	110–123†	14.7	12
	C_3-C_4 splitter	5.6	24	212	65–67‡	12	14

References:

1. Kirschbaum, *Distillier-Rektifiziertechnik*, 4th ed., Springer-Verlag, Berlin and Heidelberg, 1969.
2. Kastanek and Standart, *Sep. Sci.* **2,** 439 (1967).
3. Billet and Raichle, *Chem. Ing. Tech.*, 38, 825 (1966); **40,** 377 (1968).
4. AIChE Research Committee, *Tray Efficiency in Distillation Columns*, final report, University of Delaware, Newark, 1958.
5. Billet R., *IChemE., Symp. Ser.* **32,** p. 4:42 (1969).
6. Mayfield et al., *Ind. Eng. Chem.*, **44,** 2238 (1952).
7. Fractionation Research, Inc. "Report of Tests of Nutter Type B Float Valve Tray," July 2, 1964 from Sulzer Chem Tech.
8. Sakata and Yanagi, *IChemE., Eng. Symp.* Ser., no. 56, 3.2/21 (1979).
9. Yanagi and Sakata, *Ind. Eng. Chem. Process Des. Dec.*, **21,** 712 (1982).
10. Zuiderweg and Van der Meer, *Chem. Tech.* (Leipzig), **24,** 10 (1972).
11. Korchinsky, *Trans. I. Chem. E.*, **72,** Part A, 472 (1994).
12. Glitsch, Inc. "Glitsch Ballast Trays," Bulletin 159/160 (FRI Topical Report 15, 1958). Available from Koch-Glitsch LP, Wichita, Kans.
13. Kunesh et al., Paper presented at the AIChE Spring National Meeting, Atlanta, Ga., 1994.
14. Remesat, Chuang, and Svrcek, *Trans. I Chem. E.*, Vol. 83, Part A, p. 508, May 2005.

Notes:

*Rectangular Sulzer BDP valves.
†Glitsch V-1 round valves (Koch-Glitsch).
‡Two-pass trays, short path length.
To convert feet to meters, multiply by 0.3048; to convert inches to centimeters, multiply by 2.54; and to convert psia to kilopascals, multiply by 6.895.

Vital, Grossel and Olsen [*Hydroc. Proc.* **63,** 11, p. 147 (1984)] and Garcia and Fair [*Ind. Eng. Chem. Res.* **39,** p. 1809 (2000)] present an extensive tabulation of tray efficiency data collected from the published literature.

The *GPSA Engineering Data Book* (10th ed., Gas Processors Association, 1987) and Kaes (*Refinery Process Modeling—A Practical Guide to Steady State Modeling of Petroleum Processes Using Commercial Simulators,* Athens Printing Co., Athens, Ga., 2000) tabulate typical efficiencies in gas plant and refinery columns, respectively. Pilling (Paper presented at the 4th Topical Conference on Separations Science and Technology, November 1999, available from Sulzer Chemtech, Tulsa, Okla.) tabulated more typical efficiencies. Similar information is often available from simulation guide manuals. The quality and reliability of efficiencies from these sources vary and are generally lower than the reliability of actual measured data.

Scale-up from a Pilot- or Bench-Scale Column This is a very common scale-up. No reduction in efficiency on scale-up is expected as long as several precautions are observed. These precautions, generally relevant to pilot- or bench-scale columns, are spelled out with specific reference to the Oldershaw column.

Scale-up from Oldershaw Columns One laboratory-scale device that found wide application in efficiency investigations is the Oldershaw column [Fig. 14-44, Oldershaw, *Ind. Eng. Chem. Anal. Ed.* **13,** 265 (1941)]. This column is available from a number of laboratory supply houses and can be constructed from glass for atmospheric operation or from metal for higher pressures. Typical column diameters are 25 to 100 mm (1 to 4 in), with tray spacing the same as the column diameter.

Fair, Null, and Bolles [*Ind. Eng. Chem. Process Des. Dev.* **22,** 53 (1983)] found that efficiency measurements in Oldershaw columns closely approach the point efficiencies [Eq. (14-133)] measured in

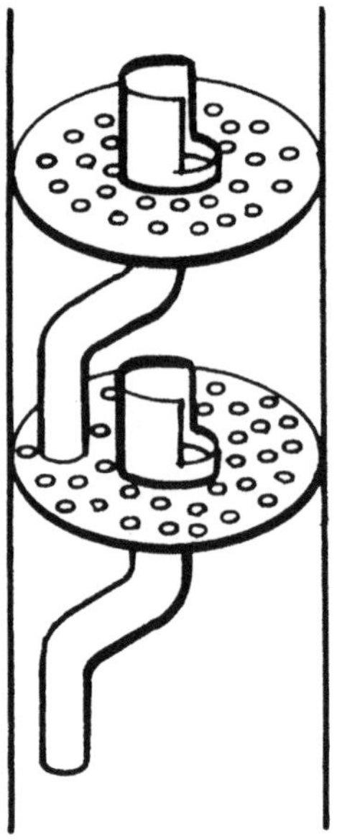

FIG. 14-44 An Oldershaw column. *(From H. Z. Kister,* Distillation Design, *copyright © 1992 by McGraw-Hill; reprinted by permission.)*

commercial sieve-tray columns (Fig. 14-45) providing (1) the systems being distilled are the same, (2) comparison is made at the same relative approach to the flood point, (3) operation is at total reflux, and (4) a standard Oldershaw device is used in the laboratory experimentation. Fair et al. compared several systems, utilizing as large-scale information the published efficiency studies of Fractionation Research, Inc. (FRI).

A mixing model can be used to convert the Oldershaw point efficiencies to overall column efficiencies. This enhances the commercial column efficiency estimates. A conservative approach suggested by Fair et al. is to apply the Oldershaw column efficiency as the estimate for the overall column efficiency of the commercial column, taking no credit for the greater plug-flow character upon scale-up. The author prefers this conservative approach, considering the poor reliability of mixing models.

Previous work with Oldershaw columns [Ellis, Barker, and Contractor, *Trans. Instn. Chem. Engnrs.* **38,** 21 (1960)], spells an additional note of caution. Cellular (i.e., wall-supported) foam may form in pilot or Oldershaw columns, but is rare in commercial columns. For a given system, higher Oldershaw column efficiencies were measured under cellular foam conditions than under froth conditions. For this reason, Gerster [*Chem. Eng. Progr.* **59**(3), 35 (1963)] warned that when cellular foam can form, scale-up from an Oldershaw column may be dangerous. The conclusions presented by Fair et al. do not extend to Oldershaw columns operating in the cellular foam regime. Cellular foam can be identified by lower pilot column capacity compared to a standard mixture that is visualized not to form cellular foam.

Heat losses are a major issue in pilot and Oldershaw columns and can lead to optimistic scale-up. Special precautions are needed to keep these at a minimum. Vacuum jackets with viewing ports are commonly used.

Uses of Oldershaw columns to less conventional systems and applications were described by Fair, Reeves, and Seibert [Topical Conference on Distillation, AIChE Spring Meeting, New Orleans, p. 27 (March 10–14, 2002)]. The applications described include scale-up in the absence of good VLE, steam stripping efficiencies, individual component efficiencies in multicomponent distillation, determining component behavior in azeotropic separation, and foam testing.

Empirical Efficiency Prediction Two empirical correlations which have been the standard of the industry for distillation tray efficiency prediction are the Drickamer and Bradford, in Fig. 14-46 [*Trans. Am. Inst. Chem. Eng.* **39,** 319 (1943)] and a modification of it by O'Connell [*Trans. Am. Inst. Chem. Eng.* **42,** 741 (1946)], in Fig. 14-47. The Drickamer-Bradford plot correlates efficiency as a function of liquid viscosity only, which makes it useful for petroleum cuts. O'Connell added the relative volatility to the x axis.

Lockett (*Distillation Tray Fundamentals,* Cambridge University Press, Cambridge, England, 1986) noted some theoretical sense in O'Connell's correlation. Higher viscosity usually implies lower diffusivity, and therefore greater liquid-phase resistance and lower efficiency. Higher relative volatility increases the significance of the liquid-phase resistance, thus reducing efficiency. Lockett expresses the O'Connell plot in equation form:

$$E_{OC} = 0.492(\mu_L\alpha)^{-0.245} \tag{14-138}$$

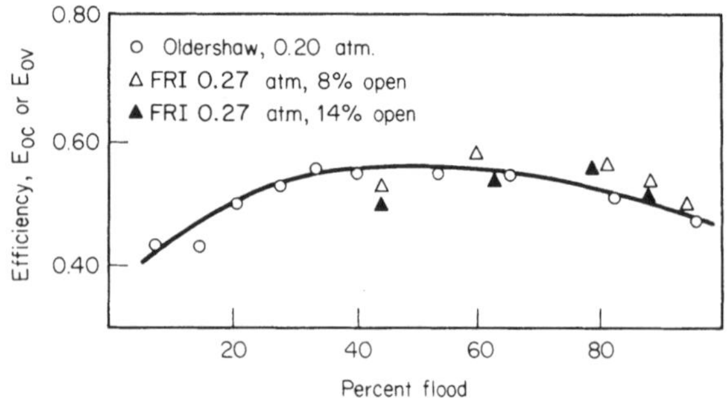

FIG. 14-45 Overall column efficiency of 25-mm Oldershaw column compared with point efficiency of 1.22-m-diameter-sieve sieve-plate column of Fractionation Research, Inc. System = cyclohexane-*n*-heptane. [*Fair, Null, and Bolles,* Ind. Eng. Chem. Process Des. Dev., **22,** *53 (1982).*]

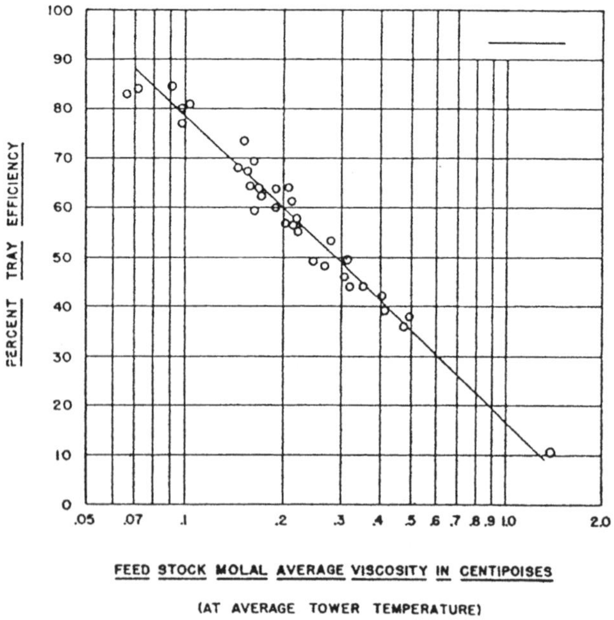

FIG. 14-46 The Drickamer and Bradford tray efficiency correlation for refinery towers. To convert centipoise to pascal-seconds, multiply by 0.0001. [*From Drickamer and Bradford,* Trans. Am. Inst. Chem. Eng. **39,** 319 *(1943). Reprinted courtesy of the AIChE.*]

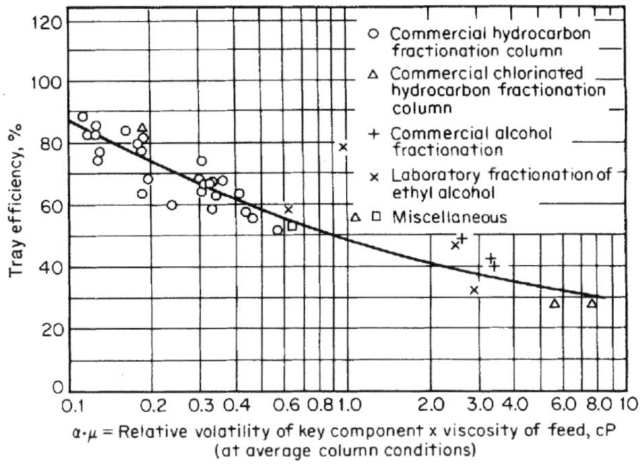

FIG. 14-47 O'Connell correlation for overall column efficiency E_{oc} for distillation. To convert centipoises to pascal-seconds, multiply by 10^{-3}. [*O'Connell,* Trans. Am. Inst. Chem. Eng., ***42,*** 741 (*1946*).]

(The viscosity is in cP and E_{OC} is fractional.) The volatility and viscosity are evaluated at the average arithmetic temperature between the column top and bottom temperatures. The relative volatility is between the key components.

The O'Connell correlation was based on data for bubble-cap trays. For sieve and valve trays, its predictions are likely to be slightly conservative.

Theoretical Efficiency Prediction Theoretical tray efficiency prediction is based on the two-film theory and the sequence of steps in Fig. 14-41. Almost all methods evolved from the AIChE model (AIChE Research Committee, *Bubble Tray Design Manual,* New York, 1958). This model was developed over 5 years in the late 1950s in three universities. Since then, several aspects of the AIChE model have been criticized, corrected, and modified. Reviews are given by Lockett (*Distillation Tray Fundamentals,* Cambridge University Press, Cambridge, England, 1986) and Chan and Fair [*Ind. Eng. Chem. Proc. Des. Dev.* **23,** 814 (1984)]. An improved version of the AIChE model, which alleviated several of its shortcomings and updated its hydraulic and mass-transfer relationships, was produced by Chan and Fair.

The Chan and Fair correlation generally gave good predictions when tested against a wide data bank, but its authors also observed some deviations. Its authors described it as "tentative until more data become available." The Chan and Fair correlation is considered the most reliable fundamental correlation for tray efficiency, but even this correlation has been unable to rectify several theoretical and practical limitations inherited from the AIChE correlation (see Kister, *Distillation Design,* McGraw-Hill, New York, 1992). Recently, Garcia and Fair (*Ind. Eng. Chem. Res.* **39,** 1818, 2000) proposed a more fundamental and accurate model that is also more complicated to apply.

The prime issue that appears to plague fundamental tray efficiency methods is their tendency to predict efficiencies of 80 to 100 percent for distillation columns larger than 1.2 m (4 ft) in diameter. In the real world, most columns run closer to 60 percent efficiency. Cai and Chen (Distillation 2003: Topical Conference Proceedings, AIChE Spring National Meeting, New Orleans, La., March 30–April 3, 2003) show that published eddy diffusivity models, which are based on small-column work, severely underestimate liquid backmixing and overestimate plug flow in commercial-scale columns, leading to optimistic efficiency predictions. Which other limitations (if any) in the theoretical methods contribute to the mismatch, and to what degree, is unknown. For this reason, the author would not recommend any currently published theoretical tray efficiency correlation for obtaining design efficiencies.

Example 12: Estimating Tray Efficiency For the column in Example 9, estimate the tray efficiency, given that at the relative volatility near the feed point is 1.3 and the viscosity is 0.25 cP.

Solution Table 14-12 presents measurements by Billet (loc. cit.) for ethylbenzene-styrene under similar pressure with sieve and valve trays. The column diameter and tray spacing in Billet's tests were close to those in Example 9. Since both have single-pass trays, the flow path lengths are similar. The fractional hole area (14 percent in Example 9) is close to that in Table 14-12 (12.3 percent for the tested sieve trays, 14 to 15 percent for standard valve trays). So the values in Table 14-12 should be directly applicable, that is, 70 to 85 percent. So a conservative estimate would be 70 percent. The actual efficiency should be about 5 to 10 percent higher.

Alternatively, using Eq. (14-138) or Fig. 14-47, $E_{OC} = 0.492(0.25 \times 1.3)^{-0.245} = 0.65$ or 65 percent. As stated, the O'Connell correlation tends to be slightly conservative. This confirms that the 70 percent above will be a good estimate.

EQUIPMENT FOR DISTILLATION AND GAS ABSORPTION: PACKED COLUMNS

Packings are generally divided into three classes:

1. *Random or dumped packings* (Figs. 14-48 and 14-49) are discrete pieces of packing, of a specific geometric shape, that are "dumped" or randomly packed into the column shell.
2. *Structured or systematically arranged packings* (Fig. 14-50) are crimped layers of corrugated sheets (usually) or wire mesh. Sections of these packings are stacked in the column.
3. *Grids.* These are also systematically arranged packings, but instead of wire mesh or corrugated sheets, these use an open-lattice structure.

Random and structured packings are common in commercial practice. The application of grids is limited primarily to heat-transfer and wash services and/or where a high fouling resistance is required. Grids are discussed in detail elsewhere (Kister, *Distillation Design,* McGraw-Hill, New York, 1992).

Figure 14-51 is an illustrative cutaway of a packed tower, depicting typical internals. This tower has a structured-packed top bed and a random-packed bottom bed. Each bed rests on a support grid or plate. The lower bed has a holddown grid at its top to restrict packing uplift. Liquid to each of the beds is supplied by a liquid distributor. An intermediate distributor, termed a *redistributor,* is used to introduce feed and/or to remix liquid at regular height intervals. The intermediate distributor in Fig. 14-51 is not self-collecting, so a chevron collector is used to collect the liquid from the bed above. An internal pipe passes this liquid to the distributor below. The collected liquid is mixed with the fresh feed (not shown) before entering the distributor. The reboiler return enters behind a baffle above the bottom sump.

As illustrated, the packing needs to be interrupted and a distributor added at each point where a feed enters or a product leaves. A simple distillation tower with a single feed will have a minimum of two beds, a rectifying bed and a stripping bed.

Packing Objectives The objective of any packing is to maximize efficiency for a given capacity, at an economic cost. To achieve these goals, packings are shaped to

1. *Maximize the specific surface area, i.e., the surface area per unit volume.* This maximizes vapor-liquid contact area, and, therefore, efficiency. A corollary is that efficiency generally increases as the random packing size is decreased or as the space between structured packing layers is decreased.
2. *Spread the surface area uniformly.* This improves vapor-liquid contact, and, therefore, efficiency. For instance, a Raschig ring (Fig. 14-48*a*) and a Pall® ring (Fig. 14-48*c*) of an identical size have identical surface areas per unit volume, but the Pall® ring has a superior spread of surface area and therefore gives much better efficiency.
3. *Maximize the void space per unit column volume.* This minimizes resistance to gas upflow, thereby enhancing packing capacity.

A corollary is that capacity increases with random packing size or with the space between structured packing layers. Comparing with the first objective, a tradeoff exists; the ideal size of packing is a compromise between maximizing efficiency and maximizing capacity.

4. *Minimize friction.* This favors an open shape that has good aerodynamic characteristics.

5. *Minimize cost.* Packing costs, as well as the requirements for packing supports and column foundations, generally rise with the weight per unit volume of packing. A corollary is that packings become cheaper as the size increases (random packing) and as the space between layers increases (structured packing).

Random Packings Historically, there were three generations of evolution in random packings. The first generation (1907 to the 1950s) produced two basic simple shapes—the Raschig ring and the Berl saddle (Fig. 14-48*a, b*) that became the ancestors of modern random packings. These packings have been superseded by more modern packing and are seldom used in modern distillation practice.

The second generation (late 1950s to the early 1970s) produced two popular geometries—the Pall® ring, which evolved from the Raschig ring, and the Intalox® saddle (Fig. 14-48*c–f*), which evolved from the Berl saddle.

BASF developed the Pall® ring by cutting windows in the Raschig ring and bending the window tongues inward. This opened up the ring, lowering the aerodynamic resistance and dramatically enhancing capacity. The bent tongues improved area distribution around the particle, giving also better efficiency. These improvements made the first generation Raschig rings obsolete for distillation.

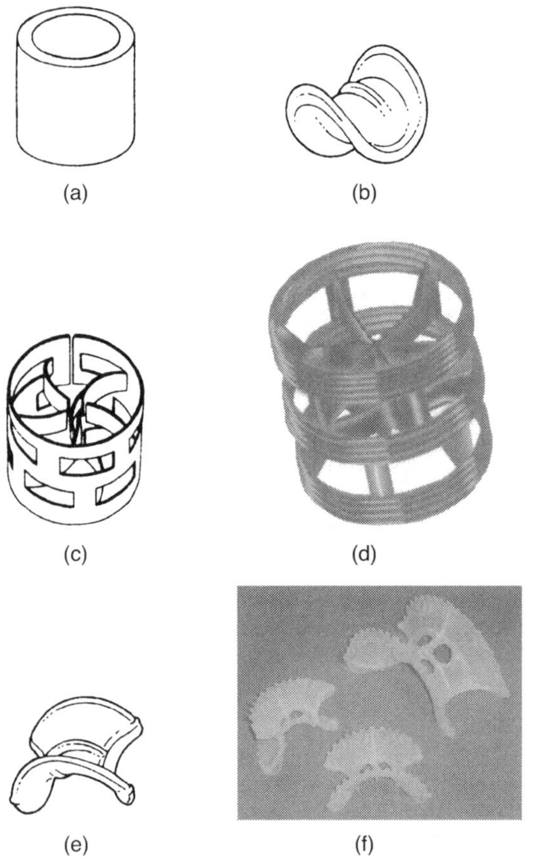

FIG. 14-48 Common first- and second-generation random packings. (*a*) Raschig ring (metal, plastic, ceramic). (*b*) Berl saddle (ceramic). (*c*) Pall ring (metal). (*d*) Pall ring (plastic). (*e*) Intalox saddle (ceramic). (*f*) Super Intalox saddle (plastic). (*Parts d, f, courtesy of Koch-Glitsch LP.*)

Berl saddles (ceramics) are still used due to their good breakage resistance.

The second-generation packings are still popular and extensively used in modern distillation practice. The third generation (the mid-1970s until present) has produced a multitude of popular geometries, most of which evolved from the Pall® ring and Intalox® saddle. Some are shown in Fig. 14-49. A more comprehensive description of the various packings is given elsewhere (Kister, *Distillation Design*, McGraw-Hill, New York, 1992).

The third generation of packing was a significant, yet not large, improvement over the second generation, so second-generation packings are still commonly used.

Structured Packings Structured packings have been around since as early as the 1940s. First-generation structured packings, such as Panapak, never became popular, and are seldom used nowadays.

The second generation of structured packings began in the late 1950s with high-efficiency wire-mesh packings such as Goodloe®, Hyperfil®, and the Sulzer® (wire-mesh) packings. By the early 1970s, these packings had made substantial inroads into vacuum distillation, where their low pressure drop per theoretical stage is a major advantage. In these services, they are extensively used today. Their high cost, high sensitivity to solids, and low capacity hindered their application outside vacuum distillation.

The corrugated-sheet packing, first introduced by Sulzer in the late 1970s, started a third generation of structured packings. With a high capacity, lower cost, and lower sensitivity to solids, while still retaining a high efficiency, these corrugated-sheet packings became competitive with conventional internals, especially for revamps. The 1980s saw an accelerated rise in popularity of structured packings, to the point of their becoming one of the most popular column internals in use today.

Corrugated structured packings are fabricated from thin, corrugated (crimped) metal sheets, arranged parallel to one another. The corrugated sheets are assembled into an element (Figs. 14-50*a, c* and 14-51). The sheets in each element are arranged at a fixed angle to the vertical. Table 14-14 contains geometric data for several corrugated packings.

Geometry (Fig. 14-52) The crimp size defines the opening between adjacent corrugated layers. Smaller B, h, and S yield narrower openings, more sheets (and, therefore, greater surface area) per unit volume, and more efficient packing, but higher resistance to gas upflow, lower capacity, and enhanced sensitivity to plugging and fouling.

The corrugations spread gas and liquid flow through a single element in a series of parallel planes. To spread the gas and liquid uniformly in all radial planes, adjacent elements are rotated so that sheets of one element are at a fixed angle to the layer below (Fig. 14-51). For good spread, element height s is relatively short (typically 200 to 300 mm, 8 to 12 in) and the angle of rotation is around 90°.

The surfaces of a few structured packings (especially those used in highly fouling environments) are smooth. Most structured packings have a roughened or enhanced surface that assists the lateral spread of liquid, promotes film turbulence, and enhances the area available for mass transfer. Texturing commonly employed is embossing and grooving (Fig. 14-50*a, b*).

The surfaces of most (but not all) structured packings contain holes that serve as communication channels between the upper and lower surfaces of each sheet. If the holes are too small, or nonexistent, both sides of a sheet will be wet only at low liquid rates. At high liquid rates, *sheeting* or *blanking* will cause liquid to run down the top surface with little liquid wetting the bottom surface [Chen and Chuang, *Hydroc. Proc.* **68**(2), 37 (1989)], which may lower efficiency. Usually, but not always, the holes are circular (Fig. 14-50*a, b*), about 4 mm in diameter. Olujic et al. (Distillation 2003: Topical Conference Proceedings, p. 523, AIChE, 2003, Spring National Meeting, New Orleans, La.) showed that the hole diameter has a complex effect, strongly dependent on packing size, on both capacity and efficiency.

Inclination Angle In each element, corrugated sheets are most commonly inclined at about 45° to the vertical (typically indicated by the letter 'Y' following the packing size). This angle is large enough for good drainage of liquid, avoiding stagnant pockets and regions of liquid accumulation, and small enough to prevent gas from bypassing the metal surfaces. In some packings, the inclination angle to the vertical is steepened to 30° (typically indicated by the letter X following the

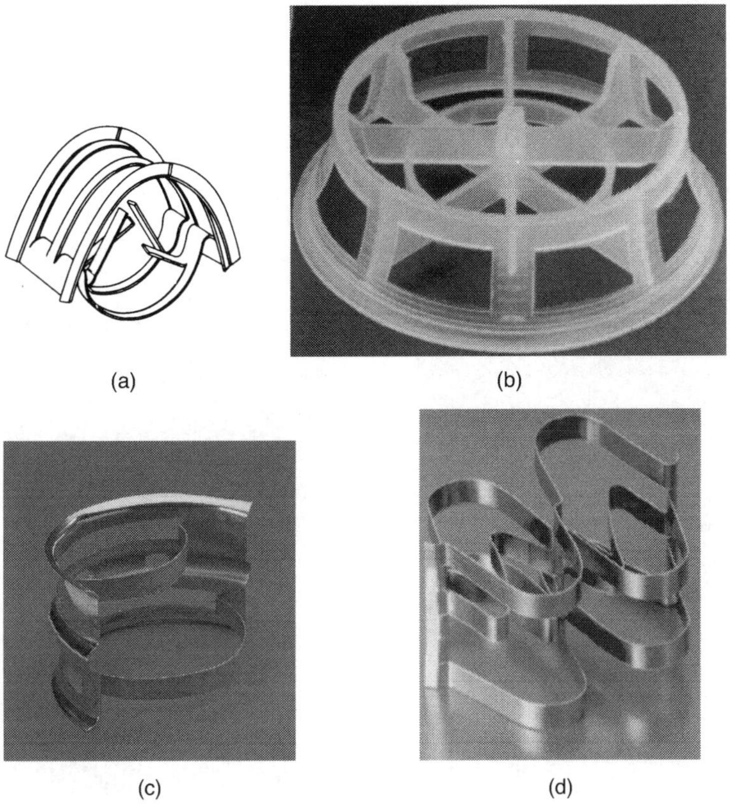

FIG. 14-49 Common third-generation random packings. (*a*) Intalox metal tower packing (IMTP). (*b*) Cascade mini-ring (CMR) (plastic). (*c*) Nutter ring (metal). (*d*) Raschig Super-Ring (metal). *(Parts a, b, courtesy of Koch-Glitsch LP; part c, courtesy of Sulzer Chemtech; part d, courtesy of Rashig AG.)*

packing size). This improves drainage, and therefore capacity, but at the expense of reduced gas-liquid contact, and therefore efficiency.

A recent development followed the realization that liquid drainage was restricted at the element-to-element transition rather than inside elements (Lockett and Billingham, *IChemE Symp. Ser.* **152,** London, 2006). This means that the liquid accumulation leading to flood initiates at the transition region. A fourth generation of structured packing started, in which the main body of each element has layers inclined at 45°, but the ends of each element are almost vertical to permit drainage at this end region (Fig. 14-50*d*; but keep in mind that successive elements are rotated 90° rather than continuous, as shown in Fig. 14-50*d*). These S-shaped or high-capacity packings offer greater capacity compared to equivalent 45° inclined packings with efficiency the same with some (Pilling and Haas, Topical Conference Proceedings, p. 132, AIChE Spring Meeting, New Orleans, March 10–14, 2002; McNulty and Sommerfeldt in "Distillation: Horizons for the New Millennium," Topical Conference Proceedings, p. 89, AIChE Spring Meeting, Houston, Tex., March 1999) and lower with others [Olujic et al., *Chem. Eng. and Proc.,* **42,** p. 55 (2003)].

PACKED-COLUMN FLOOD AND PRESSURE DROP

Pressure drop of a gas flowing upward through a packing countercurrently to liquid flow is characterized graphically in Fig. 14-53. At very low liquid rates, the effective open cross section of the packing is not appreciably different from that of dry packing, and pressure drop is due to flow through a series of variable openings in the bed. Thus, pressure drop is proportional approximately to the square of the gas velocity, as indicated in the region AB. At higher liquid rates, the effective open cross section is smaller because of the presence of liquid (region A′B′). The pressure drop is higher, but still proportional to the square of the gas velocity.

At higher gas rates, a portion of the energy of the gas stream is used to support an increasing quantity of liquid in the column. For all liquid rates, a zone is reached where pressure drop is proportional to a gas flow rate to a power distinctly higher than 2; this zone is called the *loading zone*. The increase in pressure drop is due to the liquid accumulation in the packing voids (region BC or $B'C'$)

As the liquid holdup increases, the effective orifice diameter may become so small that the liquid surface becomes continuous across the cross section of the column. Column instability occurs concomitantly with a rising continuous-phase liquid body in the column. Pressure drop shoots up with only a slight change in gas rate (condition C or C'). The phenomenon is called *flooding* and is analogous to entrainment flooding in a tray column.

Alternatively, a phase inversion occurs, and gas bubbles through the liquid. The column is not unstable and can be brought back to gas-phase continuous operation by merely reducing the gas rate. A stable operating condition beyond flooding (region CD or $C'D'$) may form with the liquid as the continuous phase and the gas as the dispersed phase [Lerner and Grove, *Ind. Eng. Chem.* **43,** 216 (1951); Teller, *Chem. Eng.* **61**(9), 168 (1954); Leung et al., *Ind. Eng. Chem. Fund.* **14** (1), 63 (1975); Buchanan, *Ind. Eng. Chem. Fund.* **15** (1), 87 (1976)].

For total-reflux distillation in packed columns, regions of loading and flooding are identified by their effects on mass-transfer efficiency, as shown in Fig. 14-54. Gas and liquid rate increase together, and a

(a)

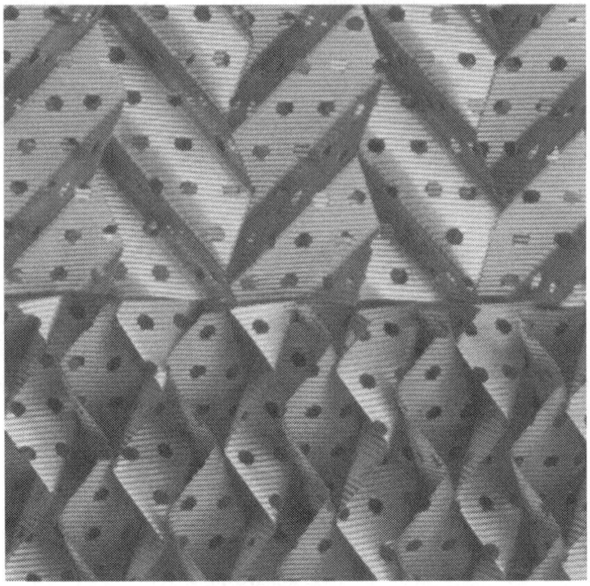

(b)

(c)

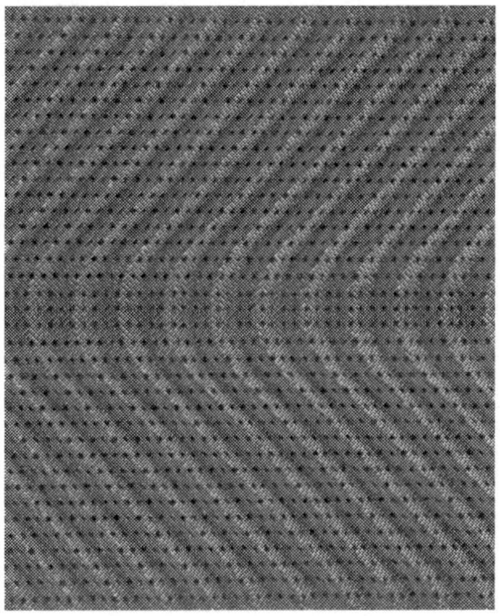

(d)

FIG. 14-50 Common structured packings. (*a*) A small element of Mellapak™ showing embossed surface, holes, and corrugated-sheet arrangement. (*b*) A closeup of the surface of Flexipac™ showing grooved surface and holes. (*c*) Fitting structured packing elements to a large-diameter tower. (*d*) Mellapak Plus™, a fourth-generation structured packing, showing a 45° inclination angle in the element and near-vertical inclination at the element-to-element transition. Note that in the tower, the successive layers will be oriented 90° to each other as in part *b*. (*Parts a, d, courtesy of Sulzer Chemtech; parts b, c, courtesy of Koch-Glitsch LP.*)

point is reached at which liquid accumulates rapidly (point *B*) and effective surface for mass transfer decreases rapidly.

Flood-Point Definition In 1966, Silvey and Keller [*Chem. Eng. Progr.* **62**(1), 68 (1966)] listed 10 different flood point definitions that have been used by different literature sources. A later survey (Kister and Gill, *Proceedings of Chemeca 92*, p. 185-2, Canberra, Australia, 1992) listed twice that many. As Silvey and Keller pointed out, the existence of so many definitions puts into question what constitutes flooding in a packed tower, and at what gas rate it occurs. Symptoms used to identify flood in these definitions include appearance of liquid on top of the bed, excessive entrainment, a sharp rise in pressure drop, a sharp rise in liquid holdup, and a sharp drop in efficiency. The survey of Kister and Gill suggests that most flood point definitions describe the point of flooding initiation (*incipient flooding*; point *C* or *C′* on Figs. 14-53 and 14-54). The different incipient flooding definitions gave surprisingly little scatter of flood point data (for a given packing under similar operating conditions). It follows that any definition describing flooding initiation should be satisfactory.

The author believes that due to the variations in the predominant symptom with the system and the packing, the use of multiple symptoms is most appropriate. The author prefers the following definition

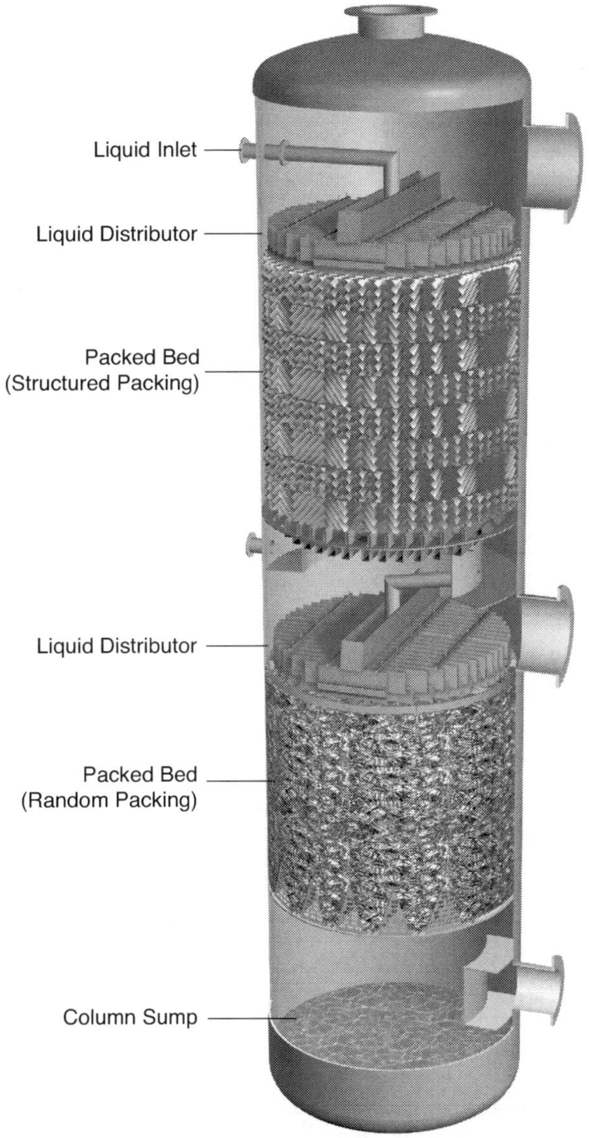

FIG. 14-51 Illustrative cutaway of a packed tower, depicting an upper bed of structured packing and a lower bed of random packing. *(Courtesy of Sulzer Chemtech.)*

by Fair and Bravo [*Chem. Eng. Symp. Ser.* **104,** A183 (1987)]: "A region of rapidly increasing pressure drop with simultaneous loss of mass transfer efficiency. Heavy entrainment is also recognized as a symptom of this region." An almost identical definition was presented earlier by Billet (*Distillation Engineering,* Chem. Publishing Co., New York, 1979).

The maximum operational capacity or throughput (often also referred to as maximum efficient capacity) is defined (Strigle, *Packed Tower Design and Applications,* 2d ed., Gulf Publishing, Houston, Tex., 1994) as the "Maximum vapor rate that provides normal efficiency of a packing" (i.e., point B in Fig. 14-54). The MOC is clear-cut in Fig. 14-54. On the other hand, locating the MOC in other cases is difficult and leaves a lot of room for subjectivity.

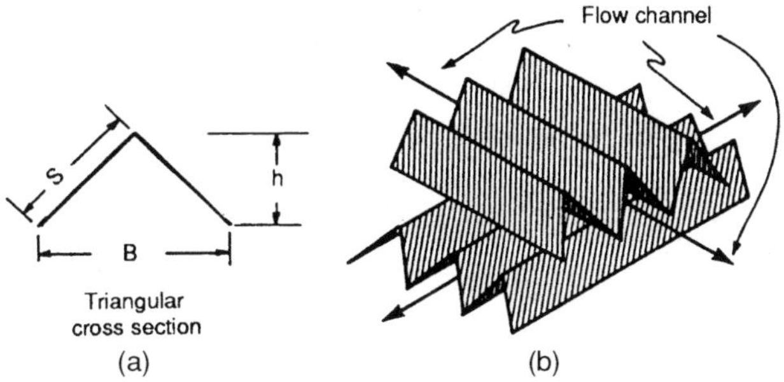

FIG. 14-52 Crimp geometry in structured packings. (*a*) Flow channel cross section. (*b*) Flow channel arrangement. *(From J. R. Fair and J. L. Bravo,* Chem. Eng. Progr., *Jan. 1990, p. 19; reproduced courtesy of the American Institute of Chemical Engineers.)*

In most cases, [Kister and Gill, *Chem. Eng. Progr.* **87**(2), 32 (1991)], the velocity at which MOC is reached is related to the flood point velocity by

$$u_{S,MOC} = 0.95\, u_{S,Fl} \tag{14-139}$$

Flood and Pressure Drop Prediction The first generalized correlation of packed-column flood points was developed by Sherwood, Shipley, and Holloway [*Ind. Eng. Chem.*, **30,** 768 (1938)] on the basis of laboratory measurements primarily on the air-water system with random packing. Later work with air and liquids other than water led to modifications of the Sherwood correlation, first by Leva [*Chem. Eng. Progr. Symp. Ser.,* **50**(1), 51 (1954)], who also introduced the pressure drop curves, and later in a series of papers by Eckert. The generalized flooding–pressure drop chart by Eckert [*Chem. Eng. Progr.* **66**(3), 39 (1970)], included in previous editions of this handbook, was modified and simplified by Strigle (*Packed Tower Design and Applications,* 2d ed., Gulf Publishing, Houston, Tex., 1994) (Fig. 14-55). It is often called the generalized pressure drop correlation (GPDC). The ordinate is a capacity parameter [Eq. (14-140)] related to the Souders-Brown coefficient used for tray columns.

$$\text{CP} = C_S F_p^{0.5} \nu^{0.05} = U_S \left(\frac{\rho_G}{\rho_L - \rho_G} \right)^{0.50} F_p^{0.5} \nu^{0.05} \tag{14-140}$$

where U_S = superficial gas velocity, ft/s
ρ_G, ρ_L = gas and liquid densities, lb/ft^3 or kg/m^3

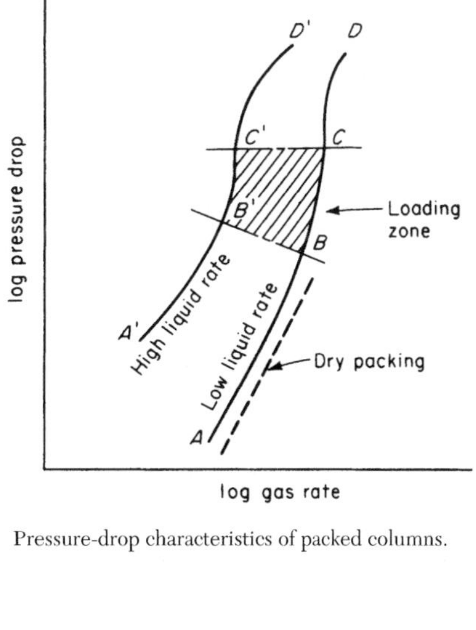

FIG. 14-53 Pressure-drop characteristics of packed columns.

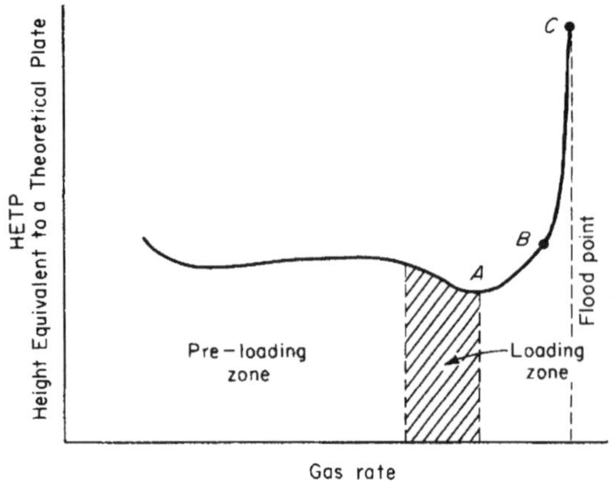

FIG. 14-54 Efficiency characteristics of packed columns (total-reflux distillation.)

F_p = packing factor, ft^{-1}
ν = kinematic viscosity of liquid, cS
C_S = C-factor, Eq. (14-77), based on tower superficial cross-sectional area, ft/s
CP = capacity factor, dimensional [units consisted with Eq. (14-140) and its symbols]

The abscissa scale term is the same flow parameter used for tray (dimensionless):

$$F_{LG} = \frac{L}{G}\left(\frac{\rho_G}{\rho_L}\right)^{0.5} \tag{14-141}$$

For structured packing, Kister and Gill [*Chem. Eng. Symp. Ser.* **128,** A109 (1992)] noticed a much steeper rise of pressure drop with flow parameter than that predicted from Fig. 14-55, and presented a modified chart (Fig. 14-56).

The GPDC charts in Figs. 14-55 and 14-56 do not contain specific flood curves. Both Strigle and Kister and Gill recommend calculating the flood point from the flood pressure drop, given by the Kister and Gill equation

$$\Delta P_{\text{flood}} = 0.12 F_p^{0.7} \tag{14-142}$$

Equation (14-142) permits finding the pressure drop curve in Fig. 14-55 or 14-56 at which incipient flooding occurs.

For low-capacity random packings, such as the small first-generation packings and those smaller than 1-in diameter (F_p > 60 ft^{-1}), calculated flood pressure drops are well in excess of the upper pressure drop curve in Fig. 14-55. For these packings only, the original Eckert flood correlation [*Chem. Eng. Prog.* **66**(3), 39 (1970)] found in pre-1997 editions of this handbook and other major distillation texts is suitable.

The packing factor F_p is empirically determined for each packing type and size. Values of F_p, together with general dimensional data for individual packings, are given for random packings in Table 14-13 (to go with Fig. 14-55) and for structured packings in Table 14-14 (to go with Fig. 14-56).

Packing flood and pressure drop correlations should always be used with caution. Kister and Gill [*Chem. Eng. Progr.,* **87**(2), 32 (1991)] showed that deviations from the GPDC predictions tend to be systematic and not random. To avoid regions in which the systematic deviations lead to poor prediction, they superimposed experimental data points for each individual packing on the curves of the GPDC. Figure 14-57 is an example. This method requires a single chart for each packing type and size. It provides the highest possible accuracy as it interpolates measured data and identifies uncertain regions. A set of charts is in Chapter 10 of Kister's book (*Distillation Design,* McGraw-Hill, New York, 1992) with

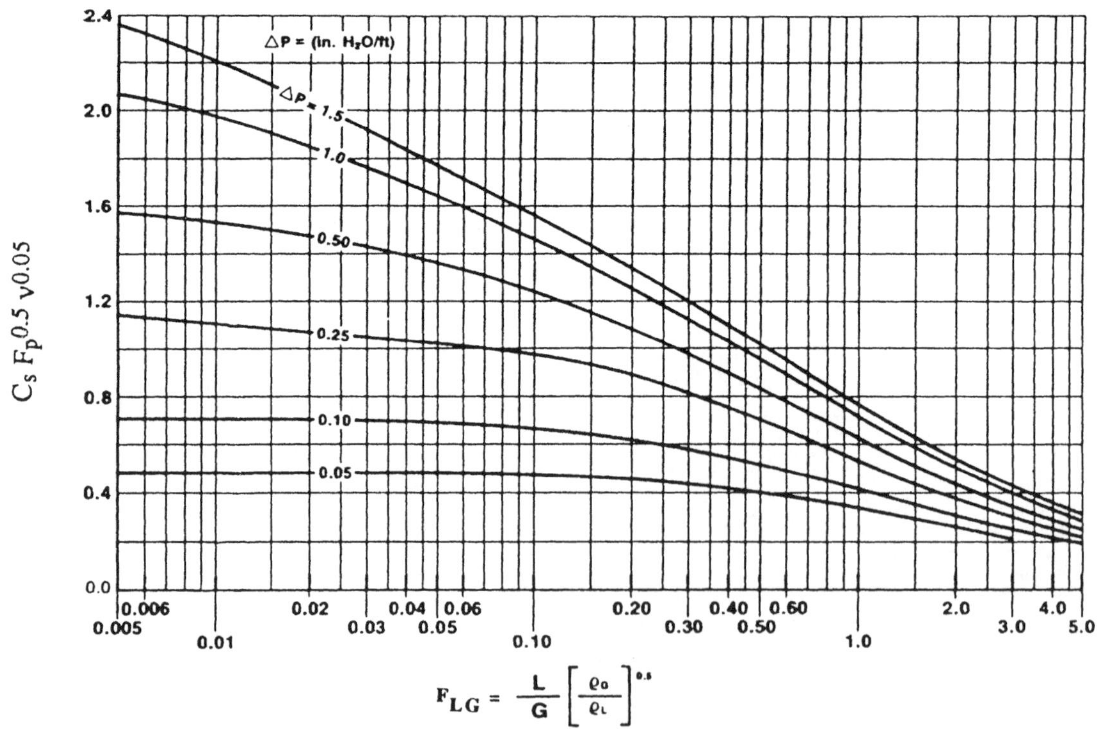

FIG. 14-55 Generalized pressure drop correlation of Eckert as modified by Strigle. To convert inches H_2O to mm H_2O, multiply by 83.31. (*From* Packed Tower Design and Applications *by Ralph E. Strigle, Jr. Copyright © 1994 by Gulf Publishing Co., Houston, Texas. Used with permission. All rights reserved.*)

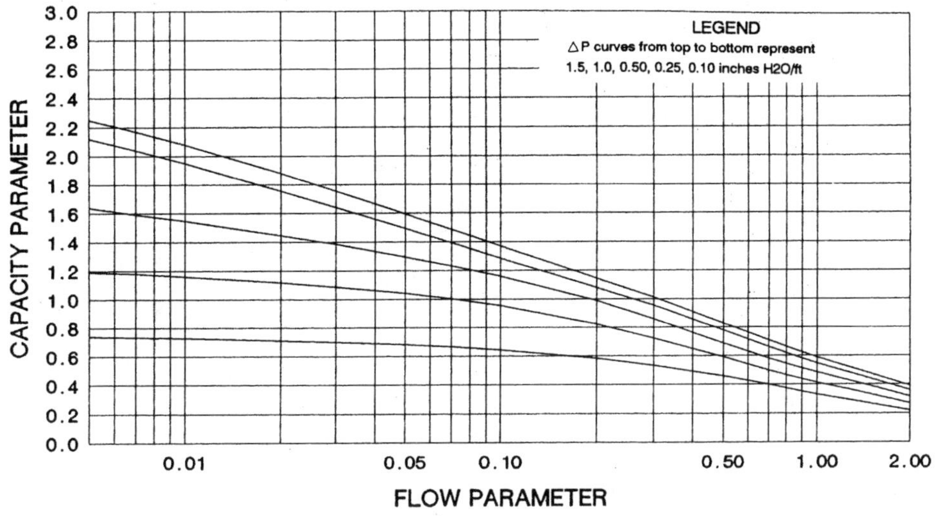

FIG. 14-56 The Kister and Gill GPDC (SP) chart for structured packings only. Abscissa and ordinate same as in Fig. 14-55. *(From Kister, H. Z., and D. R. Gill,* IChemE Symp. Ser. ***128,*** *p. A109, 1992. Reprinted courtesy of IChemE.)*

updates in Kister, Lason, and Gill, Paper presented at the AIChE Spring National Meeting, Houston, Tex., March 19–23, 1995; and in Kister, Scherffius, Afshar, and Abkar, in Distillation 2007: *Topical Conference Proceedings,* 2007 AIChE Spring National Meeting, Houston, Texas. The latter reference also discusses correct and incorrect applications of those interpolation charts.

There are many alternative methods for flood and pressure drop prediction. The Billet and Schultes [*IChemE Symp. Ser.* **104,** pp. A171 and B255 (1987)] and the Maćkowiak ("Fluiddynamik von Kolonnen mit Modernen Füllkorpern und Packungen für Gas/Flussigkeitssysteme," Otto Salle Verlag, Frankfurt am Main und Verlag Sauerländer Aarau, Frankfurt am Main, 1991) correlations are versions of the GPDC that take the liquid holdup into account. The Eiden and Bechtel correlation [*IChemE Symp. Ser.* **142,** p. 757 (1997)] is a version of the GPDC in which accuracy is improved by using constants representative of packing shape instead of packing factors. The Lockett and Billingham correlation (*IChemE Symp. Ser.* **152,** p. 400, London, 2006) uses a Wallis correlation

$$C_G^{0.5} + mC_L^{0.5} = C_{LG} \tag{14-143}$$

where
$$C_L = u_L[\rho_L/(\rho_L - \rho_G)]^{0.5} \tag{14-144}$$

and was shown to work well for high-surface-area (>400 m^2/m^3) structured packings. Here C_G is the gas C-factor, Eq. (14-77), based on the tower superficial cross-sectional area, and m and C_{LG} are constants, available from the cited reference for some packing.

A drawback of most of these correlations (except that of Eiden and Bechtel) is the unavailability of constants for many, often most, of the modern popular packings.

The above methods apply to nonfoaming systems. Foaming systems can be handled either by applying additional derating (system) factors to the flood correlation (see Table 14-9) or by limiting the calculated pressure drop to 0.25 in of water per foot of packing (Hausch, "Distillation Tools for the Practicing Engineer," Topical Conference Proceedings, p. 119, AIChE Spring Meeting, New Orleans, March 10–14, 2002).

Pressure Drop The GPDC discussed above (Figs. 14-55 and 14-56) and the Kister and Gill interpolation charts provide popular methods for calculating packing pressure drops. An alternative popular method that is particularly suitable for lower liquid loads was presented by Robbins (below).

For gas flow through dry packings, pressure drop may be estimated by use of an orifice equation. For irrigated packings, pressure drop increases because of the presence of liquid, which effectively decreases the available cross section for gas flow (Fig. 14-53). In principle, there should be a method for correcting the dry pressure drop for the presence of liquid. This approach was used by Leva [*Chem. Eng. Progr. Symp. Ser. No. 10,* **50,** 51 (1954)]. A more recent method by Robbins [*Chem. Eng. Progr.,* p. 87 (May 1991)] utilizes the same approach and is described here. The total pressure drop is

$$\Delta P_t = \Delta P_d + \Delta P_L \tag{14-145}$$

where ΔP_t = total pressure drop, inches H_2O per foot of packing

ΔP_d = dry pressure drop = $C_3 G_f^2 10^{(C_4 L_f)}$ (14-146)

ΔP_L = pressure drop due to liquid presence

= $0.4[L_f/20{,}000]^{0.1}[C_3 G_f^2 10^{(C_4 L_f)}]^4$ (14-147)

G_f = gas loading factor = $986F_s(F_{pd}/20)^{0.5}$ (14-148)

L_f = liquid loading factor = $L(62.4/\rho_L)(F_{pd}/20)^{0.5}\mu_L^{0.1}$ (14-149)

The term F_{pd} is a dry packing factor, specific for a given packing type and size. Values of F_{pd} are given in Tables 14-13 and 14-14. For operating pressures above atmospheric, and for certain packing sizes, L_f and G_f are calculated differently:

$$G_f = 986F_s(F_{pd}/20)^{0.5}10^{0.3\rho_G} \tag{14-150}$$

$$L_f = L(62.4/\rho_L)(F_{pd}/20)^{0.5}\mu_L^{0.2} \qquad F_{pd} > 200 \tag{14-151a}$$

$$L_f = L(62.4/\rho_L)(20/F_{pd})^{0.5}\mu_L^{0.1} \qquad F_{pd} < 15 \tag{14-151b}$$

The Robbins equations require careful attention to dimensions. However, use of the equations has been simplified through the introduction of Fig. 14-58. The terms L_f and G_f are evaluated, and the ΔP_L is obtained directly from the chart. Basic nomenclature for the Robbins method follows:

$C_3 = 7.4(10)^{-8}$
$C_4 = 2.7(10)^{-5}$
F_{pd} = dry packing factor, ft^{-1}
F_s = superficial F-factor for gas, $U_t\rho_g^{0.5}$, $ft/s(lb/ft^3)^{0.5}$
G = gas mass velocity, lb/hr·ft^2
G_f = gas loading factor, lb/hr·ft^2
L = liquid mass velocity, lb/hr·ft^2
L_f = liquid loading factor, lb/hr·ft^2
ΔP = pressure drop, inches H_2O/ft packing (× 83.3 = mm H_2O/m packing)

TABLE 14-13 Characteristics of Random Packings

Name	Size, mm, or no. (#)	Bed density,[*] kg/m^3	Area, m^2/m^3	% voids	Packing factor, m^{-1} Normal F_p[†]	Packing factor, m^{-1} Dry F_{pd}[‡]	Vendor
			Metals				
Pall rings (also Flexi-rings, Ballast rings, P-rings)	16	510	360	92	256	262	Various
	25	325	205	94	183	174	
	38	208	130	95	131	91	
	50	198	105	96	89	79	
	90	135	66	97	59	46	
Metal Intalox (IMTP) [also I-rings, AHPP, RSMR, MSR][§]	25	224	207	97	134	141	Koch-Glitsch [Sulzer, Amistco, Rauschert, Montz][§]
	40	153	151	97	79	85	
	50	166	98	98	59	56	
	70	141	60	98	39	—	
Nutter rings	#0.7	177	226	98	—	128	Sulzer
	#1	179	168	98	98	89	
	#1.5	181	124	98	79	66	
	#2	144	96	98	59	56	
	#2.5	121	83	98	52	49	
	#3.0	133	66	98	43	36	
Raschig Super-ring	#0.5	275	250	98	—	—	Raschig
	#0.7	185	175	98	—	—	
	#1	220	160	98	82	—	
	#1.5	170	115	98	59	—	
	#2	155	98	99	49	—	
	#3	150	80	98	36	—	
Cascade mini-rings (CMR)	#1	389	250	96	131	102	Koch-Glitsch
	#1.5	285	190	96	95	—	
	#2	234	151	97	72	79	
	#2.5	195	121	97	62	—	
	#3	160	103	98	46	43	
	#4	125	71	98	33	32	
	#5	108	50	98	26[¶]	—	
Fleximax	#300	—	141	98	85	—	Koch-Glitsch
	#400	—	85	98	56	—	
Jaeger Tripacks (Top-Pak)	#1	223	118	96	85	—	Raschig
	#2	170	75	98	46	—	
VSP	25	352	206	98	105[¶]	—	Raschig
	50	296	112	96	69	—	
Ralu-rings	25	310	215	98	157[¶]	—	Raschig
	38	260	135	97	92[¶]	—	
	50	200	105	98	66[¶]	—	
Hiflow	25	298	203	96	—	—	Rauschert
	50	175	92	98	52	—	
Hy-Pak, K-Pak, AIPR	#1	262	174	97	148	—	Koch-Glitsch, Amistco
	#1.5	180	118	98	95	—	
	#2	161	92	98	85	—	
	#3	181	69	98	52	—	
Raschig rings (1⁄16 in wall)	19	1500	245	80	722	—	Various
	25	1140	185	86	472	492	
	50	590	95	92	187	223	
	75	400	66	95	105	—	
			Ceramics				
Berl saddles	6	900	900	60	—	2950	Various
	13	865	465	62	790	900	
	25	720	250	68	360	308	
	38	640	150	71	215	154	
	50	625	105	72	150	102	
Intalox, Flexi-saddles, Torus-saddles, Novalox	6	864	984	65	—	2720	Various
	13	736	623	71	660	613	
	25	672	256	73	302	308	
	50	608	118	76	131	121	
	75	576	92	79	72	66	
Raschig rings	6	960	710	62	—	5250	Various
	13	880	370	64	1900	1705	
	25	670	190	74	587	492	
	50	660	92	74	213	230	
	75	590	62	75	121	—	
Pall ring	25	620	220	75	350	—	Raschig
	38	540	164	78	180	—	
	50	550	121	78	142	—	
	80	520	82	78	85[¶]	—	

TABLE 14-13 Characteristics of Random Packings *(Concluded)*

Name	Size, mm, or no. (#)	Bed density* kg/m³	Area, m²/m³	% voids	Packing factor, m⁻¹ Normal F_p†	Dry F_{pd}‡	Vendor
			Ceramics				
Hiflow rings	38	409	108	83	121	—	Rauschert
	50	405	89	82	95	—	
	70	333	52	86	49	—	
			Plastics				
Pall rings	15	95	350	87	320	348	Various
	25	71	206	90	180	180	
	40	70	131	91	131	131	
	50	60	102	92	85	82	
	90	43	85	95	56	39	
Super Intalox,	25	83	207	90	131	131	Various
Flexi-saddles,	50	60	108	93	92	85	
Super-Torus, Novalox	75	48	89	96	59	46	
Cascade mini-rings	#1A	53	185	97	98	92	Koch-Glitsch
(CMR)	#2A	46	115	97	59	—	
	#3A	40	74	98	39	33	
Raschig Super-ring	#0.6	62	205	96	105¶	—	Raschig
	#2	55	100	96	49	—	
Ralu-ring	15	80	320	94	230¶	—	Raschig
	25	56	190	94	135	—	
	38	65	150	95	80	—	
	50	60	110	95	55	—	
	90	40	75	96	38	—	
	125	30	60	97	30¶	—	
Snowflake	—	51	92	97	43	—	Koch-Glitsch
Nor-Pac	25	72	180	92	102	—	NSW
	38	61	144	93	69	—	
	50	53	102	94	46	—	
Tri-Packs	#1	67	158	93	53¶	—	Raschig
(Hacketten)	#2	53	125	95	39¶	43	
Ralu-Flow	#1	55	165	95	67¶	—	Raschig
	#2	54	100	95	38¶	—	
Hiflow	25	63	192	92	138	—	Rauschert
	50	59	110	93	66	—	
	90	34	65	97	30	—	
Lanpac	90	67	148	93	46	—	Lantec
Impac	#3	83	213	91	49	—	
Tellerettes	25	112	180	87	—	131	Ceilcote
	50	59	125	93	—	—	

*Values are approximate and may not be consistent. Actual number depends on wall thickness and material.

†Packing factor F_p from Kister, *Distillation Design*, McGraw-Hill, 1992; Kister, Larson, and Gill, paper presented at the Houston AIChE meeting, March 19–23, 1995; Strigle, *Packed Tower Design and Applications*, 2d ed., Gulf Publishing Co., Houston, Tex., 1994; Kister et al., in Distillation 2007. Topical Conference Proceedings, 2007 AIChE Spring National Meeting, Houston, Tex.

‡Dry packing factor F_{pd} from Robbins, *Chem. Eng. Progr.*, **87**(1), 19 (1990).

§The bracketed packings are similar to, but not the same as, the Koch-Glitsch IMTP. Some of them have holes that the IMTP do not have, and others have sizes that are different and are made by different dies.

¶Packing factor supplied by packing vendor.

Vendors: Koch-Glitsch LLP, Wichita, Kansas; Raschig GmbH, Ludwigshafen/Rhein, Germany; Sulzer Chemtech Ltd., Winterthur, Switzerland; Rauschert Verfahrenstechnik GmbH, Steinwiesen, Germany; Amistco Separation Products Inc., Alvin, Texas; Julius Montz GmbH, Hilden, Germany; Ceilcote Co., Berea, Ohio; NSW Corp., Roanoke, Virginia; Lantec Products Inc., Agoura Hills, California.

ρ_G = gas density, lb/ft³
ρ_L = liquid density, lb/ft³
μ_L = liquid viscosity, cP

The Robbins correlation applies near atmospheric pressure and under vacuum, but is not suitable above 3 bar absolute. For high (>0.3) flow parameters [Eq. (14-141)], the correlation has only been tested with air-water data.

For flood and MOC predictions, Robbins recommends his pressure drop method together with Eqs. (14-142) (flood) and (14-139) (MOC).

The GPDC and Robbins correlations are empirical. Fundamental correlations are also available. Most of these use the channel model, which attributes the pressure drop to the resistance to flow in a multitude of parallel channels. The channels may have bends, expansions, and contractions. Popular application of this approach are the Rocha et al. correlation [Rocha, Bravo, and Fair, *Ind. Eng. Chem. Res.* **32,** 641 (1993)] for structured packing and the Maćkowiak ("Fluiddynamik von Kolonnen mit Modernen Füllkorpern und Packungen für Gas/Flussigkeitssysteme," Otto Salle Verlag, Frankfurt am Main und Verlag Sauerländer Aarau, Frankfurt am Main, 1991) and Billet *(Packed Column*

TABLE 14-14 Characteristics of Structured Packings

Name	Size or number	Area, m^2/m^3	% voids*	Packing factor, m^{-1}: Normal F_p†	Packing factor, m^{-1}: Dry F_{pd}‡	Vendor
	Metals, corrugated sheets					
Mellapak	125Y	125	99	33		Sulzer
	170Y	170	99	39		
	2Y	223	99	46		
	250Y	250	98	66		
	350Y	350	98	75		
	500Y	500	98	112		
	750Y	750	97			
	125X	125	99	16		
	170X	170	99	20		
	2X	223	99	23		
	250X	250	98	26		
	500X	500	98	82		
Mellapak Plus	202Y		99			Sulzer
	252Y	250	98	39		
	352Y		98			
	452Y	350	98	69		
	752Y	500	98	131		
Flexipac	700Y	710	96			Koch-Glitsch
	500Y	495	97			
	1Y	420	98	98	(105)	
	350Y	350	98			
	1.6Y	290	98	59		
	250Y	250	99			
	2Y	220	99	49	(36)	
	2.5Y	150	99			
	3.5Y	80	99	30	(15)	
	4Y	55	99	23	(10.5)	
	1X	420	98	52		
	350X	350	98			
	1.6X	290	98	33		
	250X	250	99			
	2X	220	99	23		
	2.5X	150	99			
	3X	110	99	16		
	3.5X	80	99			
	4X	55	99			
Flexipac High-capacity	700	710	96	223		Koch-Glitsch
	500Z	495	97			
	1Y	420	98	82		
	350Y	350	98			
	1.6Y	290	99	56		
	250Y	250	99			
	2Y	220	99	43		
Intalox	1T	310	98	66		Koch-Glitsch
	1.5T	250	99			
	2T	215	99	56		
	3T	170	99	43		
	4T	135	99			
	5T	90	99			
	5TX	90	99			
Super-Pak	250	250	98	55§		Raschig
	350	350	98	70§		
Ralu-Pak	250YC	250	98	66		Raschig
Rhombopac	4M	151				Kuhni
	6M	230		59		
	9M	351				
Max-Pak	0.5-in	229	98	39		Raschig
Montz-Pak	B1-125	125	97			Montz
	B1-200	200				
	B1-250	250	95	66		
	B1-350	350	93			
	BSH-250¶	250	95			
	BSH-500¶	500	91			
	B1-250M	250		43		
	B1-350M	350				
	B1-500M	500				
	Wire Mesh					
Sulzer	AX	250	95			Sulzer
	BX	492	90	69	(52.5)	
	CY	700	85			
	BX Plus					
Wire gauze	BX	495	93			Koch-Glitsch
Montz-Pak	A3-500	500	91			Montz
Goodloe	765	1010	96			Koch-Glitsch
	773	1920	95			
	779	2640	92			
Hyperfil	2300	2300	93.6	394/230§,‖	460	Knit Mesh
	1900	1900	94.8	312/180§,‖		
	1400	1400	96.0	180/131§,‖		
	Ceramic					
Flexeramic	28	260	66	131		Koch-Glitsch
	48	160	77	79		
	88	100	86	49		
	Plastic					
Mellapak	125Y	125				Sulzer
	250Y	250		72		
	125X	125				
	250X	250				
Ralu-Pak	30/160	160	92			Raschig
Multifil plastics	P1500	1500	88.5			Knit Mesh

*% voids vary with material thickness and values may not be consistent.
†Packing factors from Kister, *Distillation Design*, McGraw-Hill, 1992; Kister, Larson, and Gill, paper presented at the Houston AIChE Meeting, March 19–23, 1995; and Kister et al., in *Distillation 2007: Proceedings of Topical Conference*, AIChE Spring Meeting, Houston, Tex., April 22–26, 2007.
‡Dry packing factors from Robbins, *Chem. Eng. Prog.*, p. 87, May 1991.
§These packing factors supplied by the packing vendor.
¶These are expanded metal packings.
‖First figure is for hydrocarbon service, second figure for aqueous service.
Vendors: Sulzer Chemtech Ltd., Winterthur, Switzerland; Koch-Glitsch LLP, Wichita, Kansas; Raschig GmbH, Ludwigshafen/Rhein, Germany; Julius Montz GmbH, Hilden, Germany; Knit Mesh/Enhanced Separation Technologies, Houston, Texas; Kuhni Ltd., Allschwil, Switzerland.

Analysis and Design, Ruhr University, Bochum, Germany, 1989) methods. Stichlmair et al. (*Distillation Principles and Practices*, Wiley, New York, 1998; *Gas Sep. Purif.* **3,** March 1989, p. 19) present alternative correlations using the particle model, that attributes packing pressure drop to friction losses due to drag of a particle. This is similar to the Ergun model for single-phase flow [*Chem. Eng. Prog.* **48**(2), 89 (1952)].

Example 13: Packed-Column Pressure Drop Air and water are flowing countercurrently through a bed of 2-inch metal Pall rings. The air mass velocity is 2.03 kg/s·m² (1500 lbs/hr·ft²), and the liquid mass velocity is 12.20 kg/s·m² (9000 lbs/hr·ft²). Calculate the pressure drop by the generalized pressure drop (GPDC, Fig. 14-55) and the Robbins methods. Properties: $\rho_G = 0.074$ lbs/ft³; $\rho_L = 62.4$ lbs/ft³, $\mu_L = 1.0$ cP, $\nu = 1.0$ cS. The packing factor $F_p = 27$ ft^{-1}. For Robbins, $F_{pd} = 24$ ft^{-1}. The flow parameter $F_{LG} = L/G\ (\rho_G/\rho_L)^{0.5} = (9000/1500)\ (0.074/62.4)^{0.5} = 0.207$. The F-factor $= F_s = U_t\rho_G^{0.5} = G/(\rho_G^{0.5} 3600) = 1500/[(0.074)^{0.5}\ (3600)] = 1.53$ ft/s(lb/ft³)$^{0.5}$.

Using the GPDC method, the capacity parameter [by Eq. (14-140)] = $U_t[\rho_G/(\rho_L - \rho_G)]^{0.5} F_p^{0.5} \nu^{0.05}$, which is roughly equivalent to

$$\frac{F_s}{\rho_L^{0.5}} F_p^{0.5} \nu^{0.05} = \frac{1.53}{62.4^{0.5}} 27^{0.5}(1.0)$$

$$= 1.01$$

Referring to Fig. 14-55, the intersection of the capacity parameter and the flow parameter lines gives a pressure drop of 0.38 inches H_2O/ft packing.

Using the Robbins method, $G_f = 986F_s(F_{pd}/20)^{0.5} = 986(1.53)(24/20)^{0.5} = 1653$. $L_f = L\ (62.4/\rho_L)(F_{pd}/20)^{0.5} \mu^{0.1} = 9000\ (1.0)(1.095)(1.0) = 9859$. $L_f/G_f = 5.96$.

From Fig. 14-58, pressure drop = 0.40 in. H_2O/ft packing.

PACKING EFFICIENCY

HETP vs. Fundamental Mass Transfer The two-film model gives the following transfer unit relationship:

$$H_{OG} = H_G + \lambda H_L \qquad (14\text{-}152)$$

where H_{OG} = height of an overall transfer unit, gas concentration basis, m

H_G = height of a gas-phase transfer unit, m

H_L = height of a liquid-phase transfer unit, m

$\lambda = m/(L_M/G_M)$ = slope of equilibrium line/slope of operating line

In design practice, a less rigorous parameter, HETP, is used to express packing efficiency. The HETP is the height of packed bed required to achieve a theoretical stage. The terms H_{OG} and HETP may be related under certain conditions:

$$\text{HETP} = H_{OG}\left[\frac{\ln \lambda}{(\lambda - 1)}\right] \qquad (14\text{-}153)$$

and since

$$Z_p = (H_{OG})(N_{OG}) = (\text{HETP})(N_t) \qquad (14\text{-}154)$$

$$N_{OG} = N_t[\ln \lambda/(\lambda - 1)] \qquad (14\text{-}155)$$

Equations (14-153) and (14-155) have been developed for binary mixture separations and hold for cases where the operating line and equilibrium line are straight. Thus, when there is curvature, the equations should be used for sections of the column where linearity can be assumed. When the equilibrium line and operating line have the same slope, HETP = H_{OG} and $N_{OG} = N_t$ (theoretical stages).

An alternative parameter popular in Europe is the NTSM (number of theoretical stages per meter) which is simply the reciprocal of the HETP.

Factors Affecting HETP: An Overview Generally, packing efficiency increases (HETP decreases) when the following occur.

- Packing surface area per unit volume increases. Efficiency increases as the particle size decreases (random packing, Fig. 14-59) or as the channel size narrows (structured packing, Fig. 14-60).
- The packing surface is better distributed around a random packing element.
- Y structured packings (45° inclination) give better efficiencies than X structured packings (60° inclination to the horizontal) of the same surface areas (Fig. 14-60).
- For constant *L/V* operation in the preloading regime, generally liquid and vapor loads have little effect on random and most corrugated sheet structured packings HETP (Figs. 14-59 and 14-60). HETP increases with loadings in some wire-mesh structured packing.
- Liquid and vapor are well distributed. Both liquid and vapor maldistribution have a major detrimental effect on packing efficiency.
- Other. These include *L/V* ratio (lambda), pressure, and physical properties. These come into play in some systems and situations, as discussed below.

HETP Prediction HETP can be predicted from mass-transfer models, rules of thumb, and data interpolation.

Mass-Transfer Models Development of a reliable mass-transfer model for packing HETP prediction has been inhibited by a lack of understanding of the complex two-phase flow that prevails in packings, by the shortage of commercial-scale efficiency data for the newer packings, and by difficulty in quantifying the surface generation in modern packings. Bennett and Ludwig (*Chem. Eng. Prog.*, p. 72, April 1994) point out that the abundant air-water data cannot be reliably used for assessing real system mass-transfer resistance due to variations in turbulence, transport properties, and interfacial areas. More important, the success and reliability of rules of thumb for predicting packing efficiency made it difficult for mass-transfer models to compete.

For random packings, the Bravo and Fair correlation [*Ind. Eng. Chem. Proc. Des. Dev.* **21,** 162 (1982)] has been one of the most popular theoretical correlations. It was shown (e.g., McDougall, *Chem SA*, p. 255, October 1985) to be better than other theoretical correlations, yet produced large discrepancies when compared to test data [Shariat and Kunesh, *Ind. Eng. Chem. Res.* **34**(4), 1273 (1995)]. For structured packings, the Bravo, Fair, and Rocha correlation [*Chem. Eng. Progr.* **86**(1), 19 (1990); *Ind. Eng. Chem. Res.* **35,** 1660 (1996)] is one of the most popular theoretical correlations. This correlation is based on the two-film theory. Interfacial areas are calculated from the packing geometry and an empirical wetting parameter.

Alternate popular theoretical correlations for random packings, structured packings, or both (e.g., Billet and Schultes, "Beitrage zur Verfahrens-und Umwelttechnik," p. 88, Ruhr Universitat, Bochum, Germany, 1991) are also available.

Rules of Thumb Since in most circumstances packing HETP is sensitive to only few variables, and due to the unreliability of even the best mass-transfer model, it has been the author's experience that rules of thumb for HETP are more accurate and more reliable than mass-transfer models. A similar conclusion was reached by Porter and Jenkins (*IChemE Symp. Ser.* **56,** Summary paper, London, 1979).

The majority of published random packing rules of thumb closely agree with one another. They are based on second- and third-generation random packings and should not be applied to the obsolete first-generation packings. Porter and Jenkins's (loc. cit.), Frank's (*Chem. Eng.*, p. 40, March 14, 1977), Harrison and France's (*Chem. Eng.*, p. 121, April 1989), Chen's (*Chem. Eng.*, p. 40, March 5, 1984), and Walas' (*Chem. Eng.*, p. 75, March 16, 1987) general rules of thumb are practically the same, have been successfully tested against an extensive data bank, and are slightly conservative, and therefore suitable for design.

For small-diameter columns, the rules of thumb presented by Frank (loc. cit.), Ludwig (*Applied Process Design for Chemical and Petrochemical Plants*, vol. 2, 2d ed., Gulf Publishing, Houston, Tex., 1979), and Vital et al. [*Hydrocarbon Processing*, **63**(12), 75 (1984)] are identical. The author believes that for small columns, the more conservative value predicted from either the Porter and Jenkins or the Frank-Ludwig-Vital rule should be selected for design. Summarizing:

$$\text{HETP} = 18D_P \qquad (14\text{-}156a)$$

$$\text{HETP} > D_T \qquad \text{for } D_T < 0.67 \text{ m} \qquad (14\text{-}156b)$$

where D_P and D_T are the packing and tower diameters, *m*, respectively, and the HETP is in meters. In high-vacuum columns (<100 mbar), and when $\lambda = mG_M/L_M$ is outside the range of 0.5 to 2.0, the above rules may be optimistic (see below).

The above rules of thumb were based on experience with Pall rings. The packing diameter may be difficult to establish for some of the modern packings, especially those of saddle or flat shape. For these,

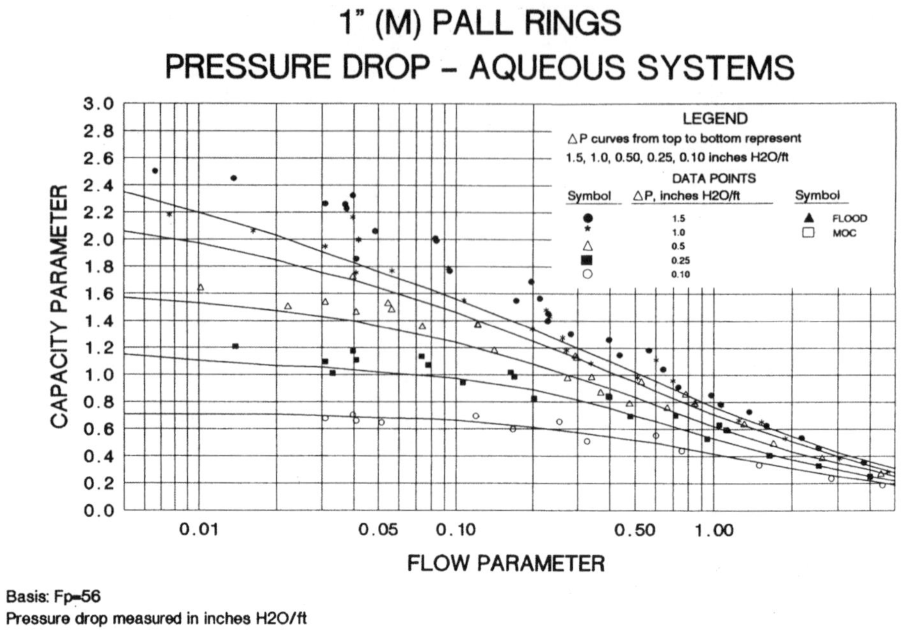

(a)

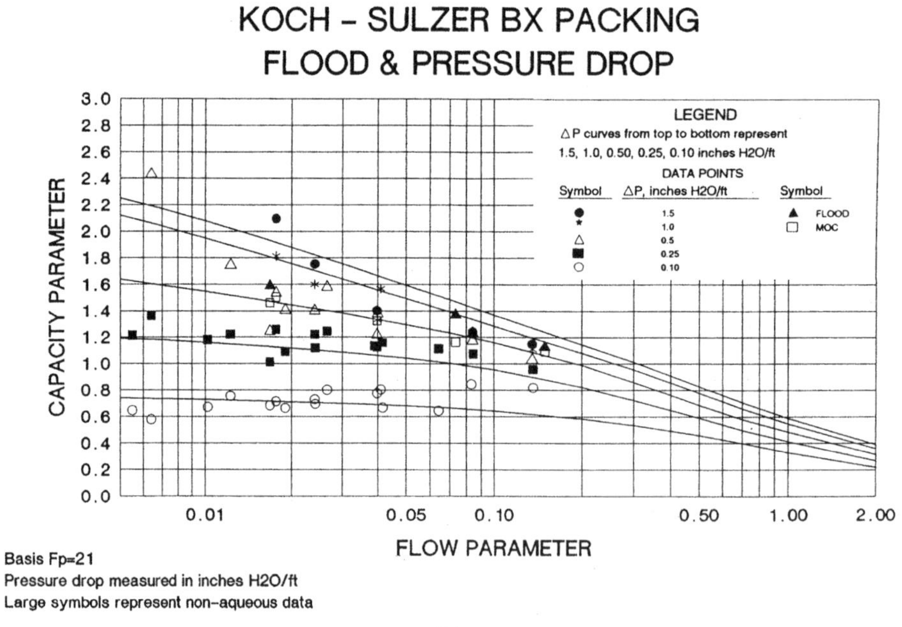

(b)

FIG. 14-57 Superimposing experimental pressure-drop data for a given packing generates a GPDC interpolation chart for this packing. (*a*) A random packing; chart is based on Eckert's GPDC, Fig. 14-55. (*b*) A structured packing; chart is based on Kister and Gill's GPDC (SP), Fig. 14-56. (*From Kister, H. Z.,* Distillation Design, *copyright © McGraw-Hill, 1992; used with permission.)*

Kister and Larson (in Schweitzer, *Handbook of Separation Techniques for Chemical Engineers,* 3d ed., McGraw-Hill, 1997) extended Eq. (14-156*a*) by expressing the packing diameter in terms of the more fundamental surface area per unit volume a_P, m²/m³. For Pall rings, it can be shown that

$$a_P = 5.2/D_P \tag{14-157}$$

and Eq. (14-156*a*) becomes

$$\text{HETP} = 93/a_P \tag{14-158}$$

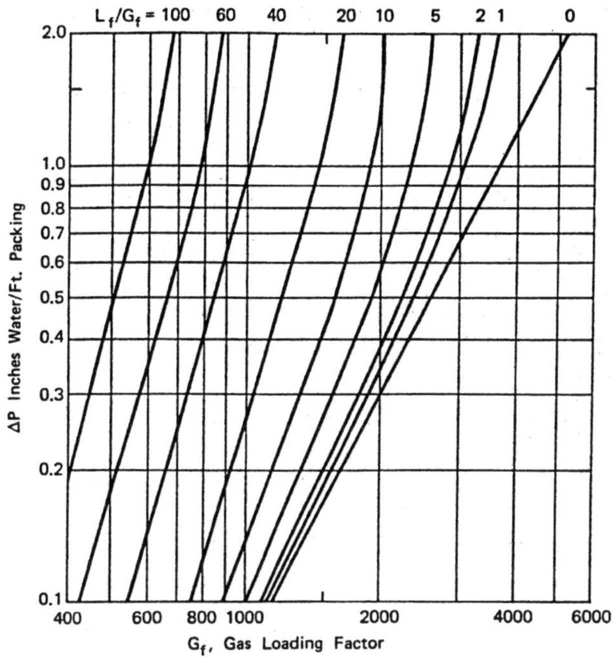

FIG. 14-58 The Robbins generalized pressure-drop correlation. (*From L. A. Robbins.* Chem Eng. Progr., *May 1991. p. 87, reprinted courtesy of the American Institute of Chemical Engineers.*)

Harrison and France (loc. cit.) presented the only published rule of thumb for structured packings efficiency as a function of packing crimp. Kister and Larson reexpressed it in terms of the surface area per unit volume to accommodate a wider range of packing geometries. The final expression is

$$\text{HETP} = 100C_{XY}/a_p + 0.10 \tag{14-159}$$

Specific surface areas are listed in Tables 14-13 and 14-14.

The above rules of thumb apply to organic and hydrocarbon systems, whose surface tensions are relatively low ($\sigma < 25$ mN/m). For higher surface tensions, the liquid does not adhere well to the packing surfaces (*underwetting*), causing higher HETPs. In a water-rich system ($\sigma = 70$ mN/m or so) HETPs obtained from Eqs. (14-156), (14-158), and (14-159) need to be doubled. For intermediate surface tension systems (some amines and glycols, whose surface tension at column conditions is 40 to 50 mN/m), HETPs obtained from Eqs. (14-156), (14-158), and (14-159) need to be multiplied by 1.5.

For random packings, Eqs. (14-156) and (14-158) apply for packings of 25-mm diameter and larger. For smaller packings, use of the a_p at 25-mm often gives a slightly conservative HETP estimate. For structured packing, C_{XY} in Eq. (14-159) reflects the effect of the inclination angle (*Fig. 14*.60) CΞΨ = 1 for Y-type, S-type, or high-capacity packings, and C_{XY} =1.45 for the larger (<300 m^2/m^3) X-type packings. There are insufficient data to determine C_{XY} for high specific area X-type packings, but Fig. 14-60 suggests it is somewhat lower than 1.45.

Compared to experimental data, the above rules of thumb are slightly conservative. Since packing data are usually measured with perfect distribution, a slight conservative bias is necessary to extend

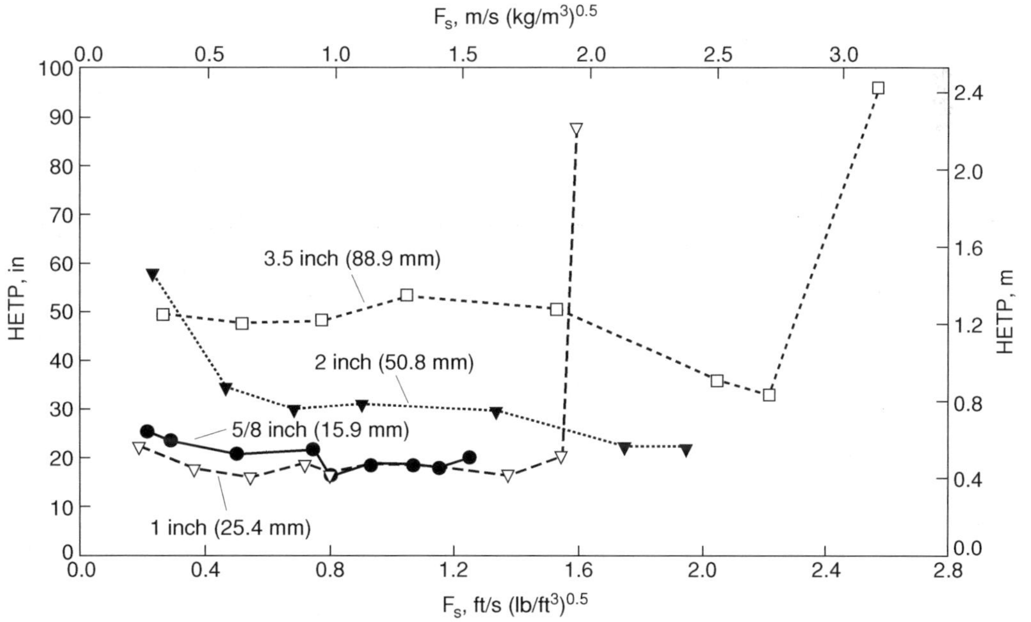

FIG. 14-59 HETP values for four sizes of metal pall rings, vacuum operation. Cyclohexane/*n*-heptane system, total reflux, 35 kPa (5.0 psia). Column diameter = 1.2 m (4.0 ft). Bed height = 3.7 m (12 ft). Distributor = tubed drip pan, 100 streams/m^2. [*Adapted from Shariat and Kunesh,* Ind. Eng. Chem. Res., ***34,*** *1273 (1995). Reproduced with permission. Copyright © 1995 American Chemical Society.*]

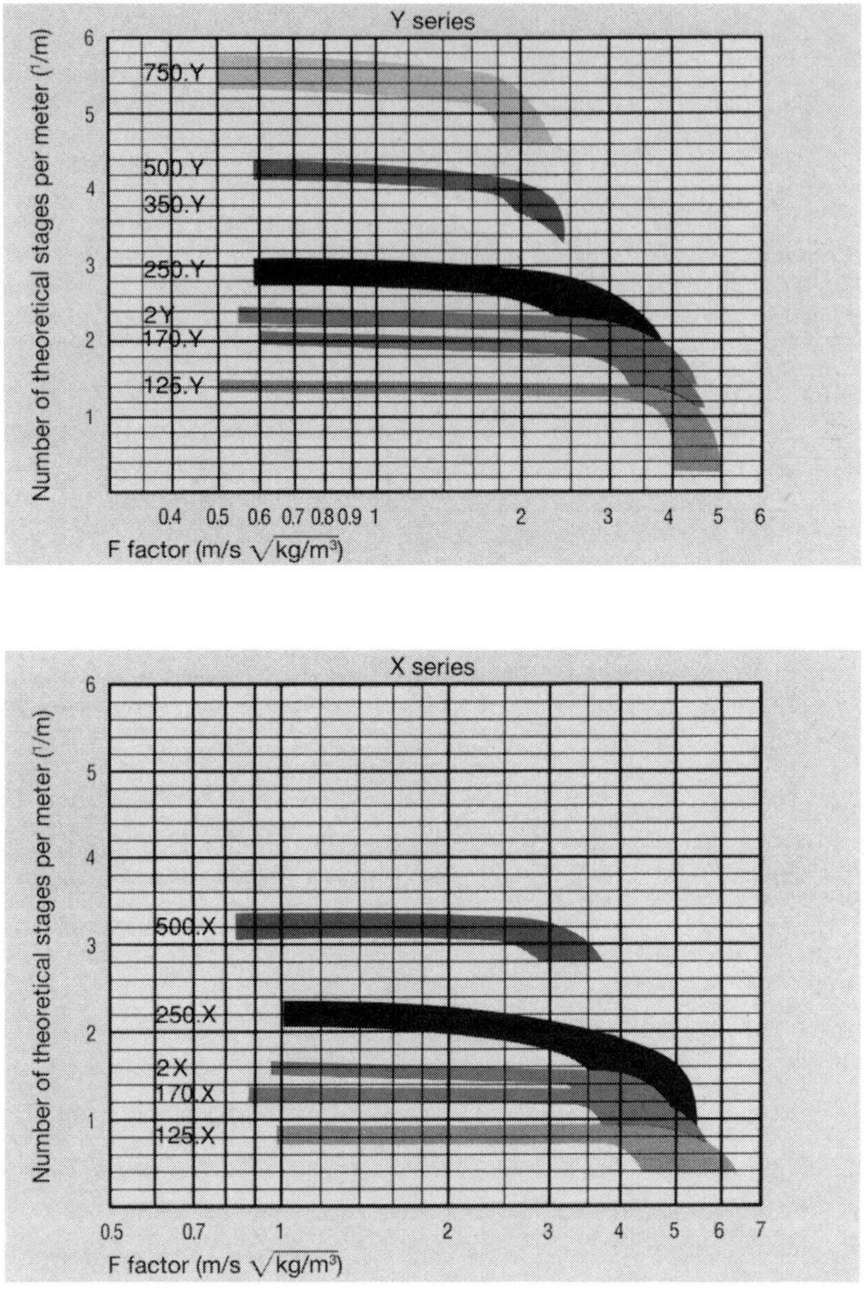

FIG. 14-60 Effect of structured packing surface areas, loads, and inclination angle on packing efficiency. Efficiency expressed as number of theoretical stages per meter, the reciprocal of HETP. Sulzer data, chlorobenzene–ethylbenzene, 100 mbar, at total reflux; 250-mm-diameter test column. (*Reprinted courtesy of Sulzer Chemtech.*)

these data to the good, yet imperfect, distributors used in the real world. For poor distributors, the above rules of thumb will usually predict well below the HETPs measured in the field.

Lockett (*Chem. Eng. Progr.*, p. 60, January 1998) simplified the fundamental Bravo-Fair-Rocha correlation [*Ind. Eng. Chem. Res.* **35**, p. 1660 (1996)] to derive an alternative rule of thumb for structured packing efficiency. This rule of thumb predicts HETPs under perfect distribution conditions. Lockett recommends caution when applying this rule of thumb for aqueous systems as it does not predict the effects of underwetting (below).

Service-Oriented Rules of Thumb Strigle (*Packed Tower Design and Applications*, 2d ed., Gulf Publishing, Houston, Tex., 1994) proposed a multitude of rules of thumb as a function of the service, column pressure, and physical properties. These rules are based on the extensive experience of Strigle and the Norton Company (now merged with Koch-Glitsch LP).

Data Interpolation Interpolation of experimental HETP data is the most reliable means of obtaining design HETP values. This is hardly surprising in an area where our understanding of the theory is so poor that rules of thumb can do better than theoretical models. The author believes that it is best to derive HETP from experimental data, and to check it against a rule of thumb.

Eckert [*Chem. Eng. Progr.* **59**(5), 76 (1963)], Chen (*Chem. Eng.* p. 40, March 5, 1984), and Vital et al. [*Hydroc. Proc.* **63**(12), 75 (1984)] tabulated experimental HETP data for various random packings. Kister (*Distillation Design,* McGraw-Hill, 1992) extended these tabulations and included published HETP data and a detailed procedure for interpolating such HETP data. A prerequisite to any interpolation of packing data is thorough familiarity with the factors that affect HETP. Overlooking any of the factors listed can easily lead to poor interpolation and grossly incorrect design. In particular, it is imperative to recognize that the quality of distribution in pilot towers is generally superior to the quality of distribution in commercial towers.

Underwetting Laboratory- and pilot-scale distillation experiments with systems that exhibit large differences in surface tension along the column such as methanol-water showed a sharp drop in efficiency at the high-surface-tension end of the column [Ponter et al., *Trans. Instn. Chem. Engineers [London],* **45,** T345 (1967)]. There appeared to be a critical methanol composition below which performance deteriorated rapidly. The poor performance at the low-methanol-concentration end appeared independent of the type and size of packing. Visual observations with disk columns attributed these effects to underwetting.

Underwetting is a packing surface phenomenon, which breaks up liquid film. The tendency of the liquid film to break (the degree of wetting) is expressed by the contact angle (Fig. 14-61). A contact angle of 0° indicates perfect wetting; an angle of 180° indicates no wetting. Mersmann and Deixler [*Chem. Ing. Tech.* **58**(1), 19 (1986)] provide a preliminary chart for estimating contact angles. The contact angle depends on both the surface and the liquid and is a strong function of composition. In systems with large surface tension gradients, both contact angles and minimum wetting rates may vary rapidly with changes of composition or surface tension. Extensive studies by Ponter et al. [loc. cit.; also, Ponter and Au-Yeung, *Chem. Ing. Tech.*, **56**(9), 701 (1984)] showed that

- Underwetting is most significant in aqueous-organic systems, and tends to occur at the high-surface-tension (aqueous) end of the composition range. Liquid viscosity may also have an effect.
- Underwetting may be alleviated by changing the material and surface roughness of the packing.
- In systems susceptible to underwetting, column efficiency can sometimes (but not always) be improved by the addition of small amounts of surfactants.

Effect of Lambda Most packed-column efficiency testing has been at total reflux. Some tests for both random and structured packings [Billet, "Packed Towers Analysis and Design," Ruhr University, Bochum, Germany, 1989; Meier, Hunkeler, and Stocker, *IChemE Symp. Ser.* **56,** 3.3/1 (1979); Eckert and Walter, *Hydroc. Proc.* **43**(2), 107 (1964)] suggest that efficiencies at finite reflux are similar to those at total reflux when lambda ($\lambda = mG_M/L_M$, which is the ratio of the slope of the equilibrium curve to the slope of the operating line) ranges between 0.5 and 2.0. This range is typical for most distillation systems.

Koshy and Rukovena [*Hydroc. Proc.,* **65**(5), 64 (1986)], experimenting with methanol-water and water-DMF using #25 IMTP packing in a pilot-scale column, observed a sharp efficiency drop when the group λ was greater than 2 or lower than 0.5. The efficiency loss escalated as λ deviated more from this range. Koshy and Rukovena recognized that surface tension gradients and underwetting may have influenced some of their findings, but argue that the lambda effect is the major cause for the efficiency differences observed in their tests. High-relative-volatility systems are those most likely to be affected by λ, because at low volatility, λ ranges from 0.5 to 2. Strigle (loc. cit.) quantified the lambda effect on HETP using the following equation:

$$\text{Actual HETP/standard HETP} = 1 + 0.278[\text{ABS}(\ln \lambda)^3] \tag{14-160}$$

For $0.5 < \lambda < 2$, Eq. (14-160) gives a ratio of less than 1.1; that is, it has a little influence on HETP.

Pressure Generally, pressure has little effect on HETP of both random and structured packing, at least above 100 mbar abs. At deep vacuum (<100 mbar), there are data to suggest that efficiency decreases as pressure is lowered for random packings [Zelvinski, Titov, and Shalygin, *Khim Tekhnol. Topl. Masel.* **12**(10) (1966)], but most of these data can also be explained by poor wetting or maldistribution.

At high pressures (>15 to 20 bar), structured packing efficiency diminishes as pressure is raised (Fitz, Shariat, and Spiegel, Paper presented at the AIChE Spring National Meeting, Houston, Tex., March 1995). Zuiderweg and Nutter [*IChemE Symp.* **128,** A481 (1992)] report the same phenomenon, but to a much smaller extent, also in random packings. They explain the efficiency reduction at higher pressure by vapor backmixing. Nooijen et al. [*IChemE Symp. Ser.* **142,** 885 (1997)] concur, bringing a supporting case study in which high-pressure distillation efficiency improved by splitting a packed bed.

With structured packings (only) FRI's high-pressure (10 to 30 bar; flow parameters >0.25) distillation tests measured maxima, termed *humps* in the HETP vs. load plot, typically at 65 to 90 percent of flood [Fitz, Shariat, and Kunesh, *IChemE Symp. Ser.* **142,** 829 (1997); Cai et al., *Trans IChemE* **81,** Part A, p. 85 (2003)]. These humps (Fig. 14-62) were not observed with lower pressure distillation (flow parameters <0.2) and appeared to intensify with higher pressure. The humps did not always occur; some tests at different distributor positioning and with larger packing showed no humps. Zuiderweg et al. [*IChemE Symp. Ser.* **142,** 865 (1997); *Trans. IChemE* **81,** Part A, p. 85 (January 2003)] and Nooijen et al. [*IChemE Symp. Ser.* **142,** 885 (1997)] explain the humps by two-phase backmixing. At the high liquid loads in high-pressure distillation, flow unevenness prematurely floods some of the packing channels, carrying vapor bubbles downward and recirculating liquid upward.

Physical Properties Data presented by a number of workers [e.g., Vital, Grossel, and Olsen, *Hydroc. Proc.* **63**(12), 75 (1984)] suggest that, generally, random packing HETP is relatively insensitive to system properties for nonaqueous systems. A survey of data in Chapter

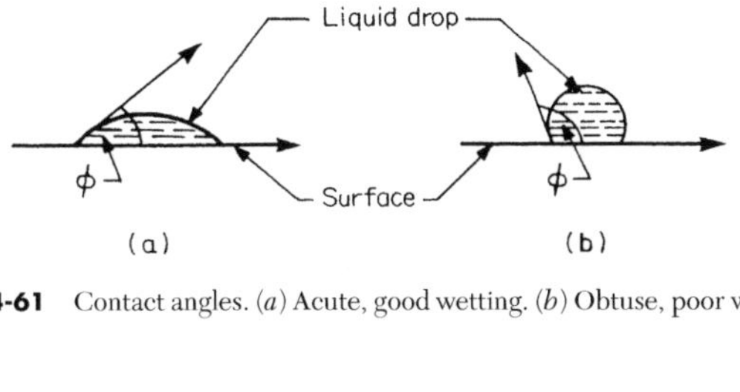

FIG. 14-61 Contact angles. (*a*) Acute, good wetting. (*b*) Obtuse, poor wetting.

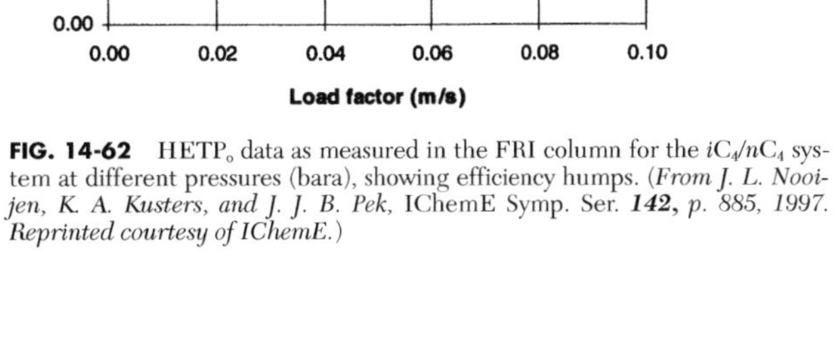

FIG. 14-62 HETP_o data as measured in the FRI column for the iC_4/nC_4 system at different pressures (bara), showing efficiency humps. (*From J. L. Nooijen, K. A. Kusters, and J. J. B. Pek,* IChemE Symp. Ser. ***142,*** *p. 885, 1997. Reprinted courtesy of IChemE.*)

11 of Kister's *Distillation Design* (McGraw-Hill, New York, 1992) leads to a similar conclusion for structured packings. For water-rich systems, packing HETPs tend to be much higher than for nonaqueous systems due to their high lambda or surface underwetting, as discussed above. High hydrogen concentrations (>30 percent or so in the gas) have also led to low packing efficiencies (Kister et al., *Proc. 4th Ethylene Producers Conference,* AIChE, New Orleans, La., p. 283, 1992), possibly due to the fast-moving hydrogen molecule dragging heavier molecules with it as it diffuses from a liquid film into the vapor.

Errors in VLE These affect packing HETP in the same way as they affect tray efficiency. The discussions and derivation earlier in this subsection apply equally to tray and packed towers.

Comparison of Various Packing Efficiencies for Absorption and Stripping In past editions of this handbook, extensive data on absorption/stripping systems were given. Emphasis was given to the following systems:

Ammonia-air-water	Liquid and gas phases contributing; chemical reaction contributing
Air-water	Gas phase controlling
Sulfur dioxide-air-water	Liquid and gas phase controlling
Carbon dioxide-air-water	Liquid phase controlling

The reader may refer to the data in the 5th edition. For the current work, emphasis will be given to one absorption system, carbon dioxide-air-caustic.

Carbon Dioxide-Air-Caustic System The vendors of packings have adopted this system as a "standard" for comparing the performance of different packing types and sizes for absorption/stripping. For tests, air containing 1.0 mol % CO_2 is passed countercurrently to a circulating stream of sodium hydroxide solution. The initial concentration of NaOH in water is 1.0 N (4.0 wt %), and as the circulating NaOH is converted to sodium carbonate it is necessary to make a mass-transfer correction because of reduced mass-transfer rate in the liquid phase. The procedure has been described by Eckert et al. [*Ind. Eng. Chem.,* **59**(2), 41 (1967); *Chem. Eng. Progr.,* **54**(1), 790 (1958)]. An overall coefficient is measured using gas-phase (CO_2) concentrations:

$$K_{OG}a_e = \frac{\text{moles } CO_2 \text{ absorbed}}{\text{time-bed volume-partial pressure } CO_2 \text{ driving force}} \tag{14-161}$$

The coefficients are usually corrected to a hydroxide conversion of 25 percent at 24°C. For other conversions, Fig. 14-14 may be used. Reported values of $K_{OG}a$ for representative random packings are given in Table 14-15. The effect of liquid rate on the coefficient is shown in Fig. 14-63.

While the carbon dioxide/caustic test method has become accepted, one should use the results with caution. The chemical reaction masks the effect of physical absorption, and the relative values in the table may not hold for other cases, where much of the resistance to mass transfer is in the gas phase. Background on this combination of physical and chemical absorption may be found earlier in the present section, under "Absorption with Chemical Reaction."

TABLE 14-15 Overall Coefficients for Representative Packings

CO_2-air-caustic system

	Nominal size, mm	Overall coefficient $K_{OG}a$, kg·moles/(hr·m³·atm)
Ceramic raschig rings	25	37.0
	50	26.1
Ceramic Intalox saddles	25	45.1
	50	30.1
Metal pall rings	25	49.6
	50	34.9
Metal Intalox saddles (IMTP®)	25	54.8
	50	39.1

NOTE: Basis for reported values: CO_2 concentration in inlet gas, 1.0 vol %; 1N NaOH solution in water, 25 percent NaOH conversion; temperature = 24°C; atmospheric pressure: gas rate = 1.264 kg/(s·m²); liquid rate = 6.78 kg/(s·m²).
SOURCE: Strigle, R. L., *Packed Tower Design and Applications,* 2d ed., Gulf Publ. Co., Houston, 1994.

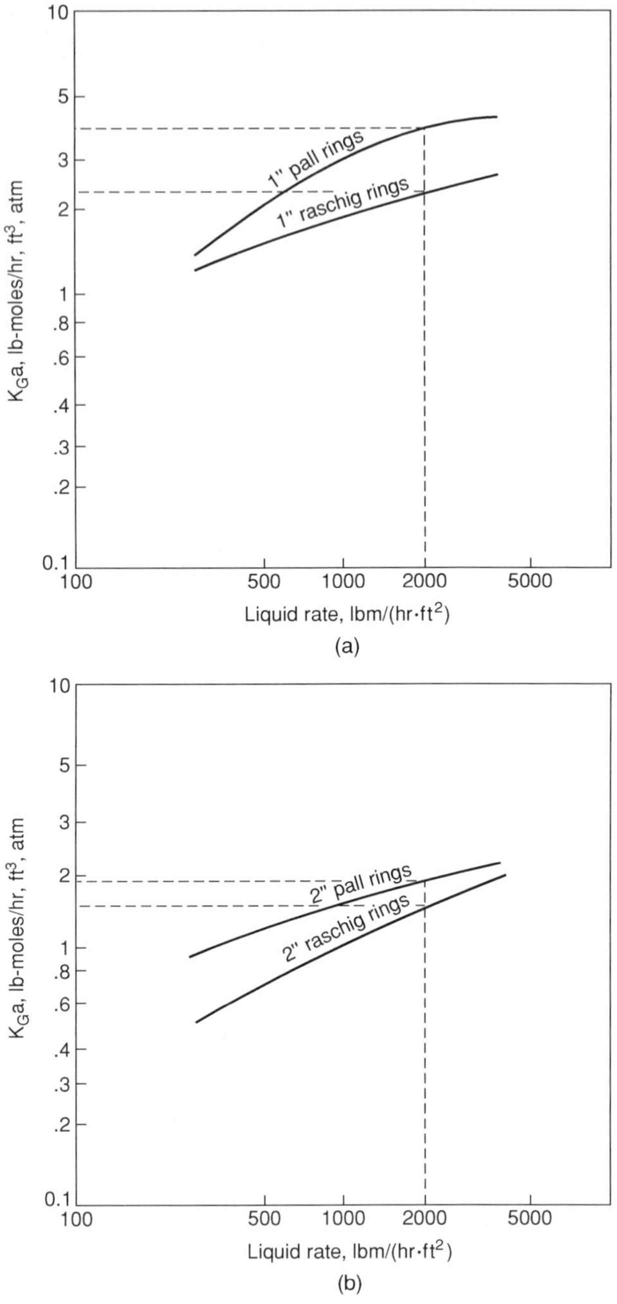

FIG. 14-63 Overall mass transfer coefficients for carbon dioxide absorbed from air by 1N caustic solution. (*a*) 1-in Pall rings and Raschig rings. (*b*) 2-in Pall rings and Raschig rings. Air rate = 0.61 kg/s·m² (450 lb/hr·ft²). To convert from lb/hr·ft² to kg/s·m², multiply by 0.00136. To convert from lb-moles/hr·ft³ atm to kg-moles/s·m³ atm, multiply by 0.0045. [*Eckert et al.,* Chem. Eng. Progr., ***54***(*1*), *70* (*1958*).]

Summary In the preloading regime, packing size, type, and distribution affect HETP. With aqueous-organic systems, HETP may be sensitive to underwetting and composition. A lambda value ($\lambda = mG_M/L_M$) outside the range of 0.5 to 2.0 causes HETP to rise, and so does a high hydrogen concentration. HETP of structured packings may also be affected by pressure (at high pressure), and vapor and liquid loads.

MALDISTRIBUTION AND ITS EFFECTS ON PACKING EFFICIENCY

Modeling and Prediction Maldistribution may drastically reduce packing efficiency. HETP may increase by a factor as high as 2 or 3 due to maldistribution. Shariat and Kunesh [*Ind. Eng. Chem. Res.*, **34**(4), 1273 (1995)] provide a good demonstration.

Early models [Mullins, *Ind. Chem. Mfr.*, **33**, 408 (1957); Manning and Cannon, *Ind. Eng. Chem.* **49**(3), 347 (1957)] expressed the effect of liquid maldistribution on packing efficiency in terms of a simple channeling model. A portion of the liquid bypasses the bed, undergoing negligible mass transfer, and then rejoins and contaminates the rest of the liquid. Huber et al. [*Chem. Ing. Tech.* **39**, 797 (1967); *Chem. Eng. Sci.* **21**, 819 (1966)] and Zuiderweg et al. [*IChemE Symp. Ser.* **104**, A217 (1987)] replaced the simple bypassing by variations in the local *L/V* ratios. The overirrigated parts have a high *L/V* ratio, the underirrigated parts a low *L/V* ratio. Regions with low *L/V* ratios experience pinching, and, therefore, produce poor separation.

Huber et al. (loc. cit.) and Yuan and Spiegel [*Chem. Ing. Tech.* **54**, 774 (1982)] added lateral mixing to the model. Lateral deflection of liquid by the packing particles tends to homogenize the liquid, thus counteracting the channeling and pinching effect.

A third factor is the nonuniformity of the flow profile through the packing. This nonuniformity was observed as far back as 1935 [Baker, Chilton, and Vernon, *Trans. Instn. Chem. Engrs.* **31**, 296 (1935)] and was first modeled by Cihla and Schmidt [*Coll. Czech. Chem. Commun.*, **22**, 896 (1957)]. Hoek (Ph.D. Thesis, The University of Delft, The Netherlands, 1983) combined all three factors into a single model, leading to the zone-stage model below.

The Zone-Stage Model Zuiderweg et al. [*IChemE Symp. Ser.* **104**, A217, A233 (1987)] extended Hoek's work combining the effects of local *L/V* ratio, lateral mixing, and flow profile into a model describing the effect of liquid maldistribution on packing efficiency. This work was performed at Fractionation Research Inc. (FRI) and at The University of Delft in The Netherlands. The model postulates that, in the absence of maldistribution, there is a "basic" (or "true" or "inherent") HETP which is a function of the packing and the system only. This HETP can be inferred from data for small towers, in which lateral mixing is strong enough to offset any pinching. For a given initial liquid distribution, the model uses a diffusion-type equation to characterize the splitting and recombining of liquid streams in the horizontal and vertical directions. The mass transfer is then calculated by integrating the liquid flow distribution at each elevation and the basic HETP. Kunesh et al. successfully applied the model to predict measured effects of maldistribution on packing efficiency. However, this model is difficult to use and has not gained industrywide acceptance.

Empirical Prediction Moore and Rukovena [*Chemical Plants and Processing* (European edition), p. 11, August 1987] proposed the empirical correlation in Fig. 14-64 for efficiency loss due to liquid maldistribution in packed towers containing Pall® rings or Metal Intalox® packing. This correlation was shown to work well for several case studies (Fig. 14-64), is simple to use, and is valuable, at least as a preliminary guide.

To quantify the quality of liquid irrigation, the correlation uses the distribution quality rating index. Typical indexes are 10 to 70 percent for most standard commercial distributors, 75 to 90 percent for intermediate-quality distributors, and over 90 percent for high-performance distributors. Moore and Rukovena present a method for calculating a distribution-quality rating index from distributor geometry. Their method is described in detail in their paper as well as in Kister's book (*Distillation Operation*, McGraw-Hill, New York, 1990).

Maximum Liquid Maldistribution Fraction f_{max}. To characterize the sensitivity of packed beds to maldistribution, Lockett and Billingham (*Trans. IChemE.* **80,** Part A, p. 373, May 2002; *Trans. IChemE.* **81,** Part A, p. 134, January 2003) modeled maldistribution as two parallel columns, one receiving more liquid $(1 + f)L$, the other receiving less $(1 - f)L$. The vapor was assumed equally split (Fig. 14-65)

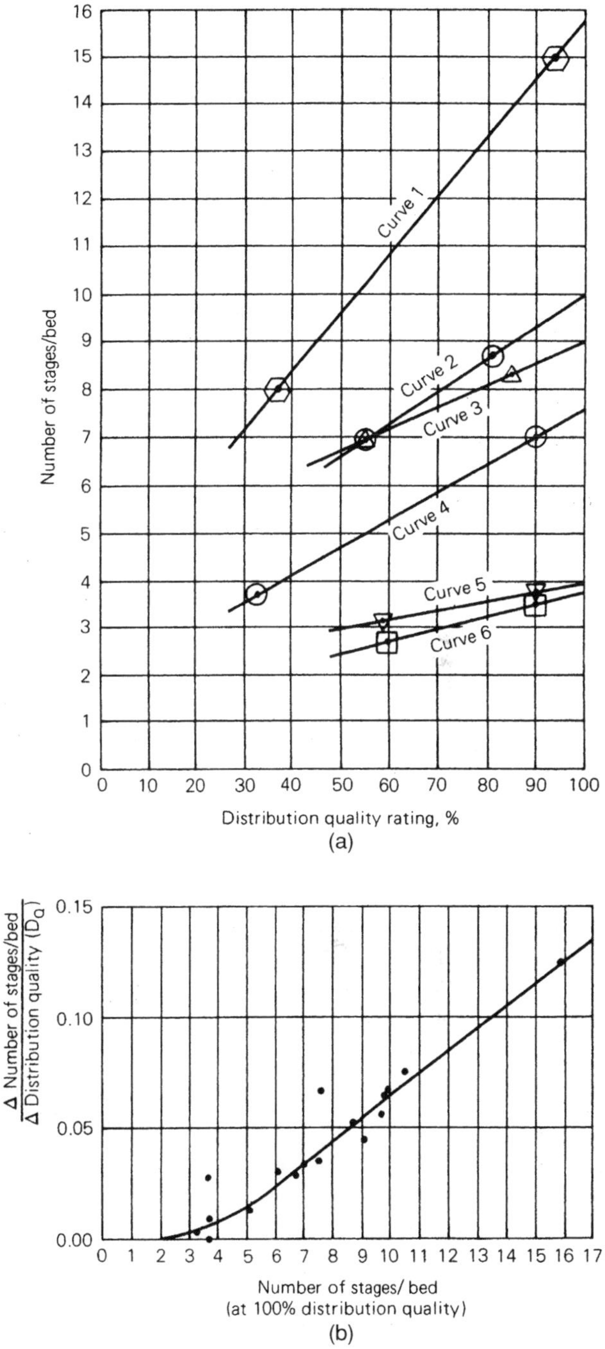

FIG. 14-64 Effect of irrigation quality on packing efficiency. (*a*) Case histories demonstrating efficiency enhancement with higher distribution quality rating. (*b*) Correlation of the effect of irrigation quality on packing efficiency. (*From F. Moore and F. Rukovena,* Chemical Plants and Processing, *Europe edition, Aug. 1987; reprinted courtesy of* Chemical Plants and Processing.)

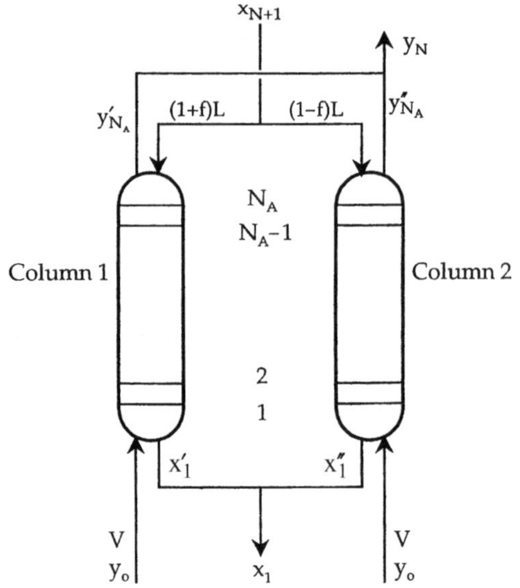

FIG. 14-65 Parallel-columns model. (*From Lockett and Billingham,* Trans. IChemE *80, Part A, p. 373, May 2002; reprinted courtesy of IChemE.*)

without lateral mixing. Because of the different L/V ratios, the overall separation is less than is obtained at uniform distribution. A typical calculated result (Fig. 14-66) shows the effective number of stages from the combined two-column system decreasing as the maldistribution fraction f increases. Figure 14-66*a* shows that the decrease is minimal in short beds (e.g., 10 theoretical stages) or when the maldistribution fraction is small. Figure 14-66*a* shows that there is a limiting fraction f_{max} which characterizes the maximum maldistribution that still permits achieving the required separation. Physically, f_{max} represents the maldistribution fraction at which one of the two parallel columns in the model becomes pinched. Figure 14-66*b* highlights the steep drop in packing efficiency upon the onset of this pinch. Billingham and Lockett derived the following equation for f_{max} in a binary system:

$$f_{max} = \frac{y_{N+1} - y_N}{y_N - y_o} + \frac{x_1 - x_o}{x_{N+1} - x_o} - \left(\frac{y_{N+1} - y_N}{y_N - y_o}\right)\left(\frac{x_1 - x_o}{x_{N+1} - x_o}\right) \quad (14\text{-}162)$$

This equation can be used to calculate f_{max} directly without the need for a parallel column model. Billingham and Lockett show that the various terms in Eq. (14-162) can be readily calculated from the output of a steady-state computer simulation. Multicomponent systems are represented as binary mixtures, either by lumping components together to form a binary mixture of pseudolight and pseudoheavy components, or by normalizing the mole fractions of the two key components. Once f_{max} is calculated, Billingham and Lockett propose the following guidelines:

- f_{max} <0.05, extremely sensitive to maldistribution. The required separation will probably not be achieved.
- 0.05 < f_{max} <0.10, sensitive to maldistribution, but separation can probably be achieved.
- 0.10 < f_{max} <0.20, not particularly sensitive to maldistribution.
- f_{max} >0.20 insensitive to maldistribution.

Figure 14-66*b* shows that shortening the bed can increase f_{max}. Relative volatility and L/V ratio also affect f_{max}. The bed length and L/V ratio can often be adjusted to render the bed less sensitive to maldistribution.

Implications of Maldistribution to Packing Design Practice These are discussed at length with extensive literature citation in Kister's book *Distillation Operation,* McGraw-Hill, New York, 1990. Below are the highlights:

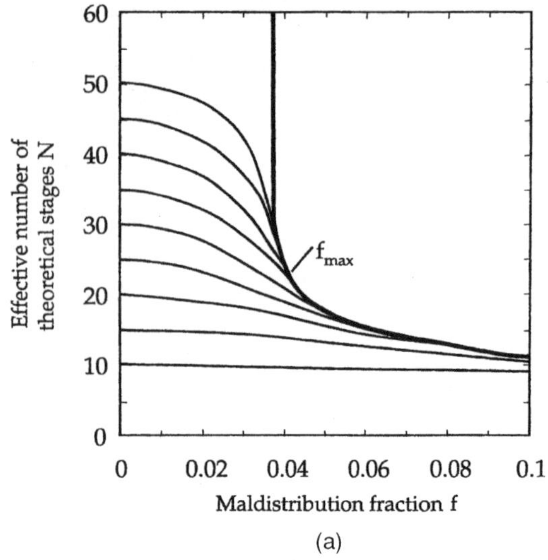

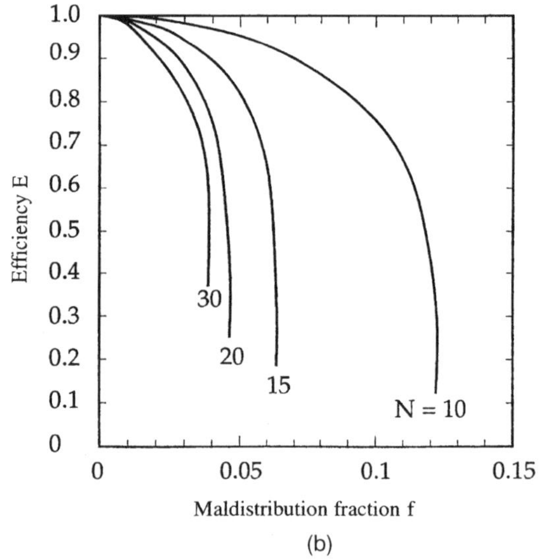

FIG. 14-66 Loss of efficiency due to maldistribution as a function of maldistribution fraction f and the number of stages per bed for a given case study. (*a*) f_{max} and reduction in number of stages. (*b*) Showing larger efficiency loss at higher number of stages per bed and higher f. The steep drops indicate pinching associated with f_{max}. (*From Lockett and Billingham,* Trans. IChemE *80, Part A, p. 373, May 2002; reprinted courtesy of IChemE.*)

1. Three factors appear to set the effect of maldistribution on efficiency:

a. Pinching. Regional changes in L/V ratio cause regional composition pinches.

b. Lateral mixing. Packing particles deflect both liquid and vapor laterally. This promotes mixing of vapor and liquid and counteracts the pinching effect.

c. Liquid nonuniformity. Liquid flows unevenly through the packing and tends to concentrate at the wall.

2. At small tower-to-packing diameter ratios (D_T/D_p < 10), the lateral mixing cancels out the pinching effect, and a greater degree of maldistribution can be tolerated without a serious efficiency loss. At high ratios (D_T/D_p > 40), the lateral mixing becomes too small to offset

the pinching effect. The effects of maldistribution on efficiency are therefore most severe in large-diameter columns and small-diameter packings.

A good design practice is to seek a packing size that gives a D_T/D_p between 10 and 40. This is often impractical, and higher ratios are common. When D_T/D_p exceeds 100, avoiding efficiency loss due to maldistribution is difficult. Either ratios exceeding 100 should be avoided, or a special allowance should be made for loss of efficiency due to maldistribution.

3. Wall flow effects become large when D_T/D_p falls below about 10. Packing diameter should be selected such that D_T/D_p exceeds 10.

4. Columns containing less than five theoretical stages per bed are relatively insensitive to liquid maldistribution. With 10 or more stages per bed, efficiency can be extremely sensitive to maldistribution (Strigle, *Packed Tower Design and Applications,* 2d ed., Gulf Publishing, Houston, Tex., 1994) (Fig. 14-66). Beds consisting of small packings or structured packings, which develop more theoretical stages per bed, are therefore more sensitive to maldistribution than equal-depth beds of larger packings. This is clearly demonstrated by FRI's experiments [Shariat and Kunesh, *Ind. Eng. Chem. Res.* **34**(4), 1273 (1995)]. Lockett and Billingham (*Trans. IChemE,* vol. 81, Part A, p. 131, January 2003) concur with these comments when their procedure (above) indicates high sensitivity to maldistribution, but allow a higher number of stages per bed when the sensitivity is low.

5. Maldistribution tends to be a greater problem at low liquid flow rates than at high liquid flow rates [Zuiderweg, Hoek, and Lahm, *I. ChemE Symp. Ser.* **104,** A217 (1987)]. The tendency to pinch and to spread unevenly is generally higher at the lower liquid flow rates.

6. A packed column has reasonable tolerance for a uniform or smooth variation in liquid distribution and for a variation that is totally random (*small-scale maldistribution*). The impact of discontinuities or zonal flow (*large-scale maldistribution*) is much more severe [Zuiderweg et al., loc. cit.; Kunesh, *Chem. Eng.*, p. 101, Dec. 7, 1987; Kunesh, Lahm, and Yanagi, *Ind. Eng. Chem. Res.* **26**(9), 1845 (1987)]. This is so because the local pinching of small-scale maldistribution is evened out by the lateral mixing, and therefore causes few ill effects. In contrast, the lateral mixing either is powerless to rectify a large-scale maldistribution or takes considerable bed length to do so (meanwhile, efficiency is lost).

Figure 14-67 shows HETPs measured in tests that simulate various types of maldistribution in FRI's 1.2-m column containing a 3.6-m bed of 1-in Pall® rings. The y axis is the ratio of measured HETP in the maldistribution tests to the HETP obtained with an excellent distributor. Analogous measurements with structured packing were reported by Fitz, King, and Kunesh [*Trans. IChemE* **77,** Part A, p. 482 (1999)]. Generally, the response of the structured packings resembled that of the Pall® rings, except as noted below.

Figure 14-67*a* shows virtually no loss of efficiency when a distributor uniformly tilts, such that the ratio of highest to lowest flow is 1.25 (i.e., a "1.25 tilt"). In contrast, an 11 percent chordal blank of a level distributor causes packing HETP to rise by 50 percent.

Figure 14-67*b* compares continuous tilts with ratios of highest to lowest flow of 1.25 and 1.5 to a situation where one-half of the distributor passes 25 percent more liquid than the other half. The latter ("zonal") situation causes a much greater rise in HETP than a "uniform" maldistribution with twice as much variation from maximum to minimum.

Figure 14-67*c* shows results of tests in which flows from individual distributor drip points were varied in a gaussian pattern (maximum/mean = 2). When the pattern was randomly assigned, there was no efficiency loss. When the variations above the mean were assigned to a "high zone," and those below the mean to a "low zone," HETP rose by about 20 percent. With structured packing, both random and zonal maldistribution caused about the same loss of efficiency at the same degree of maldistribution.

7. A packed bed appears to have a "natural distribution," which is an inherent and stable property of the packings. An initial distribution which is better than natural will rapidly degrade to it, and one that is worse will finally achieve it, but sometimes at a slow rate. If the rate is extremely slow, recovery from a maldistributed pattern may not be observed in practice (Zuiderweg et al., loc. cit.). Even though the

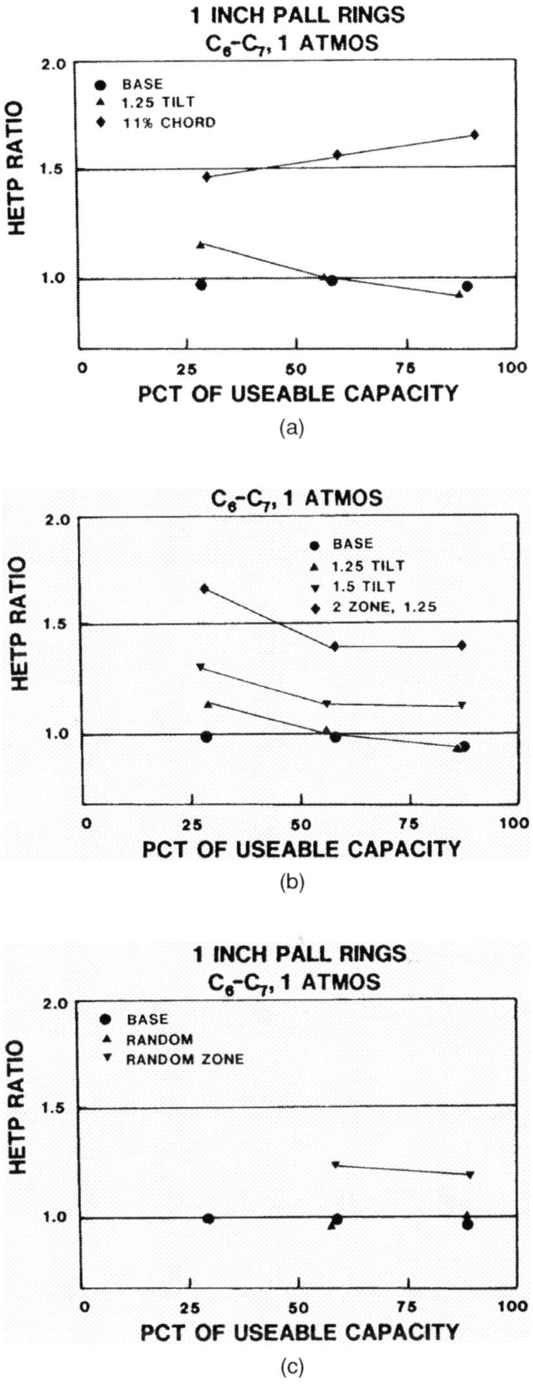

FIG. 14-67 Comparing the effects of "small-scale" and "large-scale" maldistribution on packing HETP. (*a*) Comparing the effect of a simulated continuous tilt (max/min flow ratio = 1.25) with the simulated effect of blanking a chordal area equal to 11 percent of the tower area. (*b*) Comparing the effects of simulated continuous tilts (max/min flow ratios of 1.25 and 1.5) with the effects of a situation where one-half of the distributor passes 25 percent more liquid to the other half. (*c*) Comparing the effects of random maldistribution with those of zonal maldistribution. (*Reprinted with permission from J. G. Kunesh, L. Lahm, and T. Yahagi,* Ind. Eng. Chem. Res., ***26,*** *p. 1845; copyright © 1987, American Chemical Society.*)

volumetric distribution improves along the bed, the concentration profile could have already been damaged, and pinching occurs (Bonilla, *Chem. Eng. Prog.*, p. 47, March 1993).

8. Liquid maldistribution lowers packing turndown. The 2-in Pall rings curve in Fig. 14-59 shows HETP rise upon reaching the distributor turndown limit.

9. The major source of gas maldistribution is undersized gas inlet and reboiler return nozzles, leading to the entry of high-velocity gas jets into the tower. These jets persist through low-pressure-drop devices such as packings. Installing gas distributors and improving gas distributor designs, even inlet baffles, have alleviated many of these problems. Vapor distribution is most troublesome in large-diameter columns. Strigle (*Packed Tower Design and Applications,* 2d ed., Gulf Publishing, Houston, Tex., 1994) recommends considering a gas distributing device whenever the gas nozzle *F*-factor ($F_N = u_N\rho_G^{0.5}$) exceeds 27 m/s (kg/m^3)$^{0.5}$, or the kinetic energy of the inlet gas exceeds 8 times the pressure drop through the first foot of packing, or the pressure drop through the bed is less than 0.65 mbar/m. Gas maldistribution is best tackled at the source by paying attention to the gas inlet arrangements.

10. A poor initial liquid maldistribution may cause gas maldistribution in the loading region, i.e., at high gas rates [Stoter, Olujic, and de Graauw, *IChemE Symp. Ser.* **128,** A201 (1992); Kouri and Sohlo, *IChemE Symp. Ser.* **104,** B193 (1987)]. At worst, initial liquid maldistribution may induce local flooding, which would channel the gas. The segregation tends to persist down the bed. Outside the loading region, the influence of the liquid flow on gas maldistribution is small or negligible. Similarly, in high-gas-velocity situations, the liquid distribution pattern in the bottom structured packing layers is significantly influenced by a strongly maldistributed inlet gas flow [Olujic et al., *Chem. Eng. and Processing,* **43,** 465 (2004)]. Duss [*IChemE Symp. Ser.* **152,** 418 (2006)] suggests that high liquid loads such as those experienced in high-pressure distillation also increase the susceptibility to gas maldistribution.

11. The effect of gas maldistribution on packing performance is riddled with unexplained mysteries. FRI's (Cai, Paper presented at the AIChE Annual Meeting, Reno, Nev., 2001) commercial-scale tests show little effect of gas maldistribution on both random and structured packing efficiencies. Cai et al. (*Trans IChemE* **81,** Part A, p. 85, 2003) distillation tests in a 1.2-m-diameter tower showed that blocking the central 50 percent or the chordal 30 percent of the tower cross-sectional area beneath a 1.7-m-tall bed of 250 m^2/m^3 structured packing had no effect on packing efficiency, pressure drop, or capacity. The blocking did not permit gas passage but allowed collection of the descending liquid. Simulator tests with similar blocking with packing heights ranging from 0.8 to 2.4 m (Olujic et al., *Chemical Engineering and Processing,* **43,** p. 465, 2004; *Distillation 2003: Topical Conference Proceedings,* AIChE Spring National Meeting, New Orleans, La., AIChE, p. 567, 2003) differed, showing that a 50 percent chordal blank raised pressure drop, gave a poorer gas pattern, and prematurely loaded the packing. They explain the difference by the ability of liquid to drain undisturbed from the gas in the blocked segment in the FRI tests. Olujic et al. found that while gas maldistribution generated by collectors and by central blockage of 50 percent of the cross-sectional areas was smoothed after two to three layers of structured packing, a chordal blockage of 30 to 50 percent of cross-sectional area generated maldistribution that penetrated deeply into the bed.

12. Computational fluid dynamics (CFD) has been demonstrated effective for analyzing the effects of gas inlet geometry on gas maldistribution in packed beds. Using CFD, Wehrli et al. (*Trans. IChemE* **81,** Part A, p. 116, January 2003) found that a very simple device such as the V-baffle (Fig. 14-70) gives much better distribution than a bare nozzle, while a more sophisticated vane device such as a Schoepentoeter (Fig. 14-71*c*) is even better. Implications of the gas inlet geometry to gas distribution in refinery vacuum towers was studied by Vaidyanathan et al. (Distillation 2001, Topical Conference Proceedings, AIChE Spring National Meeting, Houston, Tex., p. 287, April 22–26, 2001); Paladino et al. (Distillation 2003: Topical Conference Proceedings, AIChE Spring National Meeting, New Orleans, La., p. 241, 2003); Torres et al. (ibid., p. 284); Waintraub et al. (Distillation 2005: Topical Conference Proceedings, AIChE Spring National Meeting, Atlanta, Ga., p. 79, 2005); and Wehrli et al. (*IChemE* Symp. Ser 152, London, 2006). Vaidyanathan et al. and Torres et al. examined the effect of the geometry of a chimney tray (e.g., Fig. 14-72) above the inlet on gas distribution and liquid entrainment. Paladino et al. demonstrated that the presence of liquid in the feed affects the gas velocity profile, and must be accounted for in modeling. Paladino et al. and Waintraub et al. used their two-fluid model to study the velocity distributions and entrainment generated by different designs of vapor horns (e.g., Fig. 14-71). Wehrli et al. produced pilot-scale data simulating a vacuum tower inlet, which can be used in CFD model validation. Ali et al. (*Trans. IChemE,* vol. 81, Part A, p. 108, January 2003) found that the gas velocity profile obtained using a commercial CFD package compared well to those measured in a 1.4-m simulator equipped with structured packing together with commercial distributors and collectors. Their CFD model effectively pointed them to a collector design that minimizes gas maldistribution.

PACKED-TOWER SCALE-UP

Diameter For random packings there are many reports [Billet, *Distillation Engineering,* Chem Publishing Co., New York, 1979; Chen, *Chem. Eng.,* p. 40, March 5, 1984; Zuiderweg, Hoek, and Lahm, *IChemE. Symp. Ser.* **104,** A217 (1987)] of an increase in HETP with column diameter. Billet and Mackowiak's (Billet, Packed Column Analysis and Design, Ruhr University, Bochum, Germany, 1989) scale-up chart for Pall® rings implies that efficiency decreases as column diameter increases.

Practically all sources explain the increase of HETP with column diameter in terms of enhanced maldistribution or issues with the scale-up procedure. Lab-scale and pilot columns seldom operate at column-to-packing diameter ratios (D_T/D_p) larger than 20; under these conditions, lateral mixing effectively offsets loss of efficiency due to maldistribution pinch. In contrast, industrial-scale columns usually operate at D_T/D_p ratios of 30 to 100; under these conditions, lateral mixing is far less effective for offsetting maldistribution pinch.

To increase D_T/D_p, it may appear attractive to perform the bench-scale tests using a smaller packing size than will be used in the prototype. Deibele, Goedecke, and Schoenmaker [*IChemE Symp. Ser.* **142,** 1021 (1997)], Goedecke and Alig (Paper presented at the AIChE Spring National Meeting, Atlanta, Ga., April 1994), and Gann et al. [*Chem. Ing. Tech.,* **64**(1), 6 (1992)] studied the feasibility of scaling up from 50- to 75-mm-diameter packed columns directly to industrial columns. Deibele et al. and Gann et al. provide an extensive list of factors that can affect this scale-up, including test mixture, packing pretreatment, column structure, packing installation, snug fit at the wall, column insulation and heat losses, vacuum tightness, measurement and control, liquid distribution, reflux subcooling, prewetting, sampling, analysis, adjusting the number of stages to avoid pinches and analysis issues, evaluation procedure, and more. Data from laboratory columns can be particularly sensitive to some of these factors. Goedecke and Alig show that for wire-mesh structured packing, bench-scale efficiency tends to be better than large-column efficiency, while for corrugated-sheets structured packing, the converse occurs, possibly due to excessive wall flow. For some packings, variation of efficiency with loads at bench scale completely differs from its variation in larger columns. For one structured packing, Kuhni Rombopak 9M, there was little load effect and there was good consistency between data obtained from different sources—at least for one test mixture. Deibele et al. present an excellent set of practical guidelines to improve scale-up reliability. So, it appears that great caution is required for packing data scale-up from bench-scale columns.

Height Experimental data for random packings show that HETP slightly increases with bed depth [Billet, Distillation Engineering, Chemical Publishing Co., New York, 1979; "Packed Tower Analysis and Design," Ruhr University, Bochum, Germany, 1989; Eckert and Walter, *Hydrocarbon Processing,* **43**(2), 107 (1964)].

For structured packing, some tests with Mellapak 250Y [Meier, Hunkeler, and Stöcker, *IChemE Symp. Ser.* **56,** p. 3, 3/1 (1979)] showed no effect of bed height on packing efficiency, while others (Cai et al., *Trans IChemE,* vol. 81, Part A, p. 89, January 2003) did show a significant effect.

The effect of bed depth on packing HETP is attributed to liquid maldistribution. Zuiderweg et al. [*IChemE. Symp. Ser.* **104,** A217 (1987)] suggest that the uneven irrigation generates an uneven concentration profile and localized pinching near the bottom of the beds. The tests by Martin, Bravo, and Fair (Paper presented at the National AIChE Meeting, New Orleans, La., 1988) confirm that the problem area is near the bottom. According to the zone-stage and Lockett and Billingham models (above), as well as the empirical correlation by Moore and Rukovena (Fig. 14-64), the more stages per bed, the greater is the rise in HETP with bed depth. The presence and extent of maldistribution play an important role in determining the bed-depth effect.

As the bed depth increases, end effects (i.e., mass transfer in the region of liquid introduction and in the region where liquid drips from the packing supports) become less important. Such end effects tend to lower the HETP observed in short columns, such as pilot-plant columns.

In summary, bed depth may significantly influence HETP. This adds uncertainty to scale-up. Shallow test beds should be avoided. Most investigators use beds at least 1.5 m tall, and often more than 3 m tall. The FRI sampling technique (below) can detect maldistribution along the bed height.

Loadings For many random and corrugated-sheet structured packings, HETP is independent of vapor and liquid loadings (Figs. 14-59 and 14-60). For wire-mesh and some corrugated-sheet structured packings, HETP changes with gas and liquid loads.

Wu and Chen [*IChemE Symp. Ser.* **104,** B225 (1987)] recommend pilot testing over the entire range between the expected minimum and maximum operating rates, and taking the highest measured HETP as the basis for scale-up. The author concurs. With structured packings, the load effect may be due to liquid rather than gas loads, and the pilot tests should cover the range of liquid loads (i.e., m/s based on column cross section) that is expected in the prototype.

Wetting For operation at low liquid loads, the onset of minimum wetting can adversely affect scale-up, particularly with random packings and aqueous systems. Scale-up reliability at low liquid loads can be improved by pilot-testing at the composition range expected in the prototype, and by using identical packing materials and surface treatment in the pilot tests and in the prototype.

Underwetting At the aqueous end of aqueous-organic columns, underwetting is important. Rapid changes of concentration profiles and physical properties in organic-water separations complicate scale-up [Eiden and Kaiser, *IChemE Symp. Ser.* **142,** 757 (1997)]. Near the onset of underwetting, HETP becomes strongly dependent on composition, packing material and surface roughness, and the presence of surfactants. Scale-up reliability can be enhanced by pilot-testing at the composition range expected in the prototype, and by using identical packing material and surface treatment in the pilot tests and in the prototype.

Preflooding For one structured packing test with an aqueous system, Billet ("Packed Column Analysis and Design," Ruhr University, Bochum, Germany, 1989) measured higher efficiency for a preflooded bed compared with a non-preflooded bed. Presumably, the preflooding improved either wetting or distribution. Billet recommends preflooding the packing, both in the prototype and in the pilot column, to ensure maximum efficiency.

Sampling Fractionation Research Inc. (FRI) developed a sampling technique that eliminates the influence of "end effects" and detects a maldistributed composition profile. This technique [Silvey and Keller, *IChemE Symp. Ser.* **32,** p. 4:18 (1969)] samples the bed at frequent intervals, typically every 0.6 m or so. HETP is determined from a plot of these interbed samples rather than from the top and bottom compositions.

It is imperative that the interbed samplers catch representative samples, which are an average through the bed cross section. Caution is required when the liquid is highly aerated and turbulent (e.g., above 1300 kPa psia or above 1 m/min). The author highly recommends the FRI sampling technique for all other conditions.

Aging Billet (loc. cit.) showed that for some plastic packings in aqueous systems, the efficiency after one week's operation was almost double the efficiency of new packings. Little further change was observed after one week. Billet explains the phenomenon by improved wetting. He recommends that data for plastic packings should only be used for scale-up after being in operation for an adequately long period.

DISTRIBUTORS

Liquid Distributors A liquid distributor (or redistributor) should be used in any location in a packed column where an external liquid stream is introduced. Liquid redistributors are also used between packed beds to avoid excessive bed lengths that may impair packing efficiency or mechanical strength. It is best to have the packing supplier also supply the distributor, with the user critically reviewing the design. The user must provide the supplier with concise information about the plugging, corrosive, and foaming tendencies of the service as well as the range of liquid flow rates that it needs to handle and the physical properties of the liquid.

Olsson (*Chem. Eng. Progr.,* p. 57, October 1999) discussed the key for successful distributor design and operation. He states that it is critical to correctly evaluate the fouling potential of the service and design for it (e.g., preventing small holes, filtering the feed, etc.); to avoid gas entry into liquid distributors (e.g., no flashing feed into a liquid distributor); to systematically check the irrigation pattern using a method such as the circle analysis of Moore and Rukovena [*Chem. Plants and Process* (European ed.), p. 11, August 1987; described in detail in Kister's *Distillation Operation,* McGraw-Hill, New York, 1990]; to water-test any liquid distributor (major suppliers have dedicated test stands that are available for those purposes at cost); to ensure correct entry of a feed into a liquid distributor; and to thoroughly inspect a distributor. Kister [*Trans. IChemE.* vol. 81, Part A, p. 5 (January 2003)] found that between 80 and 90 percent of the distributor failures reported in the literature in the last 50 years could have been prevented if users and suppliers had followed Olsson's measures.

A minimum of 40 irrigation points per square meter has been recommended, with 60 to 100 per square meter being ideal [Strigle, *Packed Tower Design and Applications,* 2d ed., Gulf Publishing, Houston, Tex., 1994; Kister, *Distillation Operation,* McGraw-Hill, New York, 1990; Norton Company (now Koch-Glitsch LP), *Packed Column Internals,* Bulletin TA-80R, 1974]. Commercial-scale tests with both random and structured packings showed no improvement in packing efficiency by increasing the irrigation point density above 40 per square meter [Fitz, King, and Kunesh, *Trans. IChemE* **77,** Part A, p. 482 (1999)]. So going to larger numbers of irrigation points per square meter provides little improvement while leading to smaller holes which increases the plugging tendency. In orifice-type distributors, which are the most common type, the head-flow relationship is given by the orifice equation

$$Q = 3.96 \times 10^{-4}\, K_D n_D d_h^2\, h^{0.5} \qquad (14\text{-}163)$$

where Q is the liquid flow rate, m^3/h; K_D is the orifice discharge coefficient, with a recommended value of 0.707 (Chen, *Chem. Eng.,* p. 40, March 5, 1984); n_D is the number of holes; d_h is the hole diameter, mm; and h is the liquid head, mm. Equation (14-163) shows that at a given Q, increasing n leads to either smaller d or smaller h.

Figures 14-68 and 14-69 show common distributor types used for distillation and absorption. An excellent detailed discussion of the types of distributors and their pros and cons was given by Bonilla (*Chem. Eng. Progr.,* p. 47, March 1993). The *perforated pipe* (or *ladder pipe*) distributor (Fig. 14-68*a*) has holes on the underside of the pipes. It is inexpensive, provides a large open area for vapor flow, and does not rely on gravity. On the debit side, it is typically designed for high velocity heads, 500 to 1000 mm of water, which is 5 to 10 times more than gravity distributors, requiring [per Eq. (14-163)] either fewer irrigation points or the use of plugging-prone smaller holes. The high hole velocities make it prone to corrosion and erosion. These disadvantages make it relatively unpopular. A gravity variation of this distributor uses a liquid drum above the distributor that gravity-feeds it.

Spray distributors (Fig. 14-68*b*) are pipe headers with spray nozzles fitted on the underside. The spray nozzles are typically wide-angle (often 120°) full-cone. Spray distributors are unpopular in distillation but are common in heat transfer, washing and scrubbing services (especially in refinery towers), and in small-diameter towers where a single spray nozzle can be used. They are inexpensive and offer a large open area for vapor flow and a robustness for handling of fouling fluids when correctly designed, and the sprays themselves contribute to mass and heat transfer. On the debit side, the spray cones often generate regions of over- and underirrigation, the sprays may not be homogeneous, the spray nozzles are prone to corrosion, erosion, and damage. With highly subcooled liquids, the spray angle may collapse when pushed at high pressure drops (above 100 to 150 kPa) (Fractionation Research Inc., "A Spray Collapse Study," motion picture 919, Stillwater, Okla., 1985). The design and spray pattern are highly empirical. Sprays also generate significant entrainment to the section above [Trompiz and Fair, *Ind. Eng. Chem, Res.*, **39**(6), 1797 (2000)].

Orifice pan distributors (Fig. 14-69*a*) and *orifice tunnel* distributors (Fig. 14-69*b*) have floor holes for liquid flow and circular (Fig. 14-69*a*) or rectangular (Fig. 14-69*b*) risers for vapor passages. When they are used as redistributors, a hat is installed above each riser to prevent liquid from the bed above from dripping into the risers. Unlike the ladder pipe and spray distributors that operate by pressure drop, orifice distributors operate by gravity, and therefore use a much smaller liquid head, typically 100 to 150 mm at maximum rates. Using Eq. (14-163), the lower head translates to either more distributions points (n_D), which helps irrigation quality, or larger hole diameters, which resists plugging. However, the low liquid velocities, large residence times, and open pans (or troughs) make them more prone to plugging than the pressure distributors. A good hole pattern and avoidance of oversized risers are essential. Orifice distributors are self-collecting, a unique advantage for redistributors. Orifice distributors are one of the most popular types and are favored whenever the liquid loads are high enough to afford hole diameters large enough to resist plugging (>12 mm).

Orifice trough (or *orifice channel*) distributors (Fig. 14-69*c*–*f*) are some of the most popular types. The trough construction does away with the multitude of joints in the orifice pans, making them far more leak-resistant, a major advantage in large towers and low-liquid-rate applications. Liquid from a central parting box (Fig. 14-69*c*, *e*) or middle channel (Fig. 14-69*d*) is metered into each trough. The troughs can have floor holes, but elevating the holes above the floor (Fig. 14-69*c*–*g*) is preferred as it enhances plugging resistance. Tubes (Fig. 14-69*c*, *d*, *f*) or baffles (Fig. 14-69*e*) direct the liquid issuing from the elevated holes downward onto the packings. Orifice trough distributors are not self-collecting. When used for redistribution, they require a liquid collector to be installed above them.

Turndown of orifice distributors is constrained to about 2 : 1 by Eq. (14-163). For example, a 100-mm liquid head at the design drops to 25 mm when the liquid rate is halved. Lower heads give poor irrigation and high sensitivity to levelness. Turndown is often enhanced by using two rows of side tubes (in the Fig. 14-69*c* type) or of side holes (in the Fig. 14-69*d* or *e* types). *Perforated drip tubes* (as in Fig. 14-69*d*) are popular in either orifice trough or orifice pan distributors. The lower, smaller hole is active at low liquid rates, with the larger upper hole becoming active at higher liquid rates. Use of perforated drip tubes is not recommended when the vapor dew point is much higher than the liquid bubble point, because liquid may boil in the tubes, causing dryout underneath [Kister, Stupin, and Oude Lenferink, *IChemE. Symp. Ser.* **152,** p. 409, London (2006)].

A popular type of the orifice trough distributor is the *splash plate* distributor (Fig. 14-69*e*). The splash plates spread the issuing liquid over their lengths, making it possible to reduce the number of irrigation points. This is a special advantage with small liquid rates, as fewer irrigation points (at a given head) translate to larger, more fouling-resistant

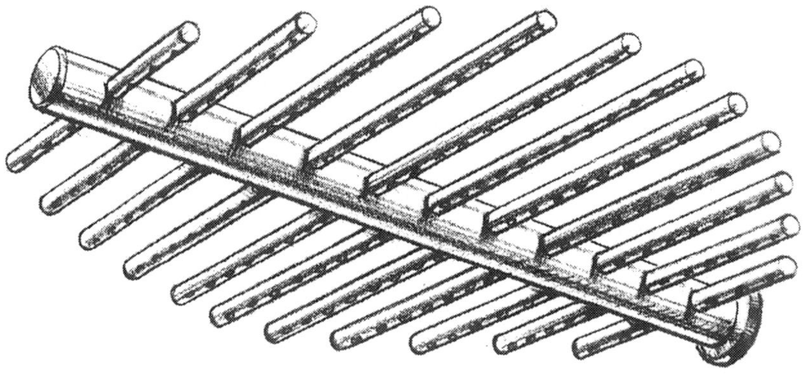

(a)

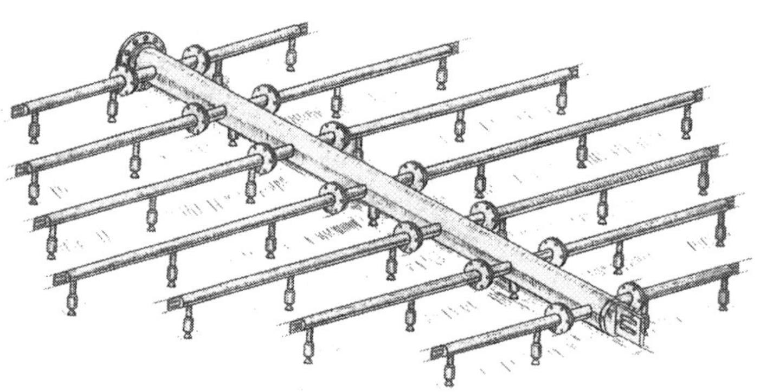

(b)

FIG. 14-68 Pressure liquid distributors. (*a*) Ladder pipe. (*b*) Spray. (*Courtesy of Koch-Glitsch LP.*)

hole diameters [Eq. (14-163)]. Lack of the drip tubes eliminates the possible in-tube boiling issue (above).

Multistage orifice trough distributors (Fig. 14-69*f*) also attempt to provide good irrigation at low liquid rates without resorting to plugging-prone small holes. The primary stage uses fewer irrigation points. Liquid from the primary stage is further split at the secondary stage. The secondary stage is of small size, so leveling and small flow variations are not of great concern. The secondary stage may use the same or a different liquid splitting principle as the primary stage. Even short layers of structured packings have been used as a secondary distribution stage.

Notched trough distributors (Fig. 14-69*g*) consist of parallel troughs with side V notches. This distributor obeys the triangular notch equation instead of the orifice equation, which makes the flow proportional to $h^{2.5}$ [instead of $h^{0.5}$ in Eq. (14-163)]. This high power renders the distributor highly sensitive to out-of-levelness and hydraulic gradients and makes it difficult to incorporate a large number of distribution points. Since the liquid issues sideways, it is difficult to predict where the liquid will hit the packings. Baffles are sometimes used to direct the liquid downward. Overall, the quality of distribution is inferior to that of orifice distributors, making notched-trough distributors unpopular. Their strength is their insensitivity to fouling and corrosive environments and their ability to handle high liquid rates at good turndown.

With any trough distributor, and especially those with V notches, excessive hydraulic gradients must be avoided. This is often achieved by using more parting boxes.

The hydraulic gradient is highest where the liquid enters the troughs, approaching zero at the end of the trough. The hydraulic gradient (between entry point and trough end) can be calculated from [Moore and Rukovena, *Chemical Plants and Processing* (European ed.), p. 11, August 1987]

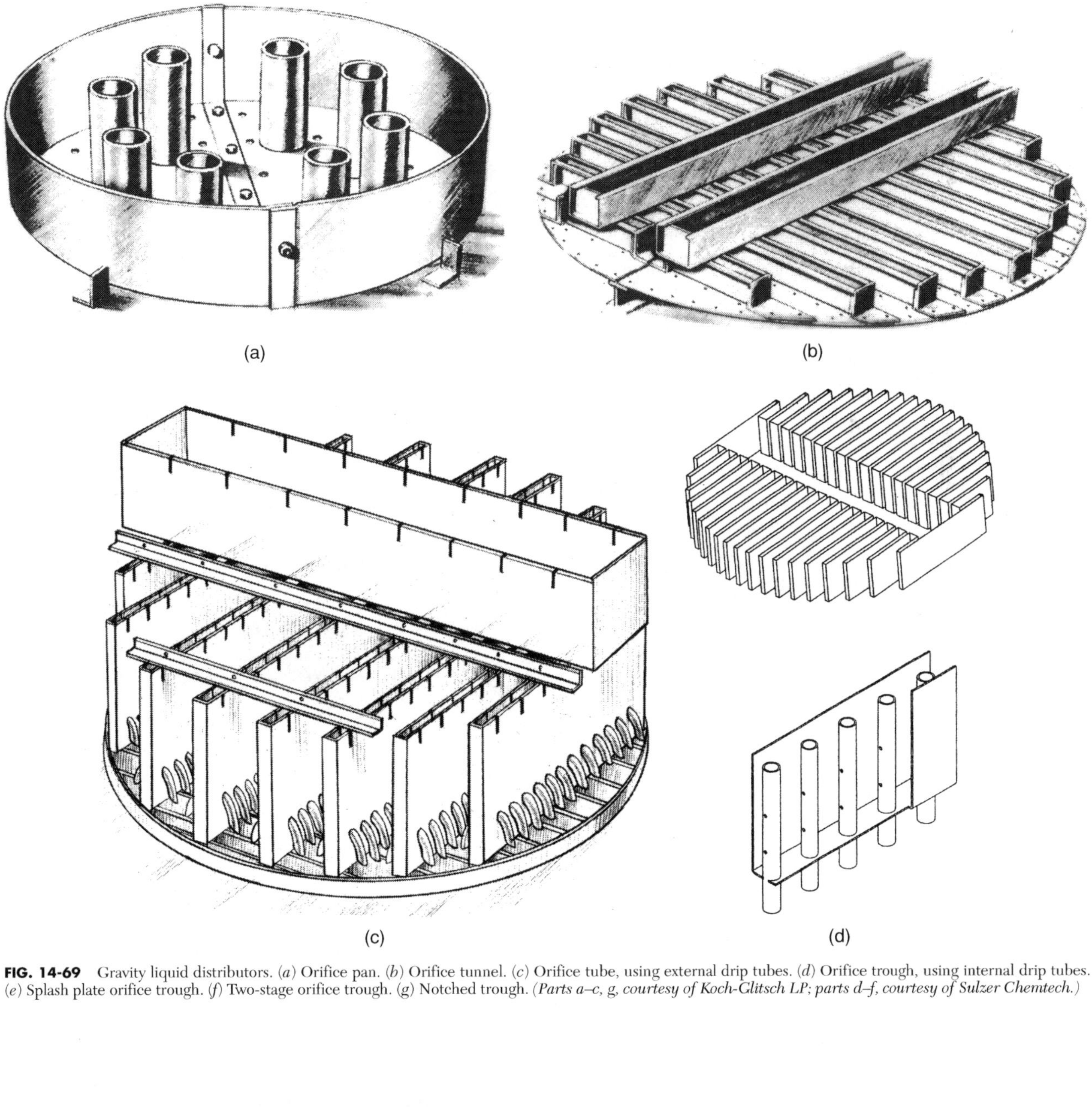

FIG. 14-69 Gravity liquid distributors. (*a*) Orifice pan. (*b*) Orifice tunnel. (*c*) Orifice tube, using external drip tubes. (*d*) Orifice trough, using internal drip tubes. (*e*) Splash plate orifice trough. (*f*) Two-stage orifice trough. (*g*) Notched trough. *(Parts a–c, g, courtesy of Koch-Glitsch LP; parts d–f, courtesy of Sulzer Chemtech.)*

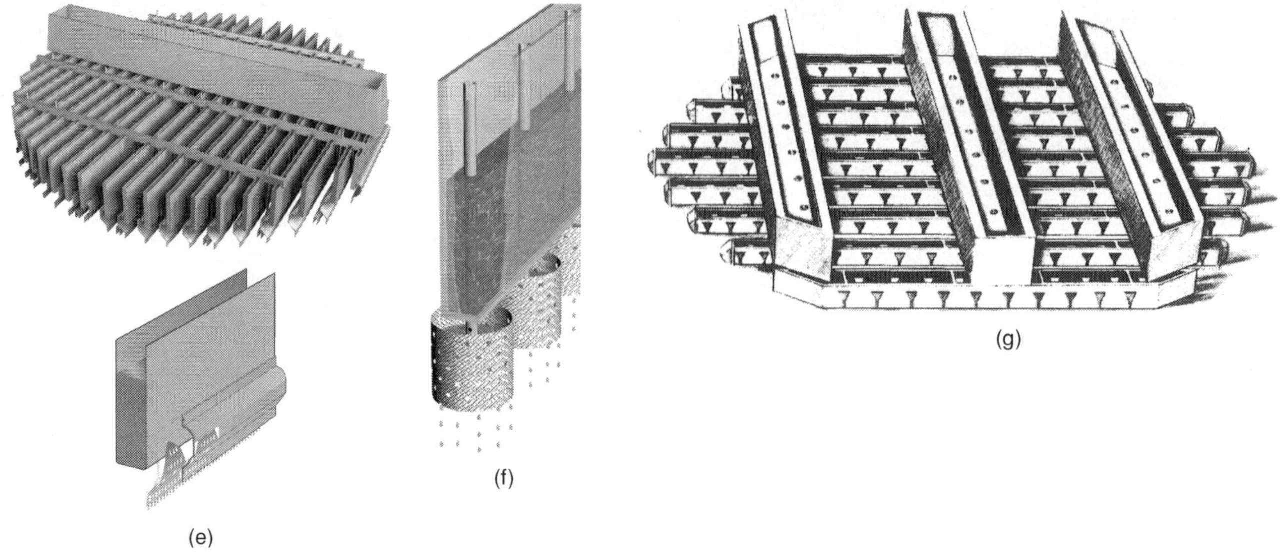

FIG. 14-69 *(Continued)*

$$h_{hg} = 51v_H^2 \tag{14-164}$$

where h_{hg} is the hydraulic gradient head, mm, and v_H is the horizontal velocity in the troughs, m/s.

Flashing Feed and Vapor Distributors When the feed or reflux is a flashing feed, the vapor must be separated out of the liquid before the liquid enters a liquid distributor. At low velocities (only), this can be achieved by a *bare nozzle* (Fig. 14-70*a*). A V *baffle* (Fig. 14-70*b*) is sometimes installed as a primitive flashing feed or vapor distributor.

For better vapor-liquid separation and distribution, with smaller-diameter towers (<1.5 m), a *flash chamber* (Fig. 14-70*c*) separates the liquid from the vapor, with the collected liquid descending via downpipes to a liquid distributor below. The flash chamber can be peripheral (Fig. 14-70*c*) or central. In larger towers, *gallery distributors* (Fig. 14-70*d*) are preferred. The flashing feed enters the peripheral section of the upper plate (the gallery) where vapor disengages and flows up, with liquid descending through holes (Fig. 14-70*d*) or down pipes onto the liquid distributor below. Alternatively, an external knockout pot is sometimes used to give separate vapor and liquid feeds.

The *vapor horn* (Fig. 14-71*a*) is unique for high-velocity feeds in which vapor is the continuous phase with liquid present as suspended drops in the feed vapor. This is common when the feed makes up the bulk of the vapor traffic in the tower section above. Typical examples are feeds to refinery vacuum and crude towers and rich solution feeds to hot carbonate regenerators. A tangential helical baffle or vapor horn, covered at the top, open at the bottom, and spiraling downward, is used at the feed entry. This baffle forces the vapor to follow the contour of the vessel as it expands and decreases in velocity. Liquid droplets, due to their higher mass, tend to collide with the tower wall, which deflects them downward, thus reducing upward entrainment. Large forces, generated by hurricane-force winds and vapor flashing, are absorbed by the entire tower wall rather than by a small area. A wear plate is required at the tower wall. Some designs have vane openings on the inside wall.

Alternatively, multivane triangular diffusers (Fig. 14-71*b, c*) such as the Schoepentoeter have been successful for high-velocity vapor-rich feeds. These are used with radial (as distinct from tangential) nozzles. The vanes knock out the liquid and direct it downward while the vapor expands to the tower diameter.

Pilot-scale tests by Fan et al. [*IChemE Symp. Ser.* **142,** 899 (1997)] compared vapor distribution and entrainment from sparger, vapor horns, and multivane triangular diffusers. Vapor horns gave the best overall performance considering vapor distribution, entrainment, and pressure drop, with multivane distributors doing well too. The designs of the inlets compared, however, were not optimized so the comparison could have reflected deviations from optimum rather than real differences between the devices.

Low-velocity vapor-only feeds often enter via bare nozzles or V baffles (above). At higher velocities, perforated vapor spargers are used. At high velocities, vapor horns and Schoepentoeters are often preferred. Alternatively or additionally, a vapor distributor may be mounted above the feed. The vapor distributor is a chimney tray (Fig. 14-72) where liquid is collected on the deck and flows via downcomers or is drawn out while vapor passes through the chimneys. To be effective as a vapor distributor, the pressure drop through the chimneys needs to be high enough to counter the maldistributed vapor. It was recommended to make the pressure drop through the chimneys at least equal to the velocity head at the tower inlet nozzle (Strigle, *Random Packings and Packed Towers,* Gulf Publishing, Houston, Tex., 1987), with common pressure drops ranging from 25 to 200 mm water.

OTHER PACKING CONSIDERATIONS

Liquid Holdup Liquid holdup is the liquid present in the void spaces of the packing. Reasonable liquid holdup is necessary for good mass transfer and efficient tower operation, but beyond that, it should be kept low. High holdup increases tower pressure drop, the weight of the packing, the support load at the bottom of the packing and tower, and the tower drainage time. Most important, when distilling thermally unstable materials, excessive holdup raises product degradation and fouling, and with hazardous chemicals, increases undesirable inventories.

The effect of liquid and gas rates on the operating holdup is shown in Figs. 14-73 and 14-74. In the preloading regime, holdup is essentially independent of gas velocity, but is a strong function of liquid flow rate and packing size. Smaller packings and high liquid rates tend to have greater holdup.

Holdup can be estimated by using Buchanan's correlation [*Ind. Eng. Chem. Fund.* **6,** 400 (1967)], as recommended in previous editions of this handbook. More recent correlations by Billet and Schultes [*IChemE. Symp. Ser.* **104,** A159 (1987)], by Maćkowiak

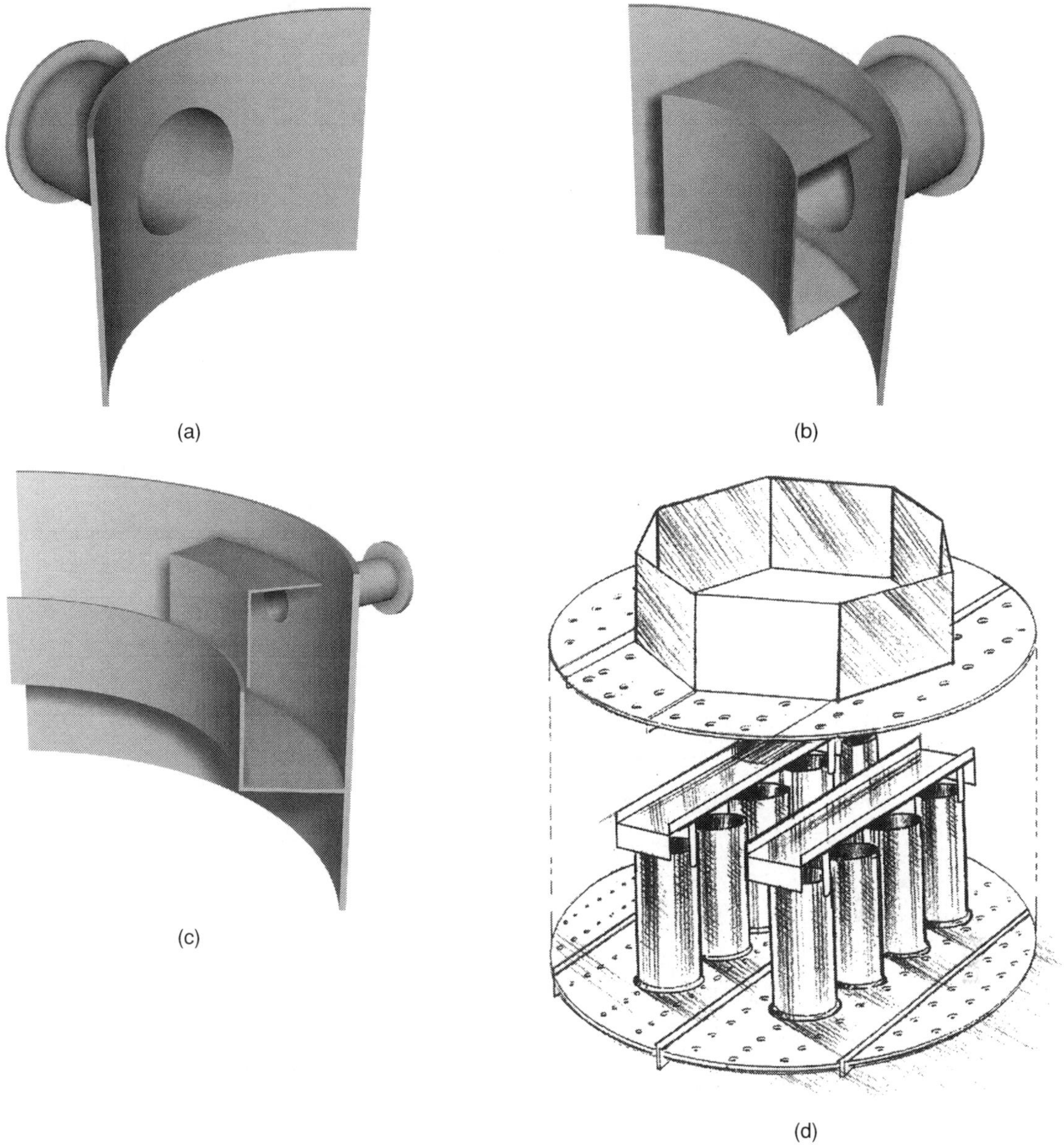

FIG. 14-70 Flashing feed and vapor distributors. (*a*) Bare nozzle. (*b*) Rounded V baffle. (*c*) Peripheral flash box—the box extends right around the tower wall, with the collected liquid descending via downpipes to a liquid distributor below. (*d*) Gallery distributor—the feed enters the gallery area (upper plate). *(Parts a–c, courtesy of Sulzer Chemtech; part d, courtesy of Koch-Glitsch LP.)*

("Fluiddynamik von Kolonnen mit Modernen Füllkörpern und Packungen für Gas/Flüssigkeitssysteme, Otto Salle Verlag, Frankfurt am Main, and Verlag Sauerländer, Aarau, 1991), and by Mersmann and Deixler [*Chem. Ing. Tech.* **58**(1), 19 (1986)] apply to second- and third-generation random packings as well as to some structured packings.

Stichlmair and Fair (*Distillation Principles and Practice*, Wiley-VCH, New York, 1998) show that liquid holdup is primarily a function of the liquid Froude number, specific surface area of the packing, and physical properties. They recommend the correction by Engel, Stichlmair, and Geipel [*Ind. Eng. Chem. Symp. Ser.* **142,** 939 (1997)].

$$h_{Lo} = 0.93\left(\frac{U_L^2 a_p}{g}\right)^{1/6}\left(\frac{\mu_L^2 a_p^3}{\rho_L^2 g}\right)^{1/10}\left(\frac{\sigma a_p^2}{1000\rho_L g}\right)^{1/8} \qquad (14\text{-}165)$$

where h_{Lo} = liquid holdup, fractional
U_L = liquid superficial velocity, m/s
a_p = packing specific surface area, m^2/m^3
g = acceleration due to gravity, m/s
μ_L = liquid viscosity, kg/(m·s)
σ = surface tension, mN/m

The Engel, Stichlmair, and Geipel correlation applies only in the preloading regime. The packing geometry is represented in the

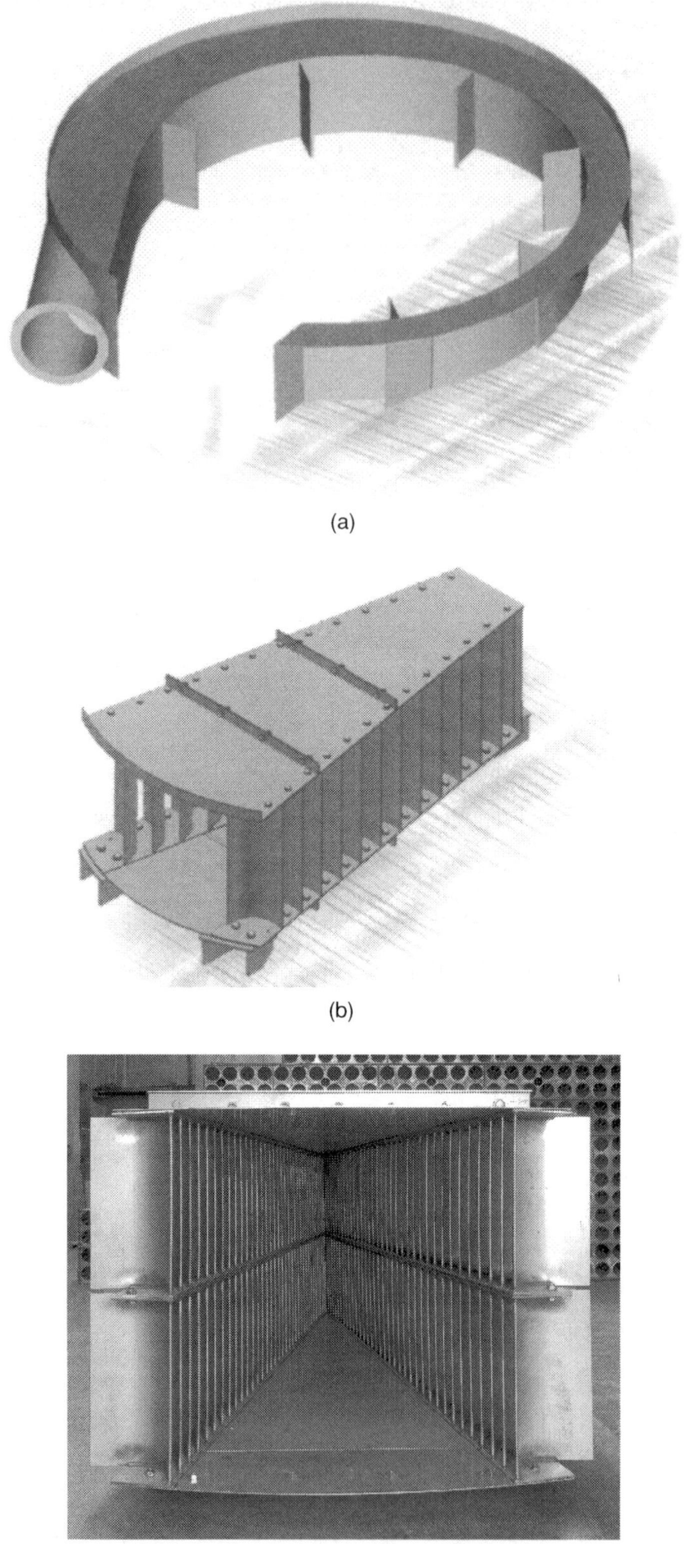

(a)

(b)

(c)

FIG. 14-71 High-velocity flashing feed and vapor distributors. (*a*) Vapor horn. (*b*) Radial vane distributor. (*c*) Schoepentoeter. *(Parts a, b, courtesy of Koch-Glitsch LP; part c, courtesy of Sulzer Chemtech.)*

FIG. 14-72 Chimney tray vapor distributor. *(Reprinted courtesy of Koch-Glitsch LP.)*

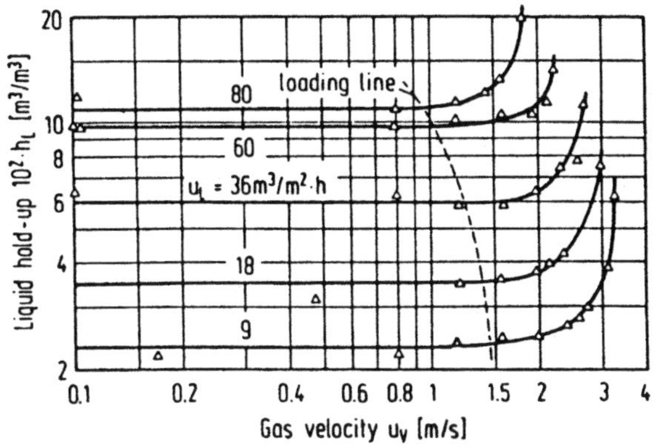

FIG. 14-74 Effect of liquid and gas rates on the operating holdup of modern random packings (25-mm NorPac®). [*From R. Billet and M. Schultes,* IChemE Symp. Ser. ***104,*** *p. A159, 1987. Reprinted courtesy of the Institution of Chemical Engineers (UK).*]

correlation solely by the readily available specific surface area (Tables 14-13 and 14-14).

Minimum Wetting Rate The minimum wetting rate (MWR) is the lower stability limit of packings. Below this liquid load the liquid film on the packing surfaces breaks up and dewetting occurs. The area available for mass transfer diminishes, and efficiency drops.

Schmidt [*IChemE Symp. Ser.* **56,** 3.1/1 (1979)] described the MWR in terms of a force balance at a dry patch along the path of a falling liquid film. While the gravity and viscous forces resist dewetting, the surface tension and vapor shear forces tend to dewet the falling film. The MWR therefore rises with an increase in surface tension and liquid density, and with a decrease in liquid viscosity. Large packing sizes and poor surface wetting characteristics also contribute to higher MWR.

Schmidt presented a fundamental correlation to predict minimum wetting for Raschig and Pall® rings. More popular have been the Glitsch rules of thumb [Table 14-16; Glitsch, Inc. (now Koch Glitsch), Bulletin 345, Dallas, Tex., 1986] for CMR® random packings with packing surface areas around 200 m²/m³. To extend these rules to other random packings, Kister (*Distillation Design*, McGraw-Hill, New York, 1992) applied Schmidt's model to give

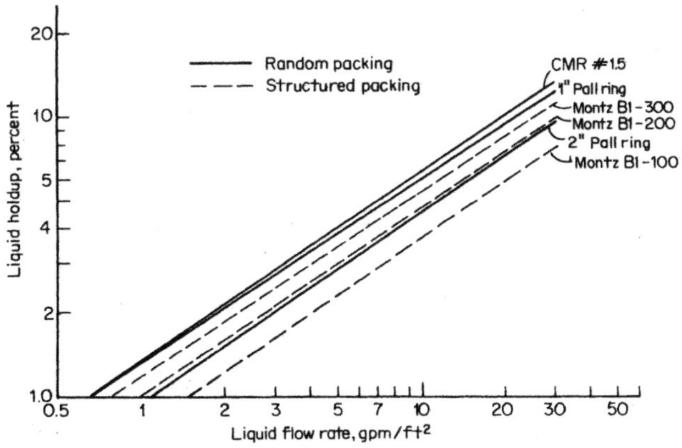

FIG. 14-73 Liquid holdup, air-water data by Billet ("Packed Column Design and Analysis," Ruhr University, Bochum, Germany), preloading regime. *(From Kister, H. Z.,* Distillation Design, *copyright © by McGraw-Hill; reprinted with permission.)*

$$Q_{MW} \approx (Q_{MW} \text{ from Table 14-16}) \times (200/a_p)^{0.5} \qquad (14\text{-}166)$$

The Glitsch brochure did not state the surface tension bases of Table 14-16. The author believes that they conservatively apply to organic and hydrocarbon systems (σ <25 mN/m). For water (σ = 70 mN/m) the author believes that double the values from Table 14-16 is reasonable.

Some surface treatments of the packing (e.g., oxidizing, sandblasting, etching) can substantially reduce the MWR. Chuang and Miller [*Can. J. Chem. Eng.* **66**(6), 377 (1988)] tested a metallic random packing with an aqueous system at low liquid rates (about 0.4 m³/hm²). They used two alternative techniques for oxidizing the packing surfaces. The packings oxidized with the more effective technique gave a column efficiency twice as high as those oxidized by the others.

Superior wetting characteristics characterize structured packings. Satisfactory performance was reported down to 0.25 and 0.12 m³/hm² in corrugated-sheet and wire-mesh metal structured packings, respectively. FRI's commercial-scale tests (Fitz and Kunesh, Paper presented at the AIChE Annual Meeting, Chicago, Nov. 1996) demonstrated good efficiencies down to 0.5 m³/hm² (no lower limit reached) with good liquid and vapor distribution.

Two Liquid Phases Two liquid phases often occur in heterogeneous distillation, in steam stripping, and in solvent recovery. Harrison (*Chem. Eng. Progr.,* p. 80, November 1990) and Meier et al. [*IChemE Symp. Ser.* **152,** 267 (2006)] saw no reason to expect unusual capacity or pressure drop limitations due to the presence of the second liquid phase, suggesting that standard correlations for these should apply.

In contrast, both expressed uncertainty in predicting packing efficiency. Harrison presented two case studies: In one, adding water to two water-insoluble organics had no effect on HETP; in another, a key

TABLE 14-16 Glitsch's Rule of Thumb for Minimum Wetting

Basis: CMR with a_p > 150 m²/m³

Material	Minimum wetting rate, m³/(m²·h)
Unglazed ceramic (chemical stoneware)	0.5
Oxidized metal (carbon steel, copper)	0.7
Surface-treated metal (etched stainless steel)	1.0
Glazed ceramic	2.0
Glass	2.5
Bright metal (stainless steel, tantalum, other alloys)	3.0
PVC-CPVC	3.5
Polypropylene	4.0
Fluoropolymers (PTFE type)	5.0

component was soluble in both liquid phases, and HETP was about 50 percent above normal. Harrison argued that a second liquid phase leads to lower efficiency only when it impairs diffusion of the key species. On this basis, Harrison expects efficiency loss also when an "inert" liquid or vapor represents a large fraction of the liquid or vapor phase. Meier et al. recommend obtaining efficiencies by scaling up laboratory-scale data using a similar type of packing.

Both Harrison and Meier et al. emphasize adequately distributing each liquid phase to the packing. Harrison noted that a well-designed ladder pipe distributor can maintain high velocities and low residence times that provide good mixing. With trough distributors that separate the phases and then distribute each to the packing, a light-to-heavy phase maldistribution may occur, especially when the phase ratio and separation vary. Meier et al. noted the existence of a cloudy two-liquid layer between the clear light and heavy liquid and recommend an additional middle distribution point for this layer. They also noticed that phase separation unevenness can have a large influence on the phase ratio irrigated to the packing.

High Viscosity and Surface Tension Bravo (Paper presented at the AIChE Spring National Meeting, Houston, Tex., 1995) studied a system that had 425-cP viscosity, 350 mN/m surface tension, and a high foaming tendency. He found that efficiencies were liquid-phase-controlled and could be estimated from theoretical HTU models. Capacity was less than predicted by conventional methods which do not account for the high viscosity. Design equations for orifice distributors extended well to the system once the orifice coefficient was calculated as a function of the low Reynolds number and the surface tension head was taken into account.

OTHER TOPICS FOR DISTILLATION AND GAS ABSORPTION EQUIPMENT

COMPARING TRAYS AND PACKINGS

Most separations can be performed either with trays or with packings. The factors below represent economic pros and cons that favor each and may be overridden. For instance, column complexity is a factor favoring trays, but gas plant demethanizers that often use one or more interreboilers are traditionally packed.

Factors Favoring Packings

Vacuum systems. Packing pressure drop is much lower than that of trays because the packing open area approaches the tower cross-sectional area, while the tray's open area is only 8 to 15 percent of the tower cross-sectional area. Also, the tray liquid head, which incurs substantial pressure drop (typically about 50 mm of the liquid per tray), is absent in packing. Typically, tray pressure drop is of the order of 10 mbar per theoretical stage, compared to 3 to 4 mbar per theoretical stage with random packings and about one-half of that with structured packings.

Consider a vacuum column with 10 theoretical stages, operating at 70-mbar top pressure. The bottom pressure will be 170 mbar with trays, but only 90 to 110 mbar with packings. The packed tower will have a much better relative volatility in the lower parts, thus reducing reflux and reboil requirements and bottom temperature. These translate to less product degradation, greater capacity, and smaller energy consumption, giving packings a major advantage.

Lower-pressure-drop applications. When the gas is moved by a fan through the tower, or when the tower is in the suction of a compressor, the smaller packing pressure drop is often a controlling consideration. This is particularly true for towers operating close to atmospheric pressure. Here excessive pressure drop in the tower increases the size of the fan or compressor (new plant), bottlenecks them (existing plant), and largely increases power consumption. Due to the compression ratio, pressure drop at the compressor discharge is far less important and seldom a controlling consideration.

Revamps. The pressure drop advantage is invaluable in vacuum column revamps, can be translated to a capacity gain, an energy gain, a separation improvement, or various combinations of these benefits. Likewise, for towers in the suction of compressors, replacing trays by packings reduces the compression ratio and helps debottleneck the compressor.

Packings also offer an easy tradeoff between capacity and separation. In the loaded sections of the tower, larger packings can overcome capacity bottlenecks at the expense of loss in separation. The separation loss can often be regained by retrofitting with smaller packings in sections of the tower that are not highly loaded. In tray towers, changing tray spacing gives similar results, but is more difficult to do.

Foaming (and emulsion). The low gas and liquid velocities in packing suppress foam formation. The large open area of the larger random packing promotes foam dispersal. Both attributes make random packing excellent for handling foams. In many cases recurrent foaming was alleviated by replacing trays by random packing, especially when tray downcomers were poorly designed.

Switching from trays to structured packing can aggravate foaming. While the low gas and liquid velocities help, the solid walls restrict lateral movement of foams and give support to the foams.

Small-diameter columns. Columns with diameter less than 1 m (3 ft) are difficult to access from inside to install and maintain the trays. "Cartridge" trays or an oversized diameter are often used. Either option is expensive. Cartridge trays also run into problems with sealing to the tower wall and matching tower to tray hardware [Sands, *Chem. Eng.*, p. 86 (April 2006)]. Packing is normally a cheaper and more desirable alternative.

Corrosive systems. The practical range of packing materials is wider. Ceramic and plastic packings are cheap and effective. Trays can be manufactured in nonmetals, but packing is usually a cheaper and more desirable alternative.

Low liquid holdup. Packings have lower liquid holdup than do trays. This is often advantageous for reducing polymerization, degradation, or the inventory of hazardous materials.

Batch distillation. Because of the smaller liquid holdup of packing, a higher percentage of the liquid can be recovered as top product.

Factors Favoring Trays

Solids. Trays handle solids much more easily than packing. Both gas and liquid velocities on trays are often an order of magnitude higher than through packing, providing a sweeping action that keeps tray openings clear. Solids tend to accumulate in packing voids. There are fewer locations on trays where solids can be deposited. Plugging in liquid distributors has been a common trouble spot. Cleaning trays is much easier than cleaning packings.

Not all trays are fouling-resistant. Floats on moving valve trays tend to "stick" to deposits on the tray deck. Fouling-resistant trays have large sieve holes or large fixed valves, and these should be used when plugging and fouling are the primary considerations.

There is much that can be done to alleviate plugging with random packing. Large, open packing with minimal pockets offers good plugging resistance. Distributors that resist plugging have large holes (> 13-mm diameter). Such large holes are readily applied with high liquid flow rates, but often not practical for small liquid flow rates.

Maldistribution. The sensitivity of packing to liquid and gas maldistribution has been a common cause of failures in packed towers. Maldistribution issues are most severe in large-diameter towers, long beds, small liquid flow rates, and smaller packing. Structured packing is generally more prone to maldistribution than random packing. While good distributor design, water testing, and inspection can eliminate most maldistribution issues, it only takes a few small details that fall through the cracks to turn success into failure. Due to maldistribution, there are far more failures experienced with packing than in trays, and it takes more trials "to get it right" than with trays. This makes trays more robust.

Complex towers. Interreboilers, intercondensers, cooling coils, and side drawoffs are more easily incorporated in trays than in packed towers. In packed towers, every complexity requires additional distribution and/or liquid collection equipment.

Feed composition variation. One way of allowing for design uncertainties and feedstock variation is by installing alternate feed points. In packed towers, every alternate feed point requires expensive distribution equipment.

Performance prediction. Due to their sensitivity to maldistribution there is greater uncertainty in predicting packed column performance.

Chemical reaction, absorption. Here the much higher liquid holdup on trays provides greater residence time for absorption or chemical reaction than does packing.

Turndown. Moving valve and bubble-cap trays normally give better turndown than packings. Unless very expensive distributors are used, packed tower turndown is usually limited by distributor turndown.

Weight. Tray towers usually weigh less than packed towers, saving on the cost of foundations, supports, and column shell.

Trays vs. Random Packings The following factors generally favor trays compared to random packings, but not compared to structured packings.

Low liquid rates. With the aid of serrated weirs, splash baffles, reverse-flow trays, and bubble-cap trays, low liquid rates can be handled better in trays. Random packings suffer from liquid dewetting and maldistribution sensitivity at low liquid rates.

Process surges. Random packings are usually more troublesome than trays in services prone to process surges (e.g., those caused by slugs of water entering a hot oil tower, relief valve lifting, compressor surges, or instability of liquid seal loops). Structured packings are usually less troublesome than trays in such services.

Trays vs. Structured Packings The following factors generally favor trays compared to structured packings, but not compared to random packings.

Packing fires. The thin sheets of structured packing (typically 0.1 mm) poorly dissipate heat away from hot spots. Also, cleaning, cooling, and washing can be difficult, especially when distributors or packing plug up. Many incidents of packing fires during turnarounds (while towers with structured packings were open to atmosphere) have been reported. Most of these fires were initiated by pyrophoric deposits, hot work (e.g., welding) above the packing, opening the tower while hot organics were still present, and packing metallurgy that was not fire-resistant. Detailed discussion can be found in Fractionation Research Inc. (FRI) Design Practices Committee, "Causes and Prevention of Packing Fires," *Chem. Eng.*, July 2007.

Materials of construction. Due to the thin sheets of structured packings, their materials of construction need to have better resistance to oxidation or corrosion. For a service in which carbon steel is usually satisfactory with trays, stainless steel is usually required with structured packings.

Column wall inspection. Due to their snug fit, structured packings are easily damaged during removal. This makes it difficult to inspect the column wall (e.g., for corrosion).

Washing and purging. Thorough removal of residual liquid, wash water, air, or process gas trapped in structured packings at start-up and shutdown is more difficult than with trays. Inadequate removal of these fluids may be hazardous.

High liquid rates. Multipass trays effectively lower the liquid load "seen" by each part of the tray. A similar trick cannot be applied with packings. The capacity of structured packings tends to rapidly fall off at high liquid rates.

Capacity and Efficiency Comparison Kister et al. [*Chem. Eng. Progr.*, **90**(2), 23 (1994)] reported a study of the relative capacity and efficiency of conventional trays, modern random packings, and conventional structured packings. They found that, for each device optimally designed for the design requirements, a rough guide could be developed on the basis of flow parameter $L/G\ (\rho_G/\rho_L)^{0.5}$ (abcissa in Figs. 14-31, 14-55, and 14-56) and the following tentative conclusions could be drawn:

Flow Parameter 0.02–0.1

1. Trays and random packings have much the same efficiency and capacity.

2. Structured packing efficiency is about 1.5 times that of trays or random packing.

3. At a parameter of 0.02, the structured packing has a 1.3–1.4 capacity advantage over random packing and trays. This advantage disappears as the parameter approaches 0.1.

Flow Parameter 0.1–0.3

1. Trays and random packings have about the same efficiency and capacity.

2. Structured packing has about the same capacity as trays and random packings.

3. The efficiency advantage of structured packing over random packings and trays decreases from 1.5 to 1.2 as the parameter increases from 0.1 to 0.3.

Flow Parameter 0.3–0.5

1. The loss of capacity of structured packing is greatest in this range.

2. The random packing appears to have the highest capacity and efficiency with conventional trays just slightly behind. Structured packing has the least capacity and efficiency.

Experience indicates that use of structured packings has capacity/efficiency disadvantages in the higher-pressure (higher-flow-parameter) region.

Zuiderweg and Nutter [*IChemE Symp. Ser.* **128,** A481 (1992)] explain the loss of capacity/efficiency by a large degree of backmixing and vapor recycle at high flow parameters, promoted by the solid walls of the corrugated packing layers.

SYSTEM LIMIT: THE ULTIMATE CAPACITY OF FRACTIONATORS

Liquid drops of various sizes form in the gas-liquid contact zones of tray or packed towers. Small drops are easily entrained upward, but their volume is usually too small to initiate excessive liquid accumulation (flooding). When the gas velocity is high enough to initiate a massive carryover of the larger drops to the tray above, or upward in a packed bed, liquid accumulation (entrainment flooding) takes place. This flood can be alleviated by increasing the tray spacing or using more hole areas on trays or by using larger, more open packings.

Upon further increase of gas velocity, a limit is reached when the superficial gas velocity in the gas-liquid contact zone exceeds the settling velocity of large liquid drops. At gas velocities higher than this, ascending gas lifts and carries over much of the tray or packing liquid, causing the tower to flood. This flood is termed *system limit* or *ultimate capacity*. This flood cannot be debottlenecked by improving packing size or shape, tray hole area, or tray spacing. The system limit gas velocity is a function only of physical properties and liquid flow rate. Once this limit is reached, the liquid will be blown upward. This is analogous to spraying water against a strong wind and getting drenched (Yangai, *Chem.* Eng., p. 120, November 1990). The system limit represents the ultimate capacity of the vast majority of existing trays and packings. In some applications, where very open packings (or trays) are used, such as in refinery vacuum towers, the system limit is the actual capacity limit.

The original work of Souders and Brown [*Ind. Eng. Chem.* **26**(1), 98 (1934), Eq. (14-80)] related the capacity of fractionators due to entrainment flooding to the settling velocity of drops. The concept of system limit was advanced by Fractionation Research Inc. (FRI), whose measurements and model have recently been published (Fitz and Kunesh, Distillation 2001: Proceedings of Topical Conference, AIChE Spring National Meeting, Houston, Tex., 2001; Stupin, FRI Topical Report 34, 1965 available through Special Collection Section, Oklahoma State University Library, Stillwater, Okla.). Figure 14-75 is a plot of FRI system limit data (most derived from tests with dual-flow trays with 29 percent hole area and 1.2- to 2.4-m tray spacing) against liquid superficial velocity for a variety of systems (Stupin, loc. cit.,

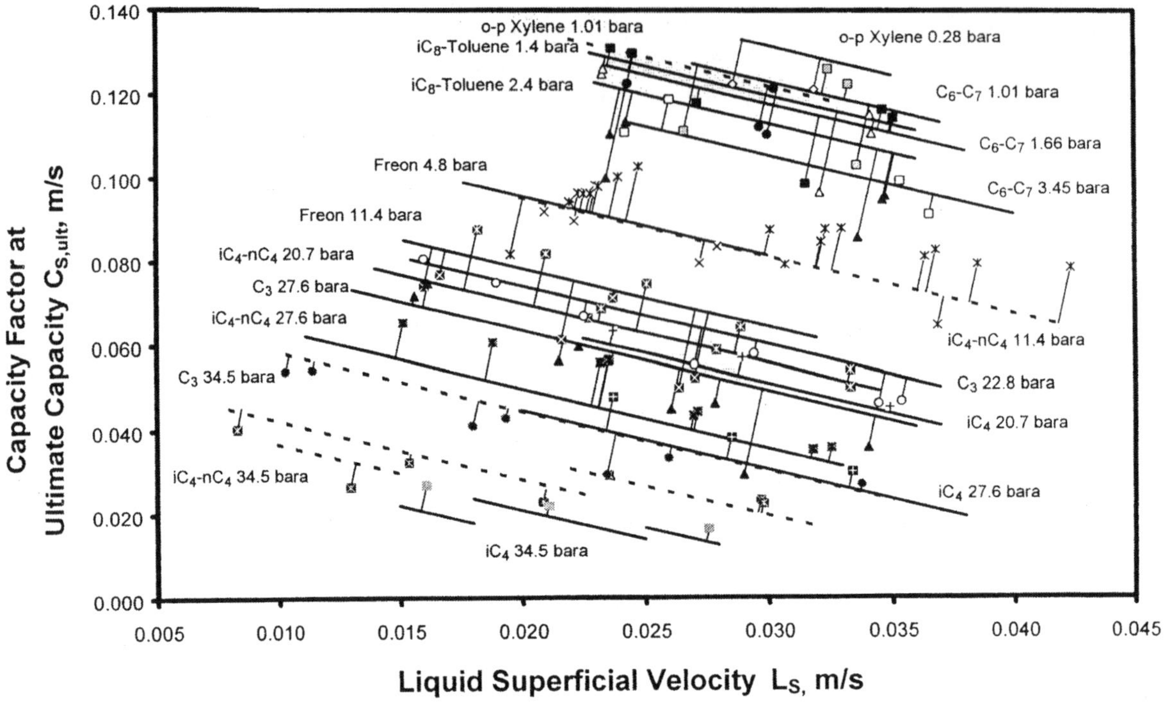

FIG. 14-75 Effect of liquid rate on ultimate capacity at higher liquid rates. *(From Stupin, W. J., and H. Z. Kister,* Trans. IChemE, *vol. 81, Part A, p. 136, January 2003. Reprinted courtesy of IChemE.)*

1965). The data show a constant-slope linear dependence of the system limit C-factor on the liquid load. There was a shortage of data at low liquid loads. Later data (Fig. 14-76) showed that as the liquid load was reduced, the system limit $C_{s,ult}$ stopped increasing and reached a limiting value. Based on this observation, Stupin and Kister [*Trans. IChemE* **81,** Part A, p. 136 (January 2003)] empirically revised the earlier Stupin/FRI correlation to give

$$C_1 = 0.445(1 - F)\left(\frac{\sigma}{\Delta\rho}\right)^{0.25} - 1.4L_S \tag{14-167}$$

$$C_2 = 0.356(1 - F)\left(\frac{\sigma}{\Delta\rho}\right)^{0.25} \tag{14-168}$$

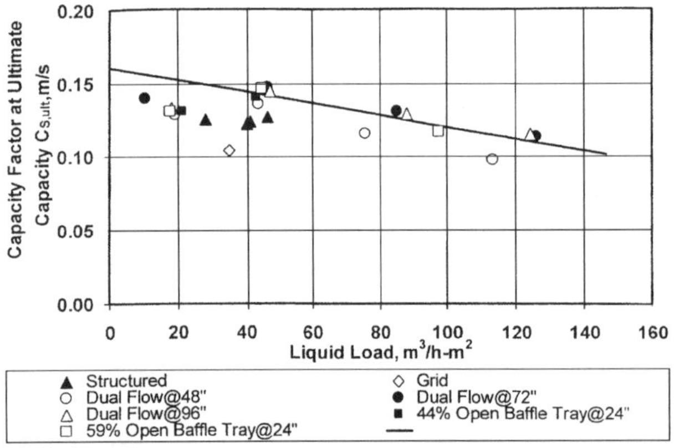

FIG. 14-76 Comparison of original ultimate capacity correlation to test data, C_6/C_7, 1.66 bar. *(From Stupin, W. J., and H. Z. Kister,* Trans. IChemE, *vol. 81, Part A, p. 136, January 2003. Reprinted courtesy of IChemE.)*

$$C_{s,ult} = \text{smaller of } C_1 \text{ and } C_2 \tag{14-169}$$

where

$$F = \frac{1}{1 + 1.4(\Delta\rho/\rho_G)^{1/2}} \tag{14-170}$$

In Eqs. (14-167) through (14-170), $C_{s,ult}$ is the system limit C-factor based on the tower superficial area [see Eq. (14-77) for C-factor definition]; L_S is the liquid superficial velocity, m/s; σ is the surface tension, mN/m; $\Delta\rho$ is the difference between the liquid and gas densities, kg/m³; and ρ_G is the gas density, kg/m³.

Stupin and Kister (loc. cit.) relate the flattening of the curve in Fig. 14-76 at low liquid loads to the formation of more, smaller, easier-to-entrain liquid drops when the liquid load is lowered beyond the limiting liquid load. It follows that devices that can restrict the formation of smaller drops may be able to approach the system limit capacity predicted by Stupin's original equation [Eq. (14-167)] even at low liquid loads.

The only devices capable of debottlenecking a tray system-limit device are those that introduce a new force that helps disentrain the vapor space. Devices that use centrifugal force (see "Centrifugal Force Deentrainment") are beginning to make inroads into commercial distillation and have achieved capacities as high as 25 percent above the system limit. Even the horizontal vapor push (see "Truncated Downcomers/Forward-Push Trays") can help settle the entrained drops, but to a much lesser extent. It is unknown whether the horizontal push alone can achieve capacities exceeding the system limit.

WETTED-WALL COLUMNS

Wetted-wall or falling-film columns have found application in mass-transfer problems when high-heat-transfer-rate requirements are concomitant with the absorption process. Large areas of open surface

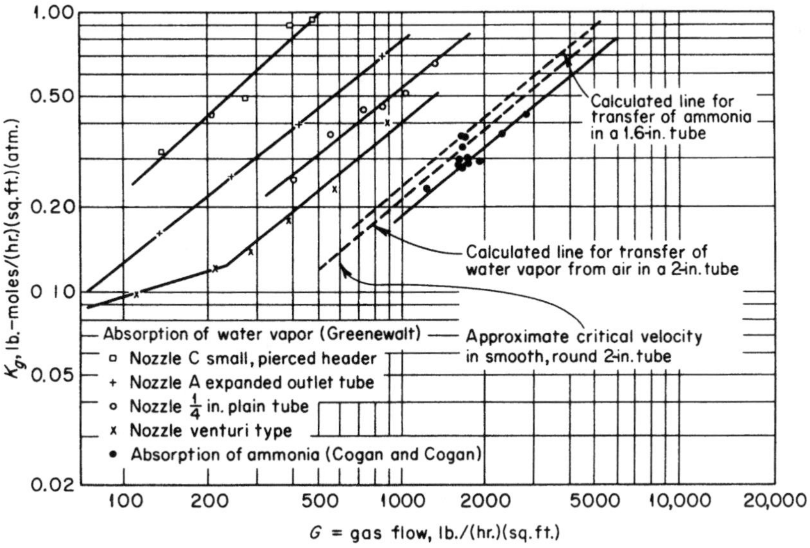

FIG. 14-77 Mass-transfer rates in wetted-wall columns having turbulence promoters. To convert pound-moles per hour-square foot-atmosphere to kilogram-moles per second-square meter-atmosphere, multiply by 0.00136; to convert pounds per hour-square foot to kilograms per second-square meter, multiply by 0.00136; and to convert inches to millimeters, multiply by 25.4. (Data of Greenewalt and Cogan and Cogan, *Sherwood, and Pigford,* Absorption and Extraction, *2d ed., McGraw-Hill, New York, 1952.*)

are available for heat transfer for a given rate of mass transfer in this type of equipment because of the low mass-transfer rate inherent in wetted-wall equipment. In addition, this type of equipment lends itself to annular-type cooling devices.

Gilliland and Sherwood [*Ind. Eng. Chem.,* **26,** 516 (1934)] found that, for vaporization of pure liquids in air streams for streamline flow,

$$\frac{k_g D_{\text{tube}}}{D_g}\,\frac{P_{BM}}{P} = 0.023 N_{\text{Re}}^{0.83} N_{\text{Sc}}^{0.44} \qquad (14\text{-}171)$$

where D_g = diffusion coefficient, ft²/h
D_{tube} = inside diameter of tube, ft
k_g = mass-transfer coefficient, gas phase, lb·mol/(h·ft²) (lb·mol/ft³)
P_{BM} = logarithmic mean partial pressure of inert gas, atm
P = total pressure, atm
N_{Re} = Reynolds number, gas phase
N_{Sc} = Schmidt number, gas phase

Note that the group on the left side of Eq. (14-171) is dimensionless. When turbulence promoters are used at the inlet-gas section, an improvement in gas mass-transfer coefficient for absorption of water vapor by sulfuric acid was observed by Greenewalt [*Ind. Eng. Chem.,* **18,** 1291 (1926)]. A falling off of the rate of mass transfer below that indicated in Eq. (14-171) was observed by Cogan and Cogan (thesis, Massachusetts Institute of Technology, 1932) when a calming zone preceded the gas inlet in ammonia absorption (Fig. 14-77).

In work with the hydrogen chloride-air-water system, Dobratz, Moore, Barnard, and Meyer [*Chem. Eng. Prog.,* **49,** 611 (1953)] using a cocurrent-flow system found that $K_g \propto G^{1.8}$ (Fig. 14-78) instead of the 0.8 power as indicated by the Gilliland equation. Heat-transfer coefficients were also determined in this study. The radical increase in heat-transfer rate in the range of $G = 30$ kg/(s·m²) [20,000 lb/(h·ft²)] was similar to that observed by Tepe and Mueller [*Chem. Eng. Prog.,* **43,** 267 (1947)] in condensation inside tubes.

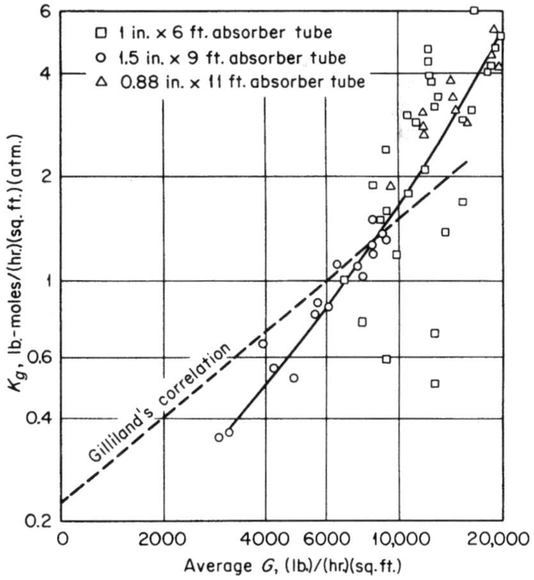

FIG. 14-78 Mass-transfer coefficients versus average gas velocity—HCl absorption, wetted-wall column. To convert pound-moles per hour-square foot-atmosphere to kilogram-moles per second-square meter-atmosphere, multiply by 0.00136; to convert pounds per hour-square foot to kilograms per second-square meter, multiply by 0.00136; to convert feet to meters, multiply by 0.305; and to convert inches to millimeters, multiply by 25.4. [*Dobratz et al.,* Chem. Eng. Prog., ***49,*** *611 (1953).*]

Gaylord and Miranda [*Chem. Eng. Prog.*, **53**, 139M (1957)] using a multitube cocurrent-flow falling-film hydrochloric acid absorber for hydrogen chloride absorption found

$$K_g = \frac{1.66(10^{-5})}{M_m^{1.75}} \left(\frac{D_{\text{tube}}G}{\mu} \right) \tag{14.172}$$

where K_g = overall mass-transfer coefficient, (kg·mol)/(s·m²·atm)
M_m = mean molecular weight of gas stream at inlet to tube
D_{tube} = diameter of tube, m
G = mass velocity of gas at inlet to tube, kg/(s·m²)
μ = viscosity of gas, Pa·s

Note that the group $D_{tube}G/\mu$ is dimensionless. This relationship also satisfied the data obtained for this system, with a single-tube falling-film unit, by Coull, Bishop, and Gaylor [*Chem. Eng. Prog.*, **45**, 506 (1949)].

The rate of mass transfer in the liquid phase in wetted-wall columns is highly dependent on surface conditions. When laminar-flow conditions prevail without the presence of wave formation, the laminar-penetration theory prevails. When, however, ripples form at the surface, and they may occur at a Reynolds number exceeding 4, a significant rate of surface regeneration develops, resulting in an increase in mass-transfer rate.

If no wave formations are present, analysis of behavior of the liquid-film mass transfer as developed by Hatta and Katori [*J. Soc. Chem. Ind.*, **37**, 280B (1934)] indicates that

$$k_\ell = 0.422 \sqrt{\frac{D_\ell \Gamma}{\rho B_F^2}} \tag{14-173}$$

where $B_F = (3u\Gamma/\rho^2 g)^{1/3}$
D_l = liquid-phase diffusion coefficient, m²/s
ρ = liquid density, kg/m³
Z = length of surface, m
k_l = liquid-film-transfer coefficient, (kg·mol)/[(s·m²)(kg·mol)/m³]
Γ = liquid-flow rate, kg/(s·m) based on wetted perimeter
μ = viscosity of liquid, Pa·s
g = gravity acceleration, 9.81 m/s²

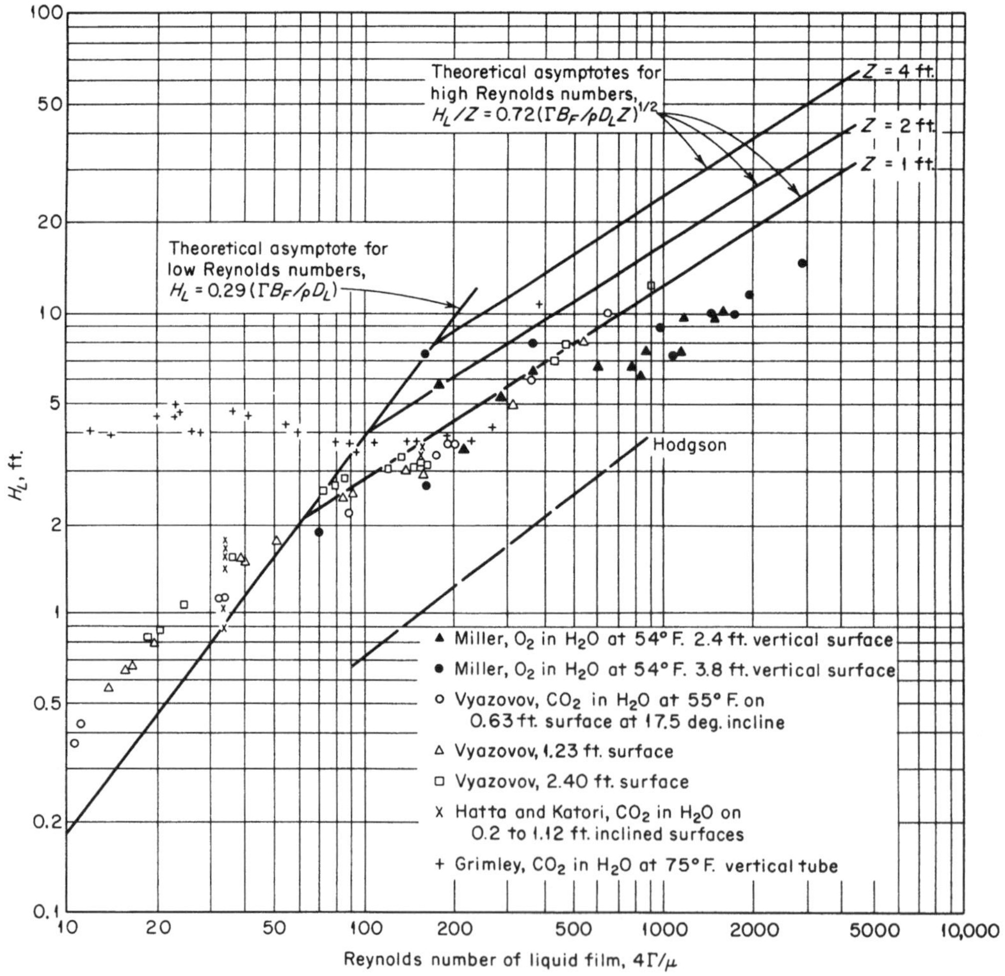

FIG. 14-79 Liquid-film resistance in absorption of gases in wetted-wall columns. Theoretical lines are calculated for oxygen absorption in water at 55°F. To convert feet to meters, multiply by 0.3048; °C = 5⁄9 (°F − 32). (*Sherwood and Pigford,* Absorption and Extraction, *2d ed., McGraw-Hill, New York, 1952.*)

When Z is large or $\Gamma/\rho B_F$ is so small that liquid penetration is complete,

$$k_\ell = 11.800\ D_\ell/B_F \tag{14-174}$$

and

$$H_\ell = 0.95\ \Gamma B_F/D_\ell \tag{14-175}$$

A comparison of experimental data for carbon dioxide absorption obtained by Hatta and Katori (op. cit.), Grimley [*Trans. Inst. Chem. Eng.*, **23,** 228 (1945)], and Vyazov [*Zh. Tekh. Fiz.* (U.S.S.R.), **10,** 1519 (1940)] and for absorption of oxygen and hydrogen by Hodgson (S.M. thesis, Massachusetts Institute of Technology, 1949), Henley (B.S. thesis, University of Delaware, 1949), Miller (B.S. thesis, University of Delaware, 1949), and Richards (B.S. thesis, University of Delaware, 1950) was made by Sherwood and Pigford (*Absorption and Extraction,* McGraw-Hill, New York, 1952) and is indicated in Fig. 14-79.

In general, the observed mass-transfer rates are greater than those predicted by theory and may be related to the development of surface rippling, a phenomenon which increases in intensity with increasing liquid path.

Vivian and Peaceman [*Am. Inst. Chem. Eng. J.*, **2,** 437 (1956)] investigated the characteristics of the CO_2-H_2O and Cl_2-HCl, H_2O system in a wetted-wall column and found that gas rate had no effect on the liquid-phase coefficient at Reynolds numbers below 2200. Beyond this rate, the effect of the resulting rippling was to increase significantly the liquid-phase transfer rate. The authors proposed a behavior relationship based on a dimensional analysis but suggested caution in its application concomitant with the use of this type of relationship. Cognizance was taken by the authors of the effects of column length, one to induce rippling and increase of rate of transfer, one to increase time of exposure which via the penetration theory decreases the average rate of mass transfer in the liquid phase. The equation is

$$\frac{k_\ell h}{D_\ell} = 0.433 \left(\frac{\mu_\ell}{\rho_\ell D_\ell}\right)^{1/2} \left(\frac{\rho_\ell^2 g h^3}{\mu_\ell^2}\right)^{1/6} \left(\frac{4\Gamma}{\mu_\ell}\right)^{0.4} \tag{14-176}$$

where D_ℓ = diffusion coefficient of solute in liquid, ft^2/h
g = gravity-acceleration constant, 4.17×10^8 ft/h^2
h = length of wetted wall, ft
k_ℓ = mass-transfer coefficient, liquid phase, ft/h
Γ = mass rate of flow of liquid, lb/(h)(ft of periphery)
μ_ℓ = viscosity of liquid, lb/(ft)(h)
ρ_ℓ = density of liquid, lb/ft^3

The equation is dimensionless.

The effect of chemical reaction in reducing the effect of variation of the liquid rate on the rate of absorption in the laminar-flow regime was illustrated by the evaluation of the rate of absorption of chlorine in ferrous chloride solutions in a wetted-wall column by Gilliland, Baddour, and White [*Am. Inst. Chem. Eng. J.*, **4,** 323 (1958)].

Flooding in Wetted-Wall Columns When gas and liquid are in counterflow in wetted-wall columns, flooding can occur at high gas rates. Methods for calculating this flood are given in "Upper Limit Flooding in Vertical Tubes." In the author's experience, Eq. (14-204) has had an excellent track record for calculating flooding in these columns.

TABLE 14-17 Relative Fabricated Cost for Metals Used in Tray-Tower Construction*

Materials of construction	Relative cost per ft² of tray area (based on carbon steel = 1)
Sheet-metal trays	
Steel	1
4–6% chrome—½ moly alloy steel	2.1
11–13% chrome type 410 alloy steel	2.6
Red brass	3
Stainless steel type 304	4.2
Stainless steel type 347	5.1
Monel	7.0
Stainless steel type 316	5.5
Inconel	8.2
Cast-iron trays	2.8

*Peters and Timmerhaus, *Plant Design and Economics for Chemical Engineers,* 4th ed., McGraw-Hill, New York, 1991. To convert cost per square foot to cost per square meter, multiply by 10.76.

COLUMN COSTS

Estimation of column costs for preliminary process evaluations requires consideration not only of the basic type of internals but also of their effect on overall system cost. For a distillation system, for example, the overall system can include the vessel (column), attendant structures, supports, and foundations; auxiliaries such as reboiler, condenser, feed heater, and control instruments; and connecting piping. The choice of internals influences all these costs, but other factors influence them as well. A complete optimization of the system requires a full-process simulation model that can cover all pertinent variables influencing economics.

Cost of Internals Installed costs of trays may be estimated from Fig. 14-80, with corrections for tray material taken from Table 14-17. For two-pass trays the cost is 15 to 20 percent higher. Approximate costs of random packing materials may be obtained from Table 14-18, but it should be recognized that, because of competition, there can be significant variations in these costs from vendor to vendor. Also, packings sold in very large quantities carry discounts. In 1995, costs of structured packings, made from sheet metal, averaged \$90 to \$110 per cubic foot, but the need for special distributors and redistributors can double the cost of structured-packings on a volumetric basis. Note

TABLE 14-18 Costs of Random Packings, Uninstalled, January 1990

Prices in dollars per ft³, 100 ft³ orders, f.o.b. manufacturing plant

	Size, in, \$/ft³			
	1	1½	2	3
Raschig rings				
Chemical porcelain	12.8	10.3	9.4	7.8
Carbon steel	36.5	23.9	20.5	16.8
Stainless steel	155	117	87.8	—
Carbon	52	46.2	33.9	31.0
Intalox saddles				
Chemical stoneware	17.6	13.0	11.8	10.7
Chemical porcelain	18.8	14.1	12.9	11.8
Polypropylene	21.2	—	13.1	7.0
Berl saddles				
Chemical stoneware	27.0	21.0	—	—
Chemical porcelain	33.5	21.5	15.6	—
Pall rings				
Carbon steel	29.3	19.9	18.2	—
Stainless steel	131	99.0	86.2	—
Polypropylene	21.2	14.4	13.1	

Peters and Timmerhaus, *Plant Design and Economics for Chemical Engineers,* 4th ed., McGraw-Hill, New York, 1991. To convert cubic feet to cubic meters, multiply by 0.0283; to convert inches to millimeters, multiply by 25.4; and to convert dollars per cubic foot to dollars per cubic meter, multiply by 35.3.

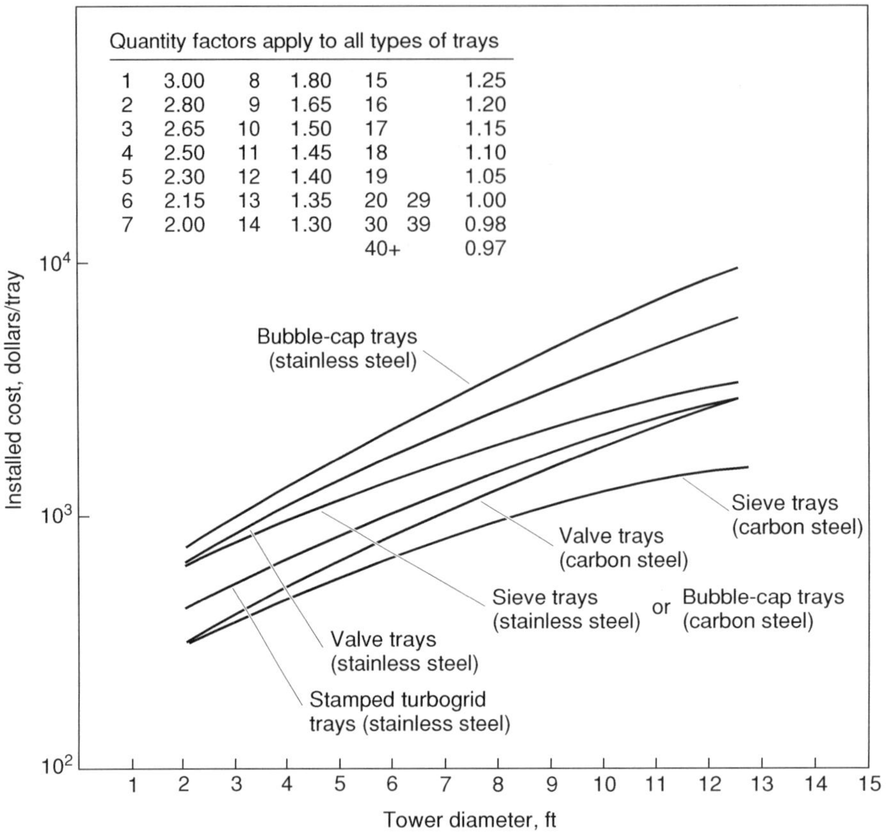

FIG. 14-80 Cost of trays. Price includes tray deck, valves, bubble caps, risers, downcomers, and structural-steel parts. The stainless steel designated is type 410 (*Peters and Timmerhaus,* Plant Design and Economics for Chemical Engineers, *4th ed., McGraw-Hill, New York, 1991*).

that for Fig. 14-80 and Table 14-17, the effective cost date is January 1990, with the Marshall and Swift cost index being taken as 904.

As indicated above, packed column internals include liquid distributors, packing support plates, redistributors (as needed), and holddown plates (to prevent movement of packing under flow conditions). Costs of these internals for columns with random packing are given in Fig. 14-81, based on early 1976 prices, and a Marshall and Swift cost index of 460.

Cost of Column The cost of the vessel, including heads, skirt, nozzles, and ladderways, is usually estimated on the basis of weight. Figure 14-82 provides early 1990 cost data for the shell and heads, and Fig. 14-83 provides 1990 cost data for connections. For very approximate estimates of complete columns, including internals, Fig. 14-84 may be used. As for Figs. 14-82 and 14-83, the cost index is 904.

PHASE DISPERSION

General References: For an overall discussion of gas-liquid breakup processes, see Brodkey, *The Phenomena of Fluid Motions,* Addison-Wesley, Reading, Massachusetts, 1967. For a discussion of atomization devices and how they work, see Masters, *Spray Drying Handbook,* 5th ed., Wiley, New York, 1991; and Lefebvre, *Atomization and Sprays,* Hemisphere, New York, 1989. A beautifully illustrated older source is Dombrowski and Munday, *Biochemical and Biological Engineering Science,* vol. 2, Academic Press, London, 1968, pp. 209–320. Steinmeyer [*Chem. Engr. Prog.,* **91**(7), 72–80 (1995)] built on Hinze's work with turbulence and showed that several atomization processes follow the common theme of breakup of large droplets due to turbulence in the gas phase. Turbulence in turn correlates with power dissipation per unit mass. In the text below, correlations referring to *power/mass* are taken from this source. For a survey on fog formation, see Amelin, *Theory of Fog Formation,* Israel Program for Scientific Translations, Jerusalem, 1967.

BASICS OF INTERFACIAL CONTACTORS

Steady-State Systems: Bubbles and Droplets Bubbles are made by injecting vapor below the liquid surface. In contrast, droplets are commonly made by atomizing nozzles that inject liquid into a vapor. Bubble and droplet systems are fundamentally different, mainly because of the enormous difference in density of the injected phase. There are situations where each is preferred. Bubble systems tend to have much higher interfacial area as shown by Example 16 contrasted with Examples 14 and 15. Because of their higher area, bubble systems will usually give a closer approach to equilibrium.

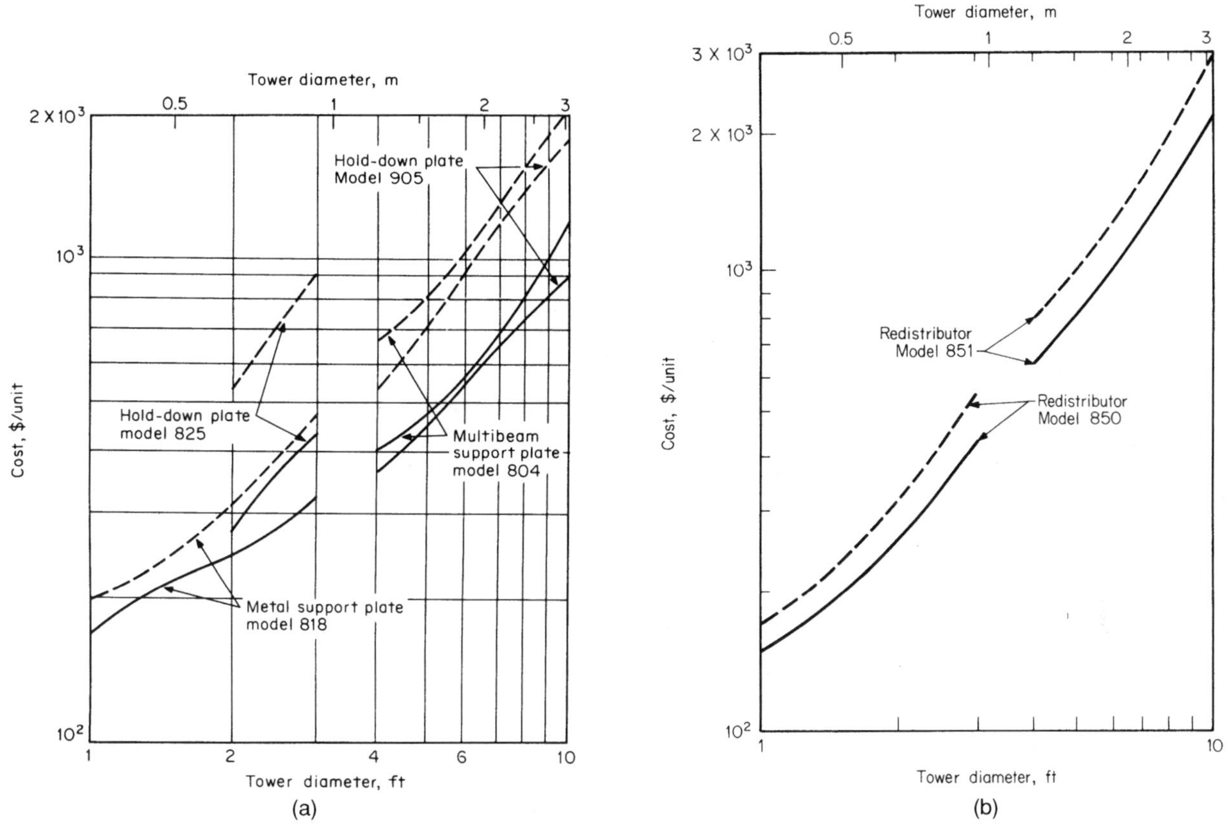

FIG. 14-81 Cost of internal devices for columns containing random packings. (*a*) Holddown plates and support plates. (*b*) Redistributors. [*Pikulik and Diaz,* Chem. Eng., *84(21), 106 (Oct. 10, 1977).*]

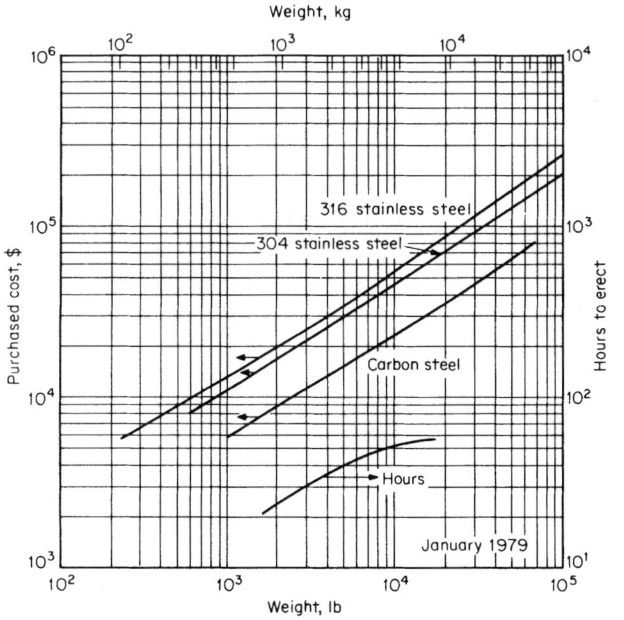

FIG. 14-82 Fabricated costs and installation time of towers. Costs are for shell with two heads and skirt, but without trays, packing, or connections. (*Peters and Timmerhaus,* Plant Design and Economics for Chemical Engineers, *4th ed., McGraw-Hill, New York, 1991.*)

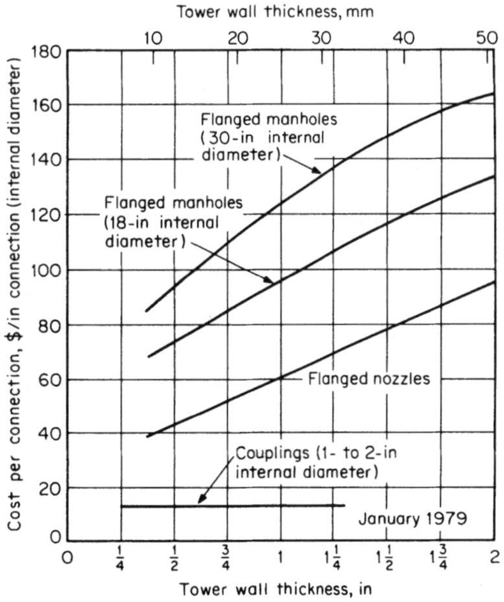

FIG. 14-83 Approximate installed cost of steel-tower connections. Values apply to 2070-kPa connections. Multiply costs by 0.9 for 1035-kPa (150-lb) connections and by 1.2 for 4140-kPa (600-lb) connections. To convert inches to millimeters, multiply by 25.4; to convert dollars per inch to dollars per centimeter, multiply by 0.394. (*Peters and Timmerhaus,* Plant Design and Economics for Chemical Engineers, *4th ed., New York, McGraw-Hill, 1991.*)

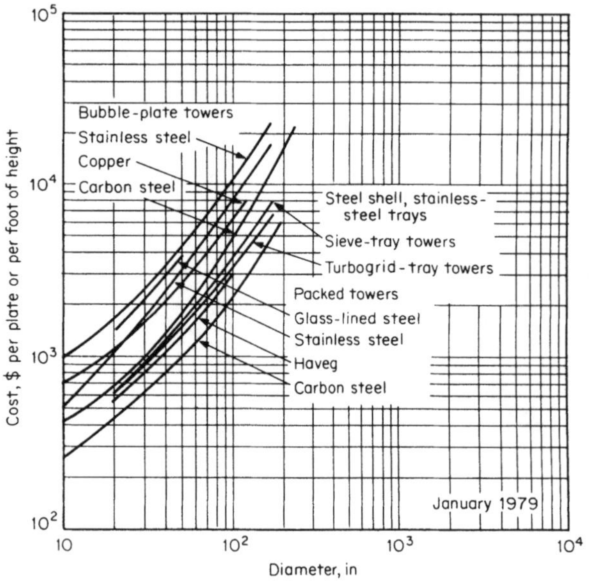

FIG. 14-84 Cost of towers including installation and auxiliaries. To convert inches to millimeters, multiply by 25.4; to convert feet to meters, multiply by 0.305; and to convert dollars per foot to dollars per meter, multiply by 3.28. (*Peters and Timmerhaus,* Plant Design and Economics for Chemical Engineers, *4th ed., McGraw-Hill, New York, 1991.*)

However, droplet systems can enable much higher energy input (via gas phase pressure drop in cocurrent systems) and, as a result, dominate applications where a quick quench is needed. See Examples 21 and 22. Conversely, droplet systems can also be designed for very low pressure drop which is advantageous in applications such as vacuum condensers.

Unstable Systems: Froths and Hollow Cone Atomizing Nozzles We usually think of interfacial contact as a steady-state system of raining droplets or rising bubbles, but some of the most efficient interfacial contactors take advantage of unstable interfacial geometry. The most common is the distillation tray which operates with a wild mix of bubbles, jets, films, and droplets. The mix is often described as *froth.* Gas pressure drop provides the energy to create the froth.

A variant on the froth contact is the reverse jet contactor (Example 22), which can be considered as an upside-down distillation tray, operated above the flooding velocity in cocurrent flow of gas and liquid. It is limited to one stage.

An entirely different unstable contactor involves the thin expanding liquid film produced by a hollow cone spray nozzle. Because of fresh surface and the thinness of the film, this can give very high transfer for liquid-limited systems. Two applications are direct contact condensation and removal of volatile components from a high-boiling residual liquid.

Surface Tension Makes Liquid Sheets and Liquid Columns Unstable Surface tension is the energy required to make an increment of interfacial surface. A sheet or column of liquid has more surface than a sphere, hence surface tension converts sheets and columns to droplets. See Fig. 14-86.

There are many different atomizers, but the underlying principle of all is the same—to first generate a flat sheet or a liquid column. Liquid sheets and columns are unstable, a small surface disturbance on either will propagate, and the liquid will reshape itself into droplets. The key property in controlling this process is surface tension. Surface tension gets a high exponent in all the atomization correlations.

Little Droplets and Bubbles vs. Big Droplets and Bubbles—Coalescence vs. Breakup When big drops are subjected to shear forces, as in falling rain, the droplets are distorted; and if the distortions are great enough, the big droplets break into little ones. This is why raindrops never exceed a certain size. A variant on this is breakup in highly turbulent systems such as in high-velocity quench systems or pneumatic nozzles. Here the droplets are distorted by the energy of the turbulent eddies.

But little droplets and bubbles have more surface per unit of liquid than big ones. Hence little droplets tend to coalesce into big ones, and will grow in size if given enough quiet time.

While the primary difficulty is estimating the interfacial area due to the unstable interface, a secondary problem is that freshly made, unstable surface gives higher transfer than older, more stable surface.

Empirical Design Tempered by Operating Data The net of these is that interfacial area is difficult to predict and interfacial contactors are difficult to design.

Prediction methods are given below but should always be tempered by operating experience.

INTERFACIAL AREA—IMPACT OF DROPLET OR BUBBLE SIZE

Transfer is aided by increased interfacial area. Interfacial area per unit volume a_D of a single droplet or bubble is inversely proportional to the diameter of the droplet or bubble D.

$$a_D = 6/D \qquad (14\text{-}177)$$

To estimate the total interfacial area in a given volume, the a_d value is multiplied by the fractional holdup of dispersed phase in the total volume.

$$a = a_D(\Phi_D) \qquad (14\text{-}178)$$

where a = interfacial area/volume and Φ_D = fraction of volume in dispersed phase = holdup. Fractional holdup in a continuous process depends on the velocities of the two phases, as if they were flowing by themselves.

$$\Phi_D = \text{(dispersed phase volume)/(volume of dispersed and continuous phases)}$$

Example 14: Interfacial Area for Droplets/Gas in Cocurrent Flow For equal mass flow of gas and liquid and with gas density 0.001 of liquid density, the gas velocity in cocurrent flow will be 1000 times the liquid velocity. This sets Φ_D.

$$\Phi_D = 1/(1 + 1000) = 0.00099$$

If the droplets are 500 μm in diameter, Eqs. (14-177) and (14-178) give

$$a = (6/0.0005)(0.00099) = 12 \text{ m}^2/\text{m}^3$$

If the droplets are 100 μm in diameter, Eqs. (14-177) and (14-178) give

$$a = (6/0.0001)(0.00099) = 60 \text{ m}^2/\text{m}^3$$

Example 15: Interfacial Area for Droplets Falling in a Vessel Droplet systems rarely exceed a Φ_D value of 0.01. At this low level, Φ_D in a low-velocity countercurrent contactor can be approximated by Eq. (14-179).

$$\Phi_D = U_L/(U_t - U_G) \qquad (14\text{-}179)$$

where U_L = liquid superficial velocity
U_t = terminal velocity of droplet
U_G = gas superficial velocity

With a gas superficial velocity of 1.5 m/s, for equal mass flow of gas and liquid, with gas density 0.001 of liquid density, and with 500-μm-diameter droplets falling at a terminal settling of 2.5 m/s, Eq. (14-179) gives a fractional holdup of liquid of

$$\Phi_D = (0.001)1.5/(2.5 - 1.5) = 0.0015$$

Equations (14-177) and (14-178) then give

$$a = (6/0.0005)(0.0015) = 18 \text{ m}^2/\text{m}^3$$

Example 16: Interfacial Area for Bubbles Rising in a Vessel For bubble systems (gases dispersed in liquids) fractional holdup can approach 0.5 as shown by Fig. 14-104. However, before reaching this holdup, the bubble systems shift to an unstable mix of bubbles and vapor jets. Hence an exact comparison to Example 14 isn't possible because at the 1.5 m/s velocity of Example 14, the system becomes a froth. But at about one-fifth the velocity of Example 14, an estimate of interfacial area is possible.

If the bubble size is 10,000 μm and fractional holdup is 0.4, Eqs. (14-177) and (14-178) give an interfacial area of

$$a = (6/0.01)(0.4) = 240\ \text{m}^2/\text{m}^3$$

Measured interfacial area in distillation trays is consistent with this high value.

Note the much higher interfacial area than in the droplet systems of Examples 14 and 15. The higher interfacial area when the gas is dispersed explains why bubbling and froth systems often give better performance than droplet systems. The big difference in interfacial area stems from the much larger volume per unit of mass of gas, i.e., lower density of the gas than the liquid.

RATE MEASURES, TRANSFER UNITS, APPROACH TO EQUILIBRIUM, AND BYPASSING

What Controls Mass/Heat Transfer: Liquid or Gas Transfer or Bypassing Either gas side or liquid side of the interface can be controlling.

Liquid-Controlled In fractionation systems with high viscosity or component relative volatility that greatly exceeds 1, the liquid side will be controlling. This is clearly illustrated by Fig. 14-47 which shows a sharp decline in efficiency with either a rise in liquid viscosity or a rise in *component relative volatility*.

Note that *high component relative volatility* means the same thing as *sparingly soluble*. Oxygen dissolving in a fermentation reactor is an example of a system being liquid-controlled due to a *sparingly soluble gas*. Another application that is liquid-controlled is the removal of *high relative volatility components* out of residual oil.

Still another case where liquid controls is in condensing a pure vapor, as in Example 23, or absorbing a pure gas, as in Example 24.

Gas-Controlled The gas side dominates in gas cooling applications. An example is the quenching of a furnace effluent with a vaporizing liquid. In this application the liquid is nearly uniform in temperature. Restated, the reduction in driving force across the liquid side of the interface is unimportant.

Other applications that are gas-side-controlled include removal of a component such as NH_3 from a gas by using an acidic liquid, or removing a component such as SO_2 from a gas with a basic liquid. See Examples 19 through 22.

Bypassing-Controlled Trayed or packed columns operate with countercurrent flow and can achieve many equilibrium stages in series by good distribution of gas and liquid, and careful control of details. Other devices such as sprays are vulnerable to bypassing and are limited to one equilibrium stage.

Rate Measures for Interfacial Processes Terminology used for reporting rate data can be confusing. Normally rate data are reported on a volumetric basis with transfer rate and effective area combined. For example, k_La denotes mass-transfer data per unit volume. The L subscript means it is referenced to the molar concentration difference between the interface and the bulk liquid. This is commonly used on data involving a sparingly soluble (high relative volatility) component. Note that the lowercase k means the data deal only with the resistance in the liquid phase.

Less commonly, data are given as k_Ga. The G subscript means it is referenced to the molar concentration difference between the interface and the gas. This might be used for data on absorbing a gas such as NH_3 by a highly acidic liquid. Note that k_Ga only deals with the resistance in the gas phase.

When one is dealing with direct contact heat transfer, the corresponding terms are h_La and h_Ga. Here the driving force is the temperature difference. The L subscript means that we are dealing with a liquid-limited process such as condensing a pure liquid. How to convert k_La data to an h_La value is illustrated by Example 23.

There are ways to combine the liquid and gas resistance to get an overall transfer rate such as K_Ga (as denoted by the uppercase K). However, data are rarely reported in this form.

Approach to Equilibrium Although rate measures such as k_Ga and h_Ga are often cited in the literature, they are often not as useful to designers as the simpler concept of approach to equilibrium. *Approach to equilibrium* compares the transfer between liquid and gas phases to the best possible that could be achieved in a single backmixed equilibrium stage.

Approach to equilibrium is easy to understand and easy to apply. Examples 17 through 23 illustrate its use.

Example 17: Approach to Equilibrium—Perfectly Mixed, Complete Exchange This would be approximated by a very long pipeline contactor where an acidic aqueous stream is injected to cool the gas and remove NH_3.

If the adiabatic saturation temperature of the gas is 70°C, at the exit of the contactor, the gas would be cooled to 70°C.

Similarly, at the exit of the contactor, the NH_3 in the gas would be zero, regardless of the initial concentration.

Example 18: Approach to Equilibrium—Complete Exchange but with 10 Percent Gas Bypassing A spray column is used, and an acidic liquid rains down on the gas of Example 17. If the initial NH_3 is 1000 ppm and 10 percent of the gas bypasses, the NH_3 in the exit gas would be

$$0.1(1000) = 100\ \text{ppm}$$

Similarly, if the gas enters at 120°C, at the exit we would find 10 percent of the differential above the adiabatic saturation temperature. For an adiabatic saturation temperature of 70°C, the exit gas temperature would be

$$70 + 0.1(120 - 70) = 75°\text{C}$$

Approach to Equilibrium—Finite Contactor with No Bypassing When there is no bypassing, the measure that sets the approach is the ratio of change to driving force. This ratio is called the *number of transfer units* N_G. It is dimensionless. For heat-transfer applications, it can be envisioned as a conventional heat exchanger where a vaporizing liquid cools a gas:

$$\text{No. of gas-phase transfer units} = \frac{T_{G,\text{out}} - T_{G,\text{in}}}{(T_G - T_L)_{\text{av}}} = N_G \qquad (14\text{-}180)$$

Where T_G = gas temperature and T_L = liquid temperature. The number of transfer units N_G can also be calculated as the capability for change divided by the thermal capacitance of the flowing streams.

$$N_G = \frac{(\text{system volume})(h_Ga)}{(\text{volumetric flow rate})\rho_Gc_G} = \frac{(\text{gas contact time})(h_Ga)}{\rho_Gc_G} \qquad (14\text{-}181)$$

where a = interfacial area per unit volume
h_G = heat-transfer coefficient from interface to gas
ρ_G = gas density
c_G = gas specific heat

Note that in the above, performance and properties all refer to the gas, which is appropriate when dealing with a gas-limited transfer process.

This leads to a way to estimate the approach to equilibrium.

$$E = 1 - e^{-N_G} \qquad (14\text{-}182)$$

where E = "approach to equilibrium" fractional removal of NH_3 or fractional approach to adiabatic liquid temperature
N_G = number of transfer units calculated relative to gas flow

Example 19: Finite Exchange, No Bypassing, Short Contactor A short cocurrent horizontal pipeline contactor gives 86 percent removal of NH_3. There is no bypassing because of the highly turbulent gas flow and injection of liquid into the center of the pipe. What would we expect the exit gas temperature to be?

Equation (14-182) says that the backcalculated N_G is 2:

$$N_G = -\ln(1 - 0.86) = 2$$

For diffusing gases of similar molecular weight, the properties that control heat transfer follow the same rules as those that control mass transfer. As a result, the NH_3 scrubbing and gas cooling processes achieve similar approaches to equilibrium.

For an entry temperature of 120°C and an adiabatic saturation temperature of 70°C, the expected outlet temperature would be

$$70 + (1 - 0.86)(120 - 70) = 77°C$$

This looks like a powerful concept, but its value is limited due to uncertainty in estimating h_Ga. Both h_G and a are difficult to estimate due to dependence on power dissipation as discussed below. The primary value of the N_G concept is in estimating an expected change from baseline data as in the comparison of Example 19 with Example 20.

Example 20: A Contactor That Is Twice as Long, No Bypassing If we double the length of the pipeline contactor, double the effective contact area, and double the number of transfer units to 4, what do we expect for performance?

For $N_G = 4$,

$$E = 1 - e^{-4} = 0.982$$

The NH_3 in the exit gas would be expected to drop to

$$(1 - 0.982)(1000) = 18 \text{ ppm}$$

and the expected outlet temperature would be

$$70 + (1 - 0.982)(120 - 70) = 70.9°C$$

If we double the length again, we increase the number of transfer units to 8 and achieve an approach of

$$E = 1 - e^{-8} = 0.9997$$

The outlet temperature would be

$$70 + (1 - 0.9997)(120 - 70) = 70.015°C$$

Similarly the NH_3 in the exit gas would be

$$(1 - 0.9997)(1000) = 0.3 \text{ ppm}$$

Note that this approximates the exit condition of Example 17.

Transfer Coefficient—Impact of Droplet Size The transfer coefficients increase as the size of droplets decreases. This is so because the transfer process is easier if it only has to move mass or heat a shorter distance (i.e., as the bubble or droplet gets smaller).

In the limiting case of quiescent small bubbles or droplets, the transfer coefficients vary inversely with average bubble or droplet diameter. For example, in heat transfer from a droplet interface to a gas, the minimum value is

$$h_{G,\min} = \text{heat transfer coefficient from interface to gas} = 2k_G/D \qquad (14\text{-}183)$$

where k_G = gas thermal conductivity and
D = droplet diameter.

IMPORTANCE OF TURBULENCE

The designer usually has control over the size of a droplet. As discussed below, several of the correlations show that droplet diameter varies with turbulent energy dissipation. For example, Eqs. (14-190) and (14-201) suggest that in droplet systems

$$D \propto \{1/(\text{gas velocity})\}^{1.2}$$

and hence from Eq. (14-178)

$$a \propto 1/D \propto (\text{gas velocity})^{1.2} \qquad (14\text{-}184)$$

However, just looking at the impact of velocity on droplet size underestimates the velocity impact because turbulence gives higher transfer than Eq. (14-183) predicts. Transfer coefficients increase as the mixing adjacent to the surface increases. This mixing depends on the energy dissipated into the phases. To a first approximation this transfer from droplets increases with local power dissipation raised to the 0.2 power.

$$h_{G,\text{turbulent}} \propto (\text{power dissipated})^{0.2}$$

and since power dissipation per unit volume increases with (velocity)^3,

$$h_{G,\text{turbulent}} \propto (\text{velocity})^{0.6} \qquad (14\text{-}185)$$

The combined effect on interfacial area and on the transfer coefficient is that the effective transfer increases greatly with gas velocity. From Eqs. (14-178) and (14-185)

$$h_Ga_{\text{turbulent}} \propto (\text{velocity})^{1.8} \qquad (14\text{-}186)$$

For quenching operations, this means that even though residence time is cut as gas velocity goes up, the effective approach to equilibrium increases. Since the volume for a given length of pipe falls with (velocity)$^{-1}$, the expected number of transfer units N_G in a given length of pipe increases with (velocity)$^{0.8}$.

$$N_{G,\text{turbulent}} \propto (h_Ga_{\text{turbulent}})(\text{volume}) \propto (\text{velocity})^{0.8} \qquad (14\text{-}187)$$

See Example 21.

EXAMPLES OF CONTACTORS

High-Velocity Pipeline Contactors High-velocity cocurrent flow can give more power input than any other approach. This is critical when extremely high rates of reaction quenching are needed.

Example 21: Doubling the Velocity in a Horizontal Pipeline Contactor—Impact on Effective Heat Transfer Velocity in pipeline quench systems often exceeds 62 m/s (200 ft/s). Note that this is far above the flooding velocity in distillation packing, distillation trays, or gas-sparged reactors. There are few data available to validate performance even though liquid injection into high-velocity gas streams has been historically used in quenching reactor effluent systems. However, the designer knows the directional impact of parameters as given by Eq. (14-187).

For example, if a 10-ft length of pipe gives a 90 percent approach to equilibrium in a quench operation, Eq. (14-182) says that the backcalculated N_G is 2.303:

$$N_G - \ln(1 - 0.9) = 2.303$$

Equation (14-187) says if we double velocity but retain the same length, we would expect an increase of N_G to 4.0.

$$N_G = 2.303(2)^{0.8} = 4$$

and

$$E = 1 - e^{-4} = 0.982$$

Restated, the approach to equilibrium rises from 90 percent to greater than 98 percent even though the contact time is cut in half.

Vertical Reverse Jet Contactor A surprisingly effective modification of the liquid injection quench concept is to inject the liquid countercurrent *upward* into a gas flowing *downward*, with the gas velocity at some multiple of the flooding velocity defined by Eq. (14-203). The reverse jet contactor can be envisioned as an upside-down distillation tray. For large gas volumes, multiple injection nozzles are used. One advantage of this configuration is that it minimizes the chance of liquid or gas bypassing. Another advantage is that it operates in the froth region which generates greater area per unit volume than the higher-velocity cocurrent pipeline quench.

The concept was first outlined in U.S. Patent 3,803,805 (1974) and was amplified in U.S. Patent 6,339,169 (2002). The 1974 patent presents data which clarify that the key power input is from the gas stream.

A more recent article discusses use of the reverse jet in refinery offgas scrubbing for removal of both SO_2 and small particles [*Hydrocarbon*

Processing, **84,**(9), 99–106 (2005)]. This article cites downward gas velocities in the range of 10 to 37 m/s and notes gas pressure drop in the range of 6 to 20 in of water. Removals of SO_2 and fine particles were both close to 99 percent. The froth produced by the contactor reverses direction, flows down, and is largely disengaged in a vessel mounted below.

Example 22: The Reverse Jet Contactor, U.S. Patent 6,339,169 This patent deals with rapid cooling and removal of NH_3 from gas exiting an acrylonitrile reactor. Liquid is injected upward. The claims suggest downward-flowing gas velocity is between 20 and 25 m/s.

Gas cooling is reported to be largely complete in 0.1 s. NH_3 removal at the exit of the contactor is reported to be greater than 99 percent. The gas is cooled by water vaporizing from the injected liquid, with total water circulated being in the range of 100 times that evaporated.

Since the gas cooling and NH_3 scrubbing move in parallel, they would be expected to achieve nearly the same approach to equilibrium as long as the pH of all the liquid stays below a key threshold. The great excess of liquid enables this.

The key is high froth interfacial area per unit volume.

Simple Spray Towers The other extreme to the pipeline and reverse jet contactors is an open vessel where spray is injected down into upflowing gas to form a rain of liquid. The advantage of simple spray towers is that they give low gas pressure drop and also tend to be nonfouling.

Even though gas velocity is well below flooding velocity, the finer droplets of the spray will be entrained. Note the wide spectrum of particle sizes shown by Fig. 14-88.

However, as shown by Examples 23 and 24, they can be extremely effective in liquid-limited systems.

Bypassing Limits Spray Tower Performance in Gas Cooling As shown by Example 18, only modest performance is achieved in gas-limited systems. The modest efficiency is due to gas bypassing. Tall spray towers are not effective countercurrent devices. Even with nominally falling droplets, there is a great deal of backmixing because there is no stabilizing pressure drop as there would be in a column filled with packing or trays. A packet of droplets weighs more than a gas-filled space. The result is that the volume that is filled with the most droplets moves down relative to all other volumes. Similarly the gas volume that has fewest droplets moves up more quickly than other volumes. This generates bypassing of liquid and gas. The flows are driven by the rain of droplets themselves. Anything less than perfect distribution of liquid and gas will induce a dodging action between the flowing streams. Most designers limit expectations for spray contactors to some fraction of a single equilibrium stage regardless of height.

One approach that has been employed to get better distribution in spray systems is to mount a single large-capacity nozzle in the center of the vessel with radial discharge of large droplets. The droplets are discharged with enough velocity to penetrate to the vessel walls.

Spray Towers in Liquid-Limited Systems—Hollow Cone Atomizing Nozzles If we follow an element of liquid leaving a hollow cone hydraulic spray nozzle, the sequence is a rapidly thinning cone followed by wave development, followed by shedding of ligaments, followed by breakage of the ligaments into droplets. See Fig. 14-86. The sequence gives high transfer for liquid-limited systems. This results from the thin sheet of the hollow cone as well as the creation of fresh surface in the breakup process.

Devolatilizers Devolatilization systems are liquid-limited due to the combination of high liquid viscosity and removal of a component with high relative volatility. Simpson and Lynn [*AIChE J.*, **23** (5), 666–673 (1977)] reported oxygen stripping from water at 98 percent complete, in less than 1 ft of contact. The concept has been employed for residual devolatilization in refineries.

Spray Towers as Direct Contact Condensers Similarly spray contactors can be highly effective for direct contact condensers, which are also liquid-limited. The high transfer rate in the initial formation of sprays is the key. Kunesh [*Ind. Engr. Chem. Res.*, **32,** 2387–2389 (1993)] reported a 97 percent approach to equilibrium in a hydrocarbon system in the 6-in space below the discharge of a row of hollow cone spray nozzles.

Other results on heat transfer in a large spray condenser are given by Waintraub et al. ("Removing Packings from Heat Transfer Sections of Vacuum Towers," AIChE 2005 Spring National Meeting, Proceedings of Topical Conference, April 10, 2005, Atlanta, p. 79). The paper highlights the importance of good gas and liquid distribution.

Converting Liquid Mass-Transfer Data to Direct Contact Heat Transfer Liquid-limited performance measures are much more commonly given for mass-transfer than for heat transfer. Often mass-transfer data are reported as k_La with units of h^{-1}. This can be converted to h_La with units of Btu/(h·°F·ft³) by Eq. (14-188).

$$h_La = 187(k_La)\,(c_L)\,(\rho_L)\,(\mu_L/T)^{0.5} \tag{14-188}$$

where μ_L = liquid, cP
T = temperature, °R
ρ_L = liquid density, lb/ft³
c_L = liquid specific heat, Btu/(lb·°F)

Calculation of transfer units for heat transfer is relatively simple. For a liquid

$$N_L = \frac{(\text{liquid contact time})(h_La)}{\rho_L c_L} \tag{14-189}$$

where ρ_L = liquid density and c_L = liquid specific heat. [See parallel gas expression, Eq. (14-181).]

Unlike gases, the liquid properties that control mass and heat transfer differ greatly. The key term is diffusivity which for liquids drops with viscosity.

The resulting values for h_La and N_L can be surprisingly large when a pure vapor such as steam is condensed. See Example 23.

Example 23: Estimating Direct Contact Condensing Performance Based on k_La Mass-Transfer Data If an aqueous system at 560°R gives a k_La of 60 h^{-1}, what does Eq. (14-188) predict for h_La in a direct contact steam condenser?

For an aqueous system

$$\mu_L = 1 \text{ cP}$$

$$\rho_L = 62 \text{ lb/ft}^3$$

$$c_L = 1 \text{ Btu/(lb·°F)}$$

and Eq. (14-189) predicts

$$h_La = 187(60)(1)(62)(1/560)^{0.5} = 29{,}400 \text{ Btu/(h·°F·ft}^3)$$

When a pure gas such as HCl is absorbed by low viscosity liquid such as water, simple spray systems can also be highly effective. See Example 24.

Example 24: HCl Vent Absorber (Kister, *Distillation Troubleshooting*, Wiley, 2006, p. 95.) A 6-in-diameter, 8-ft-tall packed bed was giving major problems due to failure of the packing support. Water was the scrubbing fluid.

The liquid distributors were replaced with carefully positioned spray nozzles, and the packing was removed. HCl in the vent was removed to a level one-fortieth of the original design.

LIQUID-IN-GAS DISPERSIONS

Liquid Breakup into Droplets There are four basic mechanisms for breakup of liquid into droplets:

- Droplets in a field of high turbulence (i.e., high power dissipation per unit mass)
- Simple jets at low velocity
- Expanding sheets of liquid at relatively low velocity
- Droplets in a steady field of high relative velocity

These mechanisms coexist, and the one that gives the smallest drop size will control. The four mechanisms follow distinctly different velocity dependencies:

1. *Breakup in a highly turbulent field* $(1/\text{velocity})^{1.2}$. This appears to be the dominant breakup process in distillation trays in the spray regime, pneumatic atomizers, and high-velocity pipeline contactors.

2. *Breakup of a low-velocity liquid jet* $(1/\text{velocity})^{0}$. This governs in special applications like prilling towers and is often an intermediate step in liquid breakup processes.

3. *Breakup of a sheet of liquid* $(1/\text{velocity})^{0.67}$. This governs drop size in most hydraulic spray nozzles.

4. *Single-droplet breakup at very high velocicty* $(1/\text{velocity})^2$. This governs drop size in free fall as well as breakup when droplets impinge on solid surfaces.

Droplet Breakup—High Turbulence This is the dominant breakup mechanism for many process applications. Breakup results from local variations in turbulent pressure that distort the droplet shape. Hinze [*Am. Inst. Chem. Eng. J.*, **1,** 289–295 (1953)] applied turbulence theory to obtain the form of Eq. (14-190) and took liquid-liquid data to define the coefficient:

$$D_{max} = 0.725(\sigma/\rho_G)^{0.6}/E^{0.4} \tag{14-190}$$

where E = (power dissipated)/mass — length²/time³
σ = surface tension — mass/time²
ρ_G = gas density — mass/length³

Note that D_{max} comes out with units of length. Since E typically varies with $(\text{gas velocity})^3$, this results in drop size dependence with $(1/\text{velocity})^{1.2}$.

The theoretical requirement for use of Eq. (14-190) is that the microscale of turbulence $<<D_{max}$. This is satisfied in most gas systems. For example, in three cases,

	(microscale of turbulence)/D_{max}
distillation tray in spray regime	0.007
pipeline @ 40 m/s and atmospheric pressure	0.012
two-fluid atomizer using 100 m/s air	0.03

Many applications involve a three-step process with high velocity first tearing wave crests away from liquid sheets, followed by breakup of ligaments into large droplets, followed by breakup of the large droplets. The prediction of final droplet size based on *power/mass* works surprisingly well, as shown by Eqs. (14-198),(14-201), (14-202), and (14-203).

Liquid-Column Breakup Because of increased pressure at points of reduced diameter, the liquid column is inherently unstable. As a result, it breaks into small drops with no external energy input. Ideally, it forms a series of uniform drops with the size of the drops set by the fastest-growing wave. This yields a dominant droplet diameter about 1.9 times the initial diameter of the jet as shown by Fig. 14-85. As shown, the actual breakup is quite close to prediction, although smaller satellite drops are also formed. The prime advantage of this type of breakup is the greater uniformity of drop size.

For high-viscosity liquids, the drops are larger, as shown by Eq. (14-191):

$$D = 1.9D_j\left[1 + \frac{3\mu_\ell}{(\sigma\rho_\ell D_j)^{1/2}}\right] \tag{14-191}$$

where D = diameter of droplet
D_j = diameter of jet
μ_ℓ = viscosity of liquid
ρ_ℓ = density of liquid
σ = surface tension of liquid

These units are dimensionally consistent; any set of consistent units can be used.

As the velocity of the jet is increased, the breakup process changes and ultimately becomes a mix of various competing effects, such as the capture of small drops by bigger ones in the slowing jet and the "turbulent breakup" of the bigger drops. The high-velocity jet is occasionally used in process applications because of the very narrow spray angle (5–20°) and the high penetration into a gas it can give. The focused stream also aids erosion of a surface.

Liquid-Sheet Breakup The basic principle of most hydraulic atomizers is to form a thin sheet that breaks via a variety of mechanisms to form ligaments of liquid which in turn yield chains of droplets. See Fig. 14-86.

For a typical nozzle, the drop size varies with $1/(\text{pressure drop})^{1/3}$. When $(\text{velocity})^2$ is substituted for pressure drop, droplet size is seen to vary with $(\text{velocity})^{-2/3}$.

Isolated Droplet Breakup—in a Velocity Field Much effort has focused on defining the conditions under which an isolated drop will break in a velocity field. The criterion for the largest stable drop size is the ratio of aerodynamic forces to surface-tension forces defined by the Weber number, N_{We} (dimensionless):

$$N_{We\ crit} = \text{constant} = [\rho_G\ (\text{velocity})^2(D_{max})/(\sigma)] \tag{14-192}$$

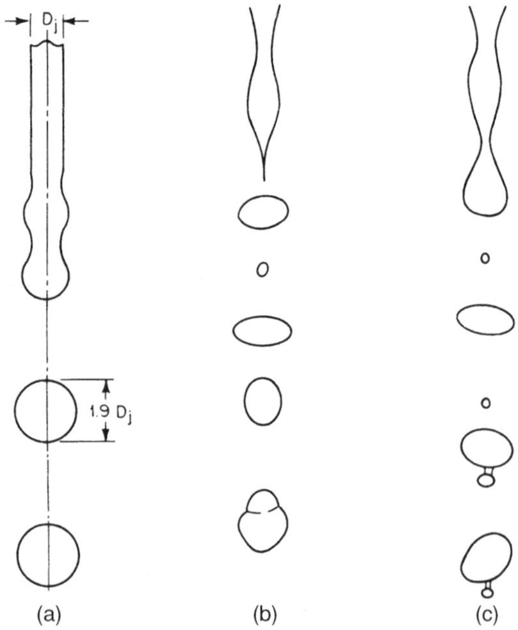

FIG. 14-85 (*a*) Idealized jet breakup suggesting uniform drop diameter and no satellites. (*b*) and (*c*) Actual breakup of a water jet as shown by high-speed photographs. [*From W. R. Marshall,* "Atomization and Spray Drying," *Chem. Eng. Prog. Monogr. Ser., no. 2 (1954).*]

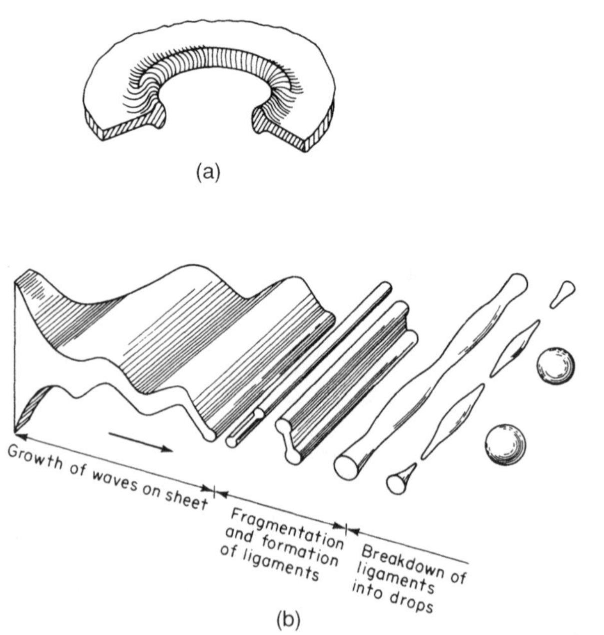

FIG. 14-86 Sheet breakup. (*a*) By perforation. [*After Fraser et al.*, Am. Inst. Chem. Eng. J., *8*(5), 672 (1962).] (*b*) By sinusoidal wave growth. [*After Dombrowski and Johns*, Chem. Eng. Sci., ***18***, *203 (1963).*]

$N_{We\ crit}$ for low-viscosity fluids commonly ranges from 10 to 20, with the larger value for a free-fall condition and the smaller for a sudden acceleration. High liquid viscosity also increases $N_{We\ crit}$.

Droplet breakup via impingement appears to follow a similar relationship, but much less data is available. This type of breakup can result from impingement on equipment walls or compressor blades. In general, there is less tendency to shatter on wetted surfaces.

Droplet Size Distribution Instead of the single droplet size implied by the discussion above, a spectrum of droplet sizes is produced. The most common ways to characterize this spectrum are:

- *Volume median (mass median)* D_{vm}. This has no fundamental meaning but is easy to determine since it is at the midpoint of a cumulative-volume plot.
- *Sauter mean* D_{32}. This has the same ratio of surface to volume as the total drop population. It is typically 70 to 90 percent of D_{vm}. It is frequently used in transport processes and is used here to characterize drop size.
- *Maximum* D_{max}. This is the largest-sized particle in the population. It is typically 3 to 4 times D_{32} in turbulent breakup processes, per Walzel [*International Chemical Engineering,* **33,** 46, (1993)]. It is the size directly calculated from the power/mass relationship. D_{32} is estimated from D_{max} by

$$D_{32} = 0.3 \cdot D_{max} \tag{14-193}$$

and D_{vm} is estimated from it by

$$D_{vm} = 0.4 \cdot D_{max} \tag{14-194}$$

However, any average drop size is fictitious, and none is completely satisfactory. For example, there is no way in which the high surface and transfer coefficients in small drops can be made available to the larger drops. Hence, a process calculation based on a given droplet size describes only what happens to that size and gives at best an approximation to the total mass.

There are a variety of ways to describe the droplet population. Figures 14-88 and 14-90 illustrate one of the most common methods, the plot of cumulative volume against droplet size on log-normal graph paper. This satisfies the restraint of not extrapolating to a negative drop size. Its other advantages are that it is easy to plot, the results are easy to visualize, and it yields a nearly straight line at lower drop sizes.

Cumulative volume over the range of 1 to 50 percent can also be shown to vary approximately as D^2. This is equivalent to finding that the number of droplets of a given size is inversely proportional to the droplet area or the surface energy of the droplet.

Atomizers The common need to disperse a liquid into a gas has spawned a large variety of mechanical devices. The different designs emphasize different advantages such as freedom from plugging, pattern of spray, small droplet size, uniformity of spray, high turndown ratio, and low power consumption.

As shown in Table 14-19, most atomizers fall into three categories:

1. Pressure nozzles (hydraulic)
2. Two-fluid nozzles (pneumatic)
3. Rotary devices (spinning cups, disks, or vaned wheels)

These share certain features such as relatively low efficiency and low cost relative to most process equipment. The energy required to produce the increase in area is typically less than 0.1 percent of the total energy consumption. This is because atomization is a secondary process resulting from high interfacial shear or turbulence. As droplet sizes decrease, this efficiency drops lower.

Other types are available that use sonic energy (from gas streams), ultrasonic energy (electronic), and electrostatic energy, but they are less commonly used in process industries. See Table 14-19 for a summary of the advantages/disadvantages of the different type units. An expanded discussion is given by Masters [*Spray Drying Handbook,* Wiley, New York (1991)].

Special requirements such as size uniformity in prilling towers can dictate still other approaches to dispersion. Here plates are drilled with many holes to develop nearly uniform columns.

Commonly, the most important feature of a nozzle is the size of droplet it produces. Since the heat or mass transfer that a given dispersion can produce is often proportional to $(1/D_d)^2$, fine drops are usually favored. On the other extreme, drops that are too fine will not settle, and a concern is the amount of liquid that will be entrained from a given spray operation. For example, if sprays are used to contact atmospheric air flowing at 1.5 m/s, drops smaller than 350 mm [terminal velocity = 1.5 m/s (4.92 ft/s)] will be entrained. Even for the relative coarse spray of the hollow-cone nozzle shown in Fig. 14-88, 7.5 percent of the total liquid mass will be entrained.

Hydraulic (Pressure) Nozzles Manufacturers' data such as shown by Fig. 14-88 are available for most nozzles for the air-water system. In Fig. 14-88, note the much coarser solid-cone spray. The coarseness results from the less uniform discharge.

Effect of Physical Properties on Drop Size Because of the extreme variety of available geometries, no attempt to encompass this variable is made here. The suggested predictive route starts with air-water droplet size data from the manufacturer at the chosen flow rate. This drop size is then corrected by Eq. (14-195) for different viscosity and surface tension:

$$\frac{D_{vm,\ system}}{D_{vm,\ water}} = \left(\frac{\sigma_{system}}{73}\right)^{0.25} \left(\frac{\mu_\ell}{1.0}\right)^{0.25} \tag{14-195}$$

where D_{vm} = volume median droplet diameter
σ = surface tension, mN/m (dyn/cm)
μ_ℓ = liquid viscosity, mPa·s (cP)

The exponential dependencies in Eq. (14-195) represent averages of values reported by a number of studies with particular weight given to Lefebvre ([*Atomization and Sprays,* Hemisphere, New York (1989)]. Since viscosity can vary over a much broader range than surface tension, it has much more leverage on drop size. For example, it is common to find an oil with 1000 times the viscosity of water, while most liquids fall within a factor of 3 of its surface tension. Liquid density is generally even closer to that of water, and since the data are not clear that a liquid density correction is needed, none is shown in Eq. (14-195). Vapor density also has an impact on dropsize but the impact is complex, involving conflicts of a number of effects, and vapor density is commonly omitted in atomizer dropsize correlations.

Effect of Pressure Drop and Nozzle Size For a nozzle with a developed pattern, the average drop size can be estimated to fall with rising ΔP (pressure drop) by Eq. (14-196):

$$\frac{D_1}{D_2} = \left(\frac{\Delta P_2}{\Delta P_1}\right)^{1/3} \tag{14-196}$$

For similar nozzles and constant ΔP, the drop size will increase with nozzle size as indicated by Eq. (14-197):

$$\frac{D_1}{D_2} = \left(\frac{\text{orifice diameter}_1}{\text{orifice diameter}_2}\right)^{1/2} \tag{14-197}$$

Once again, these relationships are averages of a number of reported values and are intended as rough guides.

The normal operating regime is well below turbulent breakup velocity. However the data of Kennedy [*J. of Engineering for Gas Turbines and Power,* **108,** 191 (1986)] at very high pressure drop in large nozzles shows a shift to a higher dependence on pressure drop. This data suggests that turbulent droplet breakup can also be governing with hydraulic spray nozzles, although this is unusual.

Spray Angle A shift to a smaller-angle nozzle gives slightly larger drops for a given type of nozzle because of the reduced tendency of the sheet to thin. Dietrich [*Proc. 1st Int. Conf. Liq. Atomization Spray Systems,* Tokyo (1978)] shows the following:

Angle	25°	50°	65°	80°	95°
D_{vm}, μm	1459	1226	988	808	771

In calculating the impact point of spray, one should recognize that the spray angle closes in as the spray moves away from the nozzle. This is caused by loss of momentum of the spray to the gas.

TABLE 14-19 Atomizer Summary

Types of atomizer	Design features	Advantages	Disadvantages
Pressure.	Flow $\alpha(\Delta P/\rho_\ell)^{1/2}$. Only source of energy is from fluid being atomized.	Simplicity and low cost.	Limited tolerance for solids; uncertain spray with high-viscosity liquids; susceptible to erosion. Need for special designs (e.g., bypass) to achieve turndown.
1. Hollow cone.	Liquid leaves as conical sheet as a result of centrifugal motion of liquid. Air core extends into nozzle.	High atomization efficiency.	Concentrated spray pattern at cone boundaries.
a. Whirl chamber (see Fig. 14-87*a*).	Centrifugal motion developed by tangential inlet in chamber upstream of orifice.	Minimum opportunity for plugging.	
b. Grooved core.	Centrifugal motion developed by inserts in chamber.	Smaller spray angle than 1*a* and ability to handle flows smaller than 1*a*.	
2. Solid cone (see Fig. 14-87*b*).	Similar to hollow cone but with insert to provide even distribution.	More uniform spatial pattern than hollow cone.	Coarser drops for comparable flows and pressure drops. Failure to yield same pattern with different fluids.
3. Fan (flat) spray.	Liquid leaves as a flat sheet or flattened ellipse.	Flat pattern is useful for coating surfaces and for injection into streams.	Small clearances.
a. Oval or rectangular orifice (see Fig. 14-87*c*). Numerous variants on cavity and groove exist.	Combination of cavity and orifice produces two streams that impinge within the nozzle.		
b. Deflector (see Fig. 14-87*d*).	Liquid from plain circular orifice impinges on curved deflector.	Minimal plugging.	Coarser drops.
c. Impinging jets (see Fig. 14-87*e*).	Two jets collide outside nozzle and produce a sheet perpendicular to their plane.	Different liquids are isolated until they mix outside of orifice. Can produce a flat circular sheet when jets impinge at 180°.	Extreme care needed to align jets.
4. Nozzles with wider range of turndown.			
a. Spill (bypass) (see Fig. 14-87*f*).	A portion of the liquid is recirculated after going through the swirl chamber.	Achieves uniform hollow cone atomization pattern with very high turndown (50:1).	Waste of energy in bypass stream. Added piping for spill flow.
b. Poppet (see Fig. 14-87*g*).	Conical sheet is developed by flow between orifice and poppet. Increased pressure causes poppet to move out and increase flow area.	Simplest control over broad range.	Difficult to maintain proper clearances.
Two-fluid (see Fig. 14-87*h*).	Gas impinges coaxially and supplies energy for breakup.	High velocities can be achieved at lower pressures because the gas is the high-velocity stream. Liquid-flow passages can be large, and hence plugging can be minimized.	Because gas is also accelerated, efficiency is inherently lower than pressure nozzles.
Sonic.	Gas generates an intense sound field into which liquid is directed.	Similar to two-fluid but with greater tolerance for solids.	Similar to two-fluid.
Rotary wheels (see Fig. 14-87*i*) disks, and cups.	Liquid is fed to a rotating surface and spreads in a uniform film. Flat disks, disks with vanes, and bowl-shaped cups are used. Liquid is thrown out at 90° to the axis.	The velocity that determines drop size is independent of flow. Hence these can handle a wide range of rates. They can also tolerate very viscous materials as well as slurries. Can achieve very high capacity in a single unit; does not require a high-pressure pump.	Mechanical complexity of rotating equipment. Radial discharge.
Ultrasound.	Liquid is fed over a surface vibrating at a frequency > 20 kHz.	Fine atomization, small size, and low injection velocity.	Low flow rate and need for ultrasound generator.

At some low flow, pressure nozzles do not develop their normal pattern but tend to approach solid streams. The required flow to achieve the normal pattern increases with viscosity.

Two-Fluid (Pneumatic) Atomizers This general category includes such diverse applications as venturi atomizers and reactor-effluent quench systems in addition to two-fluid spray nozzles. Depending on the manner in which the two fluids meet, several of the breakup mechanisms may be applicable, but the final one is high-level turbulent rupture.

As shown by Table 14-20, empirical correlations for two-fluid atomization show dependence on high gas velocity to supply atomizing energy, usually to a power dependence close to that for turbulent breakup. In addition, the correlations show a dependence on the ratio of gas to liquid and system dimension.

Further differences from hydraulic nozzles (controlled by sheet and ligament breakup) are the stronger increase in drop size with increasing surface tension and decreasing gas density.

The similarity of these dependencies to Eq. (14-190) led to a reformulation with two added terms that arise naturally from the theory of power dissipation per unit mass. The result is Eq. (14-198) which is labeled *power/mass* in Table 14-20.

$$D_{32} = 0.29\left(\frac{\sigma}{\rho_G}\right)^{0.6}(1/\text{velocity})^{1.2}\left(1+\frac{L}{G}\right)^{0.4}\left(D_{\text{nozzle}}\right)^{0.4} \quad (14\text{-}198)$$

where σ = surface tension
ρ_G = gas density
L/G = mass ratio of liquid flow to gas flow
D_{nozzle} = diameter of the air discharge

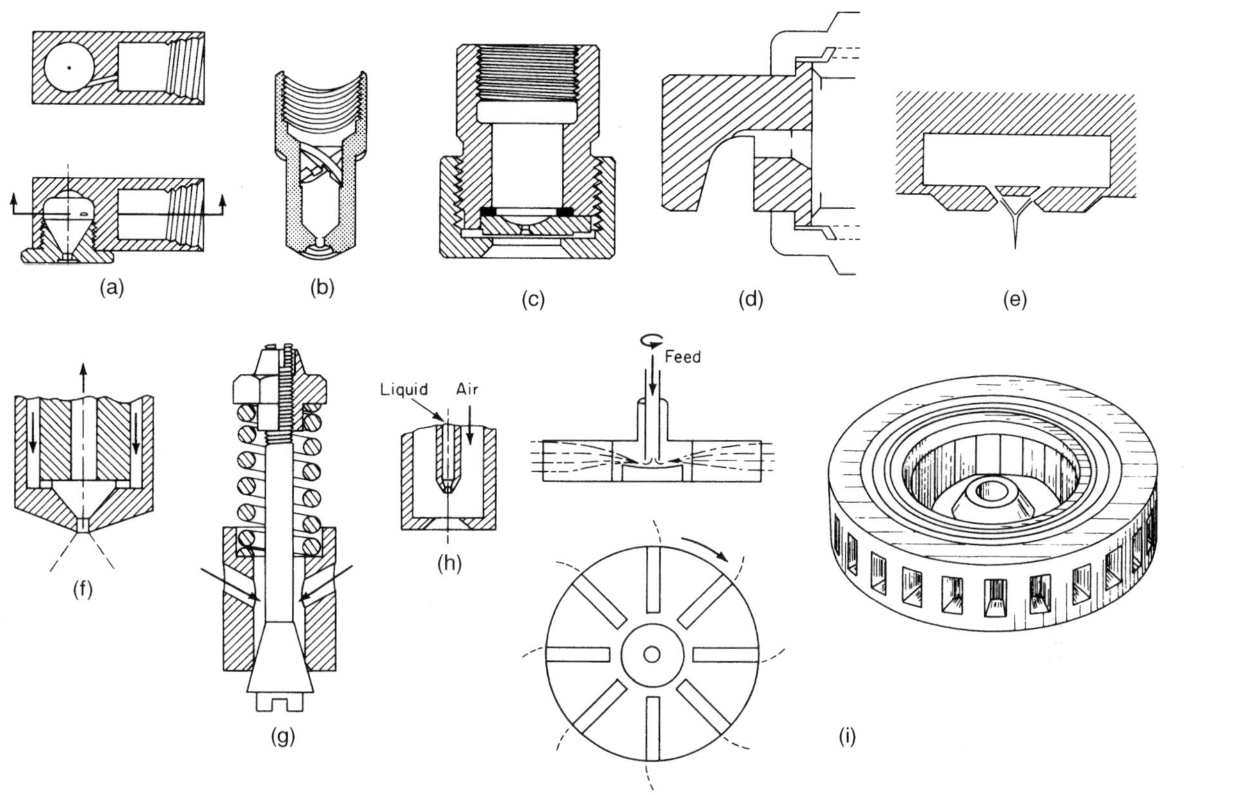

FIG. 14-87 Charactersitic spray nozzles. (*a*) Whirl-chamber hollow cone. (*b*) Solid cone. (*c*) Oval-orifice fan. (*d*) Deflector jet. (*e*) Impinging jet. (*f*) Bypass. (*g*) Poppet. (*h*) Two-fluid. (*i*) Vaned rotating wheel.

This is remarkably similar to the empirical two-fluid atomizer relationships of El-Shanawany and Lefebvre [*J. Energy,* **4,** 184 (1980)] and Jasuja [*Trans. Am. Soc. Mech. Engr.,* **103,** 514 (1981)]. For example, El-Shanawany and Lefebvre give a relationship for a prefilming atomizer:

$$D_{32} = 0.0711(\sigma/\rho_G)^{0.6}(1/\text{velocity})^{1.2}(1 + L/G)(D_{\text{nozzle}})^{0.4}(\rho_L/\rho_G)^{0.1} + 0.015[(\mu_L)^2/(\sigma \times \rho_L)]^{0.5}(D_{\text{nozzle}})^{0.5}(1 + L/G) \quad (14\text{-}199)$$

where μ_L is liquid viscosity.

According to Jasuja,

$$D_{32} = 0.17(\sigma/\rho_G)^{0.45}(1/\text{velocity})^{0.9}(1 + L/G)^{0.5}(D_{\text{nozzle}})^{0.55} + \text{viscosity term} \quad (14\text{-}200)$$

[Eqs. (14-198), (14-199), and (14-200) are dimensionally consistent; any set of consistent units on the right-hand side yields the droplet size in units of length on the left-hand side.]

The second, additive term carrying the viscosity impact in Eq. (14-199) is small at viscosities around 1 cP but can become controlling as viscosity increases. For example, for air at atmospheric pressure atomizing water, with nozzle conditions

$$D_{\text{nozzle}} = 0.076 \text{ m (3 in)}$$

$$\text{velocity} = 100 \text{ m/s (328 ft/s)}$$

$$L/G = 1$$

El-Shanaway measured 70 μm and his Eq. (14-199) predicted 76 μm. The *power/mass* correlation [Eq. (14-198)] predicts 102 μm. The agreement between both correlations and the measurement is much better than normally achieved.

Rotary Atomizers For rotating wheels, vaneless disks, and cups, there are three regimes of operation. At low rates, the liquid is shed directly as drops from the rim. At intermediate rates, the liquid leaves the rim as threads; and at the highest rate, the liquid extends from the edge as a thin sheet that breaks down similarly to a fan or hollow-cone spray nozzle. As noted in Table 14-19, rotary devices have many unique advantages such as the ability to handle high viscosity and slurries and produce small droplets without high pressures. The prime applications are in spray drying. See Masters [*Spray Drying Handbook,* Wiley, New York (1991)] for more details.

Pipeline Contactors The correlation for droplet diameter based on power/mass is similar to that for two-fluid nozzles. The dimensionless correlation is

$$D_{32} = 0.8(\sigma/\rho_G)^{0.6}(1/\text{velocity})^{1.2}(D_{\text{pipe}})^{0.4} \quad (14\text{-}201)$$

(The relation is dimensionally consistent; any set of consistent units on the right-hand side yields the droplet size in units of length on the left-hand side.)

The relationship is similar to the empirical correlation of Tatterson, Dallman, and Hanratty [*Am. Inst. Chem. Eng. J.,* **23**(1), 68 (1977)]

$$D_{32} \sim \left(\frac{\sigma}{\rho_G}\right)^{0.5}(1/\text{velocity})^1(D_{\text{pipe}})^{0.5}$$

Predictions from Eq. (14-201) align well with the Tatterson data. For example, for a velocity of 43 m/s (140 ft/s) in a 0.05-m (1.8-inch) equivalent diameter channel, Eq. (14-201) predicts D_{32} of 490 microns, compared to the measured 460 to 480 microns.

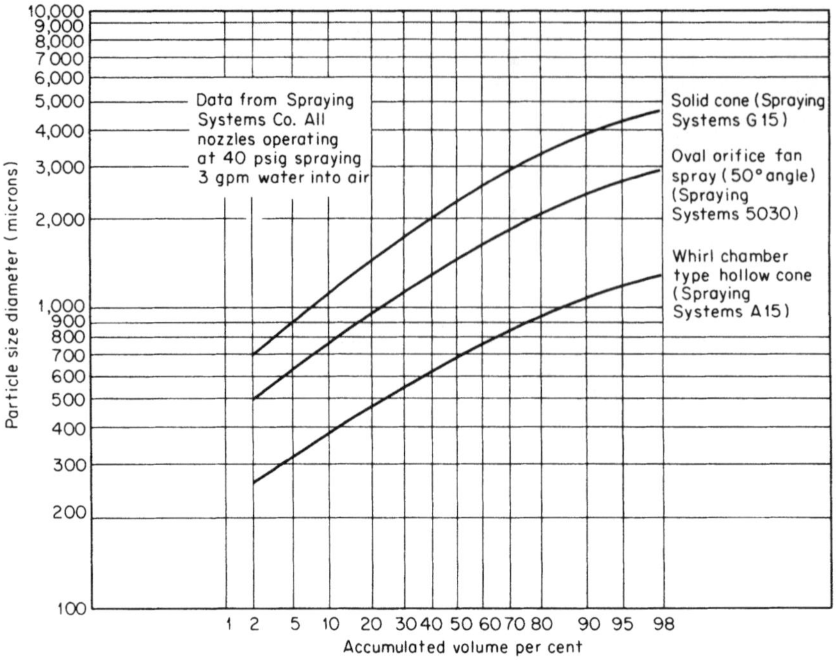

FIG. 14-88 Droplet-size distribution for three different types of nozzles. To convert pounds per square inch gauge to kilopascals, multiply by 6.89; to convert gallons per minute to cubic meters per hour, multiply by 0.227. (*Spraying Systems Inc.*)

Entrainment due to Gas Bubbling/Jetting through a Liquid Entrainment generally limits the capacity of distillation trays and is commonly a concern in vaporizers and evaporators. Fortunately, it is readily controllable by simple inertial entrainment capture devices such as wire mesh pads in gravity separators.

In distillation towers, entrainment lowers the tray efficiency, and 1 pound of entrainment per 10 pounds of liquid is sometimes taken as the limit for acceptable performance. However, the impact of entrainment on distillation efficiency depends on the relative volatility of the component being considered. Entrainment has a minor impact on close separations when the difference between vapor and liquid concentration is small, but this factor can be dominant for systems where the liquid concentration is much higher than the vapor in equilibrium with it (i.e., when a component of the liquid has a very low volatility, as in an absorber).

As shown by Fig. 14-90, entrainment droplet sizes span a broad range. The reason for the much larger drop sizes of the upper curve is the short disengaging space. For this curve, over 99 percent of the entrainment has a terminal velocity greater than the vapor velocity. For contrast, in the lower curve the terminal velocity of the largest particle reported is the same as the vapor velocity. For the settling velocity to limit the maximum drop size entrained, at least 0.8 m (30 in) disengaging space is usually required. Note that even for the lower curve, less than 10 percent of the entrainment is in drops of less than 50 μm. The coarseness results from the relatively low power dissipation per mass on distillation trays. This means that it is relatively easy to remove by a device such as a wire mesh pad. Over 50 percent is typically captured by the underside of the next higher tray or by a turn in the piping leaving an evaporator. Conversely, though small on a mass basis, the smaller drops are extremely numerous. On a number basis, more than one-half of the drops in the lower curve are under 5 μm. These can serve as nuclei for fog condensation in downstream equipment.

Entrainment E is inherent in the bubbling process and can stem from a variety of sources, as shown by Fig. 14-89. However, the biggest practical problem is entrainment generated by the kinetic energy of the flowing vapor rather than the bubbling process. As vapor velocity approaches the flooding limit [Eq. (14-168)], the entrainment rises approximately with $(\text{velocity})^8$.

Pinczewski and Fell [*Trans. Inst. Chem Eng.*, **55,** 46 (1977)] show that the velocity at which vapor jets onto the tray sets the droplet size, rather than the superficial tray velocity. The power/mass correlation predicts an average drop size close to that measured by Pinczewski and Fell. Combination of this prediction with the estimated fraction of the droplets entrained gave a relationship for entrainment, Eq. (14-202). The dependence of entrainment with the eighth power of velocity even approximates the observed velocity dependence, as flooding is approached.

$$E = \frac{\text{constant}(\text{velocity})^8(\rho_G)^4}{(A_f)^3(\rho_L - \rho_G)^{2.5}\sigma^{1.5}} \qquad (14\text{-}202)$$

(Here E is the mass of entrainment per mass of vapor and A_f is the fractional open area on the tray.)

When flooding is defined as the condition that gives E of 1, the flood velocity is estimated by Eq. (14-203).

TABLE 14-20 Exponential Dependence of Drop Size on Different Parameters in Two-Fluid Atomization

	Relative velocity	Surface tension	Gas density	$1 + L/G$	Atomizer dimension
Jasuja (empirical for small nozzle)	−0.9	0.45	−0.45	0.5	0.55
El-Shanawany and Lefebvre (empirical for small nozzle)	−1.2	0.6	−0.7	1	0.40
Tatterson, Dallman, and Hanratty (pipe flow)	−1	0.5	−0.5		0.5
Power/mass	−1.2	0.6	−0.6	0.4	0.4

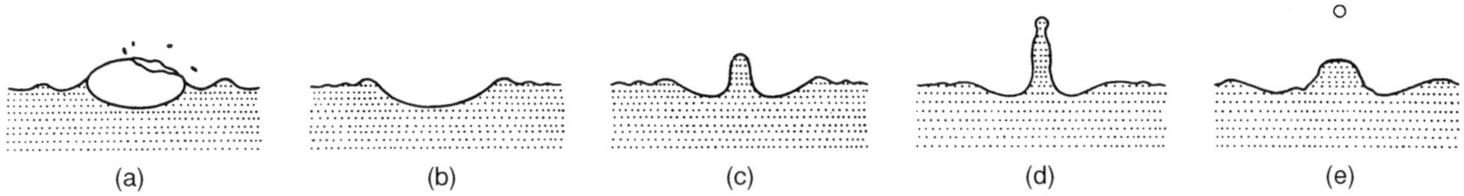

FIG. 14-89 Mechanism of the burst of an air bubble on the surface of water. [*Newitt, Dombrowski, and Knellman,* Trans. Inst. Chem. Eng., **32,** 244 (*1954*).]

$$U_{\text{flood}} = \frac{1.25(A_f)^{3/8}[g(\rho_L - \rho_G)]^{0.3125}\,\sigma^{0.1875}}{\rho_G^{0.5}} \tag{14-203}$$

The relationship is dimensionally consistent; any set of consistent units on the right-hand side yields velocity units on the left-hand side. It is similar in form to Eq. (14-168) and provides a conceptual framework for understanding the ultimate distillation column capacity concept.

"Upper Limit" Flooding in Vertical Tubes If, instead of a gas jet being injected into a liquid as in distillation, the liquid runs down the walls and the gas moves up the center of the tube, higher velocities can be achieved than shown by Eq. (14-168) or (14-203). This application is important in the design of vertical condensers.

Maharudrayya and Jayanti [*AIChE J.,* **48**(2), 212–220 (2002)] show that peak pressure drop in a 25-mm vertical tube occurs at a value close to that predicted by Eq. (14-168) or (14-203). At this velocity about 20 percent of the injected liquid is being entrained out the top of the tube. However, the condition where essentially all liquid was entrained didn't occur until a velocity over twice the value estimated from Eq. (14-168) and Eq. (14-203).

The higher velocities at modest entrainment observed by Maharudrayya and Jayanti were obtained with special smooth entry of gas (and exit of liquid) at the bottom of the tube. Hewitt (*Handbook of Heat Exchanger Design,* pp. 2.3.2-23, 1992) suggests that these values should be derated by at least 35 percent for more typical sharp heat exchanger tube entry. Similar to the smooth entry effect, other data suggest that countercurrent capacity can be increased by providing an extension of the tube below the tube sheet, with the bottom of the extension cut on a steep angle (>60°) to the horizontal. The tapered extension facilitates drainage of liquid.

An extensive data bank correlated by Diehl and Koppany [*Chem. Eng. Prog. Symp. Ser.,* **65,** 77–83 (1965)] also gave higher allowable entry velocities than Eq. (14-168) or (14-203). Dielhl and Koppany's correlation [Eq. (14-204)] is dimensional, and appears to give a much higher dependence on σ than the more recent work. However, for many fluids, $\sigma^{0.5}$ is essentially the same as the combination $\sigma^{0.1875}(\rho_l - \rho_g)^{0.3125}$ that appears in Eq. (14-203). Hence Eq. (14-204) gives a similar physical property dependence.

$$U_f = F_1 F_2(\sigma/\rho_g)^{0.5} \tag{14-204}$$

where U_f = flooding gas velocity, m/s
F_1 = 1.22 when $3.2\,d_i/\sigma > 1.0$
= $1.22\,(3.2\,d_i/\sigma)^{0.4}$ when $3.2\,d_i/\sigma < 1.0$
$F_2 = (G/L)^{0.25}$
G/L = gas-liquid mass ratio
d_i = inside diameter of column, mm
σ = surface tension, mN/m (dyn/cm)
ρ_g = gas density, kg/m^3

The primary reason for citing Eq. (14-204) is the large successful experience base in practical applications. Note that the reduction in allowable gas velocity for small diameters given by the F_1 factor is conceptually the same as the effect of using smaller-diameter packing in distillation. Note also that over the range of G/L between 1 and 0.1, the Maharudrayya and Jayanti data show a similar reduction in allowable gas rate to the F_2 factor in Eq. (14-204). The phenomenon behind this is that a thicker liquid film on the tube wall is more easily entrained.

While the limiting phenomenon of upper limit flooding in a vertical pipe is similar to *ultimate capacity* in distillation, there is a distinct difference. *Upper limit* in a vertical pipe applies to a design where a conscious effort should be made to minimize gas-liquid contact. Carried to extremes, it would involve separate tubes for liquid flowing down and vapor going up. In contrast, *ultimate capacity* in a distillation column corresponds to the condition where effective mass transfer disappears due to high entrainment. One could force more vapor up through the contactor, but fractionation would be poor.

Fog Condensation—The Other Way to Make Little Droplets For a variety of reasons, a gas or vapor can become supersaturated with a condensable component. Surface tension and mass transfer impose barriers on immediate condensation, so growth of fog particles lags behind what equilibrium predicts. Droplets formed by fog condensation are usually much finer (0.1 to 10 µm) than those formed by mechanical breakup and hence more difficult to collect. Sometimes fog can be a serious problem, as in the atmospheric discharge of a valuable or a hazardous material. More commonly, fog is a curiosity rather than a dominating element in chemical processing.

Fog particles grow because of excess saturation in the gas. Usually this means that the gas is supersaturated (i.e., it is below its dew point). Sometimes, fog can also grow on soluble foreign nuclei at partial pressures below saturation. Increased saturation can occur through a variety of routes:

1. Mixing of two saturated streams at different temperatures. This is commonly seen in the plume from a stack. Since vapor pressure is an exponential function of temperature, the resultant mixture of two saturated streams will be supersaturated at the mixed temperature. Uneven flow patterns and cooling in heat exchangers make this route to supersaturation difficult to prevent.
2. Increased partial pressure due to reaction. An example is the reaction of SO_3 and H_2O to yield H_2SO_4, which has much lower vapor pressure than its components.
3. Isoentropic expansion (cooling) of a gas, as in a steam nozzle.
4. Cooling of a gas containing a condensable vapor. Here the problem is that the gas cools faster than condensable vapor can be removed by mass transfer.

These mechanisms can be observed in many common situations. For example, fog via mixing can be seen in the discharge of breath on a cold day. Fog via adiabatic expansion can be seen in the low-pressure area over the wing of an airplane landing on a humid summer day; and fog via condensation can be seen in the exhaust from an automobile air conditioner (if you follow closely enough behind another car to pick up the ions or NO molecules needed for nucleation). All of these occur at a very low supersaturation and appear to be keyed to an abundance of foreign nuclei. All of these fogs also quickly dissipate as heat or unsaturated gas is added.

The supersaturation in condensers arises for two reasons. First, the condensable vapor is generally of higher molecular weight than the noncondensable gas. This means that the molecular diffusivity of the vapor will be much less than the thermal diffusivity of the gas. Restated, the ratio of N_{Sc}/N_{Pr} is greater than 1. The result is that a condenser yields more heat-transfer units $dT_g/(T_g - T_i)$ than mass-transfer units $dY_g/(Y_g - Y_i)$. Second, both transfer processes derive their driving force from the temperature difference between the gas T_g and the interface T_i. Each incremental decrease in interface temperature

TABLE 14-21 Simulation of Three Heat Exchangers with Varying Foreign Nuclei

	1	2	3
Weight fraction, noncondensable			
Inlet	0.51	0.42	0.02
Outlet	0.80	0.80	0.32
Molecular weight			
Inert	28	29	29
Condensable	86	99	210
Temperature difference between gas and liquid interface, K			
Inlet	14	24	67
Outlet	4	10	4
Percent of liquid that leaves unit as fog			
Nuclei concentration in inlet particles/cm^3			
100	0.05	1.1	2.2
1,000	0.44	5.6	3.9
10,000	3.2	9.8	4.9
100,000	9.6	11.4	5.1
1,000,000	13.3	11.6	
10,000,000	14.7		
∞	14.7	11.8	5.1
Fog particle size based on 10,000 nuclei/cm^3 at inlet, μm	28	25	4

yields the same relative increase in temperature driving force. However, the interface vapor pressure can only approach the limit of zero. Because of this, for equal molecular and thermal diffusivities a saturated mixture will supersaturate when cooled. The tendency to supersaturate generally increases with increased molecular weight of the condensable, increased temperature differences, and reduced initial superheating. To evaluate whether a given condensing step yields fog requires rigorous treatment of the coupled heat-transfer and mass-transfer processes through the entire condensation. Steinmeyer [*Chem. Eng. Prog.*, **68**(7), 64 (1972)] illustrates this, showing the impact of foreign-nuclei concentration on calculated fog formation. See Table 14-21. Note the relatively large particles generated for cases 1 and 2 for 10,000 foreign nuclei per cm^3. These are large enough to be fairly easily collected. There have been very few documented problems with industrial condensers despite the fact that most calculate to generate supersaturation along the condensing path. The explanation appears to be a limited supply of foreign nuclei.

Ryan et al. [*Chem. Eng. Progr.*, **90**(8), 83 (1994)] show that separate mass and heat transfer-rate modeling of an HCl absorber predicts 2 percent fog in the vapor. The impact is equivalent to lowering the stage efficiency to 20 percent.

Spontaneous (Homogeneous) Nucleation This process is quite difficult because of the energy barrier associated with creation of the interfacial area. It can be treated as a kinetic process with the rate a very steep function of the supersaturation ratio (S = partial pressure of condensable per vapor pressure at gas temperature). For water, an increase in S from 3.4 to 3.9 causes a 10,000-fold increase in the nucleation rate. As a result, below a critical supersaturation (S_{crit}), homogeneous nucleation is slow enough to be ignored. Generally, S_{crit} is defined as that which limits nucleation to one particle produced per cubic centimeter per second. It can be estimated roughly by traditional theory (*Theory of Fog Condensation,* Israel Program for Scientific Translations, Jerusalem, 1967) using the following equation:

$$S_{crit} = \exp\left[0.56 \frac{M}{\rho_l}\left(\frac{\sigma}{T}\right)^{3/2}\right] \quad (14\text{-}205)$$

where σ = surface tension, mN/m (dyn/cm)
ρ_l = liquid density, g/cm^3
T = temperature, K
M = molecular weight of condensable

Table 14-22 shows typical experimental values of S_{crit} taken from the work of Russel [*J. Chem. Phys.*, **50,** 1809 (1969)]. Since the critical supersaturation ratio for homogeneous nucleation is typically greater than 3, it is not often reached in process equipment. However, fog formation is typically found in steam turbines. Gyarmathy [*Proc. Inst. Mech. E., Part A: J. Power and Energy* **219**(A6), 511–521 (2005)] reports fog in the range 3.5 to 5 percent of total steam flow, with average fog diameter in the range of 0.1 to 0.2 μm.

TABLE 14-22 Experimental Critical Supersaturation Ratios

	Temperature, K	S_{crit}
H_2O	264	4.91
C_2H_5OH	275	2.13
CH_4OH	264	3.55
C_6H_6	253	5.32
CCl_4	247	6.5
$CHCl_3$	258	3.73
C_6H_5Cl	250	9.5

Growth on Foreign Nuclei As noted above, foreign nuclei are often present in abundance and permit fog formation at much lower supersaturation. For example,

1. *Solids.* Surveys have shown that air contains thousands of particles per cubic centimeter in the 0.1-μm to 1-μm range suitable for nuclei. The sources range from ocean-generated salt spray to combustion processes. The concentration is highest in large cities and industrial regions. When the foreign nuclei are soluble in the fog, nucleation occurs at S values very close to 1.0. This is the mechanism controlling atmospheric water condensation. Even when not soluble, a foreign particle is an effective nucleus if wet by the liquid. Thus, a 1-μm insoluble particle with zero contact angle requires an S of only 1.001 in order to serve as a condensation site for water.

2. *Ions.* Amelin [*Theory of Fog Condensation,* Israel Program for Scientific Translations, Jerusalem, (1967)] reports that ordinary air contains even higher concentrations of ions. These ions also reduce the required critical supersaturation, but by only about 10 to 20 percent, unless multiple charges are present.

3. *Entrained liquids.* Production of small droplets is inherent in the bubbling process, as shown by Fig. 14-90. Values range from near zero to 10,000/cm^3 of vapor, depending on how the vapor breaks through the liquid and on the opportunity for evaporation of the small drops after entrainment.

As a result of these mechanisms, most process streams contain enough foreign nuclei to cause some fogging. While fogging has been reported in only a relatively low percent of process partial condensers, it is rarely looked for and volunteers its presence only when yield losses or pollution is intolerable.

Dropsize Distribution Monodisperse (nearly uniform droplet size) fogs can be grown by providing a long retention time for growth. However, industrial fogs usually show a broad distribution, as in Fig. 14-91. Note also that for this set of data, the sizes are several orders of magnitude smaller than those shown earlier for entrainment and atomizers.

The result, as discussed in a later subsection, is a demand for different removal devices for the small particles.

While generally fog formation is a nuisance, it can occasionally be useful because of the high surface area generated by the fine drops. An example is insecticide application.

GAS-IN-LIQUID DISPERSIONS

GENERAL REFERENCES: Design methods for agitated vessels are presented by Penney in Couper et al., *Chemical Process Equipment, Selection and Design*, Chap. 10, Gulf Professional Publishing, Burlington, Mass., 2005. A comprehensive review of all industrial mixing technology is given by Paul, Atemo-Obeng, and Kresta, *Handbook of Industrial Mixing,* Wiley, Hoboken, N.J., 2004. Comprehensive treatments of bubbles or foams are given by Akers, *Foams: Symposium 1975*, Academic Press, New York, 1973; Bendure, *Tappi*, **58,** 83 (1975); Benfratello, *Energ Elettr.,* **30,** 80, 486 (1953); Berkman and Egloff, *Emulsions and Foams,* Reinhold, New York, 1941, pp. 112–152; Bikerman, *Foams,* Springer-Verlag, New York, 1975; *Kirk-Othmer Encyclopedia of Chemical Technology*, 4th ed., Wiley, New York, 1993, pp. 82–145; Haberman and Morton, *Report 802,* David W. Taylor Model Basin, Washington, 1953; Levich, *Physicochemical Hydrodynamics,* Prentice-Hall, Englewood Cliffs, NJ, 1962; and Soo, *Fluid Dynamics of Multiphase Systems,* Blaisdell, Waltham, Massachusetts,

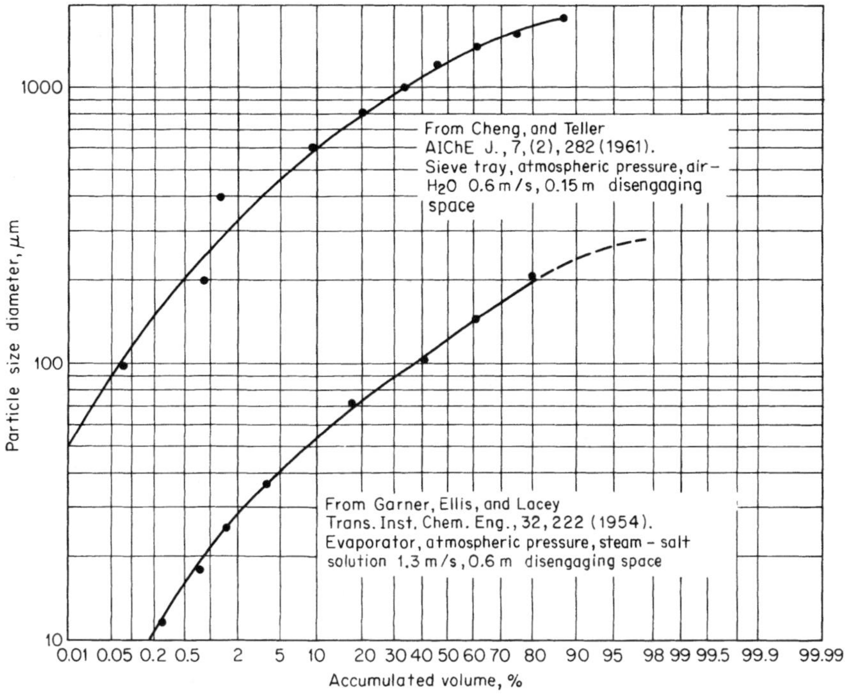

FIG. 14-90 Entrainment droplet-size distribution. To convert meters per second to feet per second, multiply by 3.28, to convert meters to feet multiply by 3.28.

1967. The formation of bubbles is comprehensively treated by Clift, Grace, and Weber, *Bubbles, Drops and Particles,* Academic, New York, 1978; Kumar and Kuloor, *Adv. Chem. Eng.* **8,** 255–368 (1970); and Wilkinson and Van Dierendonck, *Chem. Eng Sci.,* **49,** 1429–1438 (1994). Design methods for units operation in bubble columns and stirred vessels are covered by Atika and Yoshida, *Ind. Eng Chem. Process Des. Dev.,* **13,** 84 (1974); Calderbank, *The Chem. Eng.* (London), CE209 (October, 1967); and Mixing, vol. II, Academic, New York, 1967, pp. 1–111; Fair, *Chem. Eng.* **74,** 67 (July 3, 1967); Jordan, *Chemical Process Dev.,* Interscience, New York, 1968, part 1, pp. 111–175; Mersmann, *Ger. Chem. Eng.* **1,** 1 (1978); Resnick and Gal-Or, *Adv. Chem. Eng.,* **7,** 295–395 (1968); Valentin, *Absorption in Gas-Liquid Dispersions,* E. & F. N. Spon, London, 1967; Tatterson, *Fluid Mixing and Gas Dispersion in Agitated Tanks,* McGraw-Hill, 1991; and Deckwer and Schumpe, *Chem. Eng. Sci.,* **48,** 889–991 (1993).

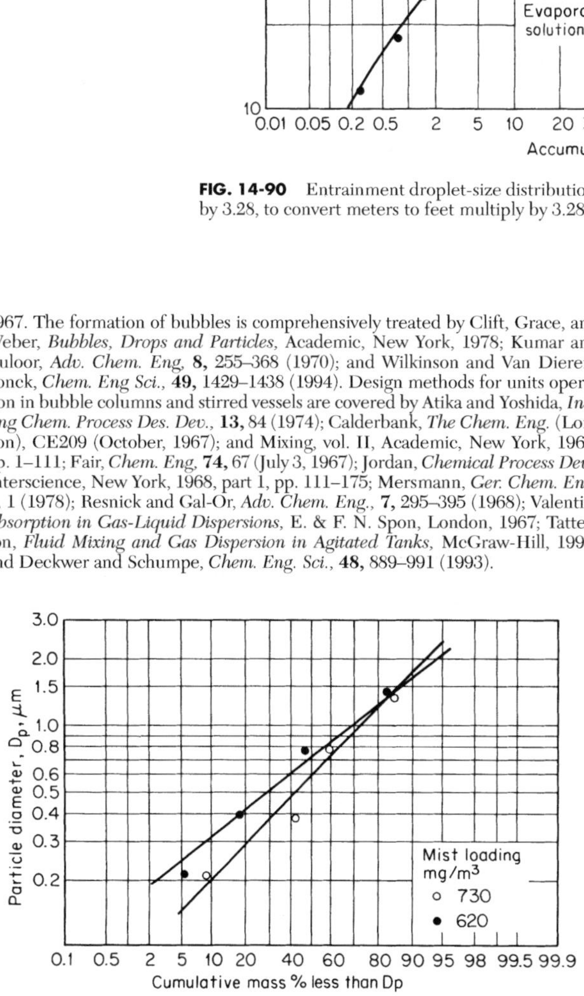

FIG. 14-91 Particle-size distribution and mist loading from absorption tower in a contact H_2SO_4 plant. [*Gillespie and Johnstone,* Chem. Eng. Prog., ***51***(2), 74 (1955).]

Review of foam rheology is given by Herzhaft [*Oil & Gas Sci. & Technol.* **54,** 587 (1999)] and Heller and Kuntamukkula [*Ind. Eng. Chem. Res.,* **26,** 318 (1987)]. The influence of surface-active agents on bubbles and foams is summarized in selected passages from Schwartz and Perry, *Surface Active Agents,* vol. 1, Interscience, New York, 1949; and from Schwartz, Perry, and Berch, *Surface Active Agents and Detergents,* vol. 2, Interscience, New York, 1958. See also Elenkov, *Theor. Found Chem. Eng.,* **1,** 1, 117 (1967); and Rubel, *Antifoaming and Defoaming Agents,* Noyes Data Corp., Park Ridge, NJ, 1972.

A review of foam stability also is given by de Vries, Meded, *Rubber Sticht. Delft.* No. 328, 1957. Foam-separation methodology is discussed by Aguoyo and Lemlich, *Ind. Eng. Chem. Process Des. Dev.,* **13,** 153 (1974); and Lemlich, *Ind. Eng Chem.,* **60,** 16 (1968). The following reviews of specific applications of gas-to-liquid dispersions are recommended: Industrial fermentations Aiba, Humphrey, and Millis, *Biochemical Engineering,* Academic, New York, 1965. Finn, *Bacteriol. Rev.,* **18,** 254 (1954). Oldshue, "Fermentation Mixing Scale-Up Techniques," in *Biotechnology and Bioengineering,* vol. 8, 1966, pp. 3–24. Aerobic oxidation of wastes: Eckenfelder and McCabe, *Advances in Biological Waste Treatment,* Macmillan, New York, 1963. Eckenfelder and O'Connor, *Biological Waste Treatment,* Pergamon, New York, 1961. McCabe and Eckenfelder, *Biological Treatment of Sewage and Industrial Wastes,* vol. 1, Reinhold, New York, 1955. Proceedings of Industrial Waste Treatment Conference, Purdue University, annually. Zlokarnik, *Adv. Biochem. Eng.,* **11,** 158–180 (1979). *Cellular elastomers:* Fling, *Natural Rubber Latex and Its Applications: The Preparation of Latex Foam Products,* British Rubber Development Board, London, 1954. Gould, in *Symposium on Application of Synthetic Rubbers,* American Society for Testing and Materials, Philadelphia, 1944, pp. 90–103. *Firefighting foams:* Perri, in Bikerman, op. cit., Chap. 12. Ratzer, *Ind. Eng. Chem.,* **48,** 2013 (1956). *Froth-flotation methods and equipment:* Booth, in Bikerman, op. cit., Chap. 13. Gaudin, *Flotation,* McGraw-Hill, New York, 1957. Taggart, *Handbook of Mineral Dressing,* Wiley, New York, 1945, Sec. 12, pp. 52–81. Tatterson, *Fluid Mixing and Gas Dispersion in Agitated Tanks,* McGraw-Hill, New York, 1991.

Objectives of Gas Dispersion The dispersion of gas as bubbles in a liquid or in a plastic mass is effected for one of the following purposes: (1) gas-liquid contacting (to promote absorption or stripping, with or without chemical reaction), (2) agitation of the liquid phase, or

(3) foam or froth production. Gas-in-liquid dispersions also may be produced or encountered inadvertently, sometimes undesirably.

Gas-Liquid Contacting Usually this is accomplished with conventional columns or with spray absorbers (see preceding subsection "Liquid-in-Gas Dispersions"). For systems containing solids or tar likely to plug columns, absorptions accomplished by strongly exothermic reactions, or treatments involving a readily soluble gas or a condensable vapor, however, bubble columns or agitated vessels may be used to advantage.

Agitation Agitation by a stream of gas bubbles (often air) rising through a liquid is often employed when the extra expense of mechanical agitation is not justified. Gas spargers may be used for simple blending operations involving a liquid of low volatility or for applications where agitator shaft sealing is difficult.

Foam Production This is important in froth-flotation separations; in the manufacture of cellular elastomers, plastics, and glass; and in certain special applications (e.g., food products, fire extinguishers). Unwanted foam can occur in process columns, in agitated vessels, and in reactors in which a gaseous product is formed; it must be avoided, destroyed, or controlled. Berkman and Egloff (*Emulsions and Foams*, Reinhold, New York, 1941, pp. 112–152) have mentioned that foam is produced only in systems possessing the proper combination of interfacial tension, viscosity, volatility, and concentration of solute or suspended solids. From the standpoint of gas comminution, foam production requires the creation of small bubbles in a liquid capable of sustaining foam.

Theory of Bubble and Foam Formation A bubble is a globule of gas or vapor surrounded by a mass or thin film of liquid. By extension, globular voids in a solid are sometimes called bubbles. Foam is a group of bubbles separated from one another by thin films, the aggregation having a finite static life. Although nontechnical dictionaries do not distinguish between foam and froth, a technical distinction is often made. A highly concentrated dispersion of bubbles in a liquid is considered a froth even if its static life is substantially nil (i.e., it must be dynamically maintained). Thus, all foams are also froths, whereas the reverse is not true. The term *lather* implies a froth that is worked up on a solid surface by mechanical agitation; it is seldom used in technical discussions. The thin walls of bubbles comprising a foam are called *laminae* or *lamellae*.

Bubbles in a liquid originate from one of three general sources: (1) They may be formed by desupersaturation of a solution of the gas or by the decomposition of a component in the liquid; (2) They may be introduced directly into the liquid by a bubbler or sparger or by mechanical entrainment; and (3) They may result from the disintegration of larger bubbles already in the liquid.

Generation Spontaneous generation of gas bubbles within a homogeneous liquid is theoretically impossible (Bikerman, *Foams: Theory and Industrial Applications*, Reinhold, New York, 1953, p. 10). The appearance of a bubble requires a gas nucleus as a void in the liquid. The nucleus may be in the form of a small bubble or of a solid carrying adsorbed gas, examples of the latter being dust particles, boiling chips, and a solid wall. A void can result from cavitation, mechanically or acoustically induced. Basu, Warrier, and Dhir [*J. Heat Transfer*, **124**, 717 (2002)] have reviewed boiling nucleation, and Blander and Katz [*AIChE J.*, **21**, 833 (1975)] have thoroughly reviewed bubble nucleation in liquids.

Theory permits the approximation of the maximum size of a bubble that can adhere to a submerged horizontal surface if the contact angle between bubble and solid (angle formed by solid-liquid and liquid-gas interfaces) is known [Wark, *J. Phys. Chem.*, **37**, 623 (1933); Jakob, *Mech. Eng.*, **58**, 643 (1936)]. Because the bubbles that actually rise from a surface are always considerably smaller than those so calculated and inasmuch as the contact angle is seldom known, the theory is not directly useful.

Formation at a Single Orifice The formation of bubbles at an orifice or capillary immersed in a liquid has been the subject of much study, both experimental and theoretical. Kulkarni and Joshi [*Ind. Eng. Chem. Res.*, **44**, 5873 (2005)] have reviewed bubble formation and rise. Bikerman (op. cit., Secs. 3 to 7), Valentin (op. cit., Chap. 2), Jackson (op. cit.), Soo (op. cit., Chap. 3), Fair (op. cit.), Kumer et al. (op. cit.), Clift et al. (op. cit.) and Wilkinson and Van Dierendonck [*Chem. Eng. Sci.*, **49**, 1429 (1994)] have presented reviews and analyses of this subject.

There are three regimes of bubble production (Silberman in *Proceedings of the Fifth Midwestern Conference on Fluid Mechanics*, Univ. of Michigan Press, Ann Arbor, 1957, pp. 263–284): (1) single-bubble, (2) intermediate, and (3) jet.

Single-Bubble Regime Bubbles are produced one at a time, their size being determined primarily by the orifice diameter d_o, the interfacial tension of the gas-liquid film σ, the densities of the liquid ρ_L and gas ρ_G, and the gravitational acceleration g according to the relation

$$d_b/d_o = [6\sigma/gd_o^2(\rho_L - \rho_G)]^{1/3} \qquad (14\text{-}206)$$

where d_b is the bubble diameter.

$$f = Q/(\pi d_b^3/6) = Qg(\rho_L - \rho_G)/(\pi d_o\sigma) \qquad (14\text{-}207)$$

where f is the frequency of bubble formation and Q is the volumetric rate of gas flow in consistent units.

Equations (14-206) and (14-207) result from a balance of bubble buoyancy against interfacial tension. They include no inertia or viscosity effects. At low bubbling rates (<1/s), these equations are quite satisfactory. Van Krevelen and Hoftijzer [*Chem. Eng. Prog.*, **46**, 29 (1950)], Guyer and Peterhaus [*Helv. Chim. Acta*, **26**, 1099 (1943)] and Wilkinson (op. cit.) report good agreement with Eq. (14-206) for water, transformer oil, ether, and carbon tetrachloride for vertically oriented orifices with $0.004 < D < 0.95$ cm. If the orifice diameter becomes too large, the bubble diameter will be smaller than the orifice diameter, as predicted by Eq. (14-206), and instability results; consequently, stable, stationary bubbles cannot be produced. Kulkarni et al. (op. cit.) have discussed much more detailed models for predicting bubble size; however, the models are very difficult and tedious to use in practice. These more sophisticated models need to be considered if the added accuracy warrants the extra effort.

For bubbles formed in water, the orifice diameter that permits bubbles of about its own size is calculated as 0.66 cm. Davidson and Amick [*AIChE J.*, **2**, 337 (1956)] confirmed this estimate in their observation that stable bubbles in water were formed at a 0.64-cm orifice but could not be formed at a 0.79-cm orifice.

For very thin liquids, Eqs. (14-206) and (14-207) are expected to be valid up to a gas-flow Reynolds number of 200 (Valentin, op. cit., p. 8). For liquid viscosities up to 100 cP, Datta, Napier, and Newitt [*Trans. Inst. Chem. Eng.*, **28**, 14 (1950)] and Siems and Kauffman [*Chem. Eng. Sci.*, **5**, 127 (1956)] have shown that liquid viscosity has very little effect on the bubble volume, but Davidson and Schuler [*Trans. Instn. Chem. Eng.*, **38**, 144 (1960)] and Krishnamurthi et al. [*Ind. Eng. Chem. Fundam.*, **7**, 549 (1968)] have shown that bubble size increases considerably over that predicted by Eq. (14-206) for liquid viscosities above 1000 cP. In fact, Davidson et al. (op. cit.) found that their data agreed very well with a theoretical equation obtained by equating the buoyant force to drag based on Stokes' law and the velocity of the bubble equator at break-off:

$$d_b = \left(\frac{6}{\pi}\right)\left(\frac{4\pi}{3}\right)^{1/4}\left(15 \times \frac{\nu Q}{2g}\right)^{3/4} \qquad (14\text{-}208)$$

where ν is the liquid kinematic viscosity and Q is the gas volumetric flow rate. This equation is dimensionally consistent. The relative effect of liquid viscosity can be obtained by comparing the bubble diameters calculated from Eqs. (14-206) and (14-208). If liquid viscosity appears significant, one might want to use the long and tedious method developed by Krishnamurthi et al. (op. cit.) and the review by Kulkarni et al. (op. cit.) that considers both surface-tension forces and viscous-drag forces.

Intermediate Regime This regime extends approximately from a Reynolds number of 200 to one of 2100. As the gas flow through a

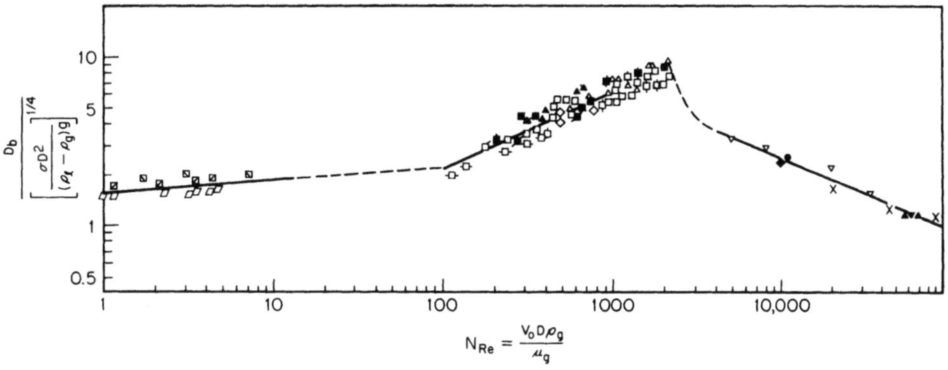

FIG. 14-92 Bubble-diameter correlation for air sparged into relatively inviscid liquids. D_b = bubble diameter, D = orifice diameter, V_o = gas velocity through sparging orifice, P = fluid density, and μ = fluid viscosity. [*From* Can. J. Chem. Eng., **54,** 503 (*1976*).]

submerged orifice increases beyond the limit of the single-bubble regime, the frequency of bubble formation increases more slowly, and the bubbles begin to grow in size. Between the two regimes there may indeed be a range of gas rates over which the bubble size decreases with increasing rate, owing to the establishment of liquid currents that nip the bubbles off prematurely. The net result can be the occurrence of a minimum bubble diameter at some particular gas rate [Mater, *U.S. Bur. Mines Bull.* 260 (1927) and Bikerman, op. cit., p. 4]. At the upper portion of this region, the frequency becomes very nearly constant with respect to gas rate, and the bubble size correspondingly increases with gas rate. The bubble size is affected primarily by (1) orifice diameter, (2) liquid-inertia effects, (3) liquid viscosity, (4) liquid density, and (5) the relationship between the constancy of gas flow and the constancy of pressure at the orifice.

Kumar et al. have done extensive experimental and theoretical work reported in *Ind. Eng. Chem. Fundam.*, **7,** 549 (1968); *Chem. Eng. Sci*, **24,** part 1, 731; part 2, 749; part 3, 1711 (1969) and summarized in *Adv. Chem. Eng.*, **8,** 255 (1970). They, along with other investigators—Swope [*Can. J Chem. Eng.*, **44,** 169 (1972)], Tsuge and Hibino [*J. Chem. Eng. Japan,* **11,** 307 (1972)], Pinczewski [*Chem. Eng. Sci.*, **36,** 405 (1981)], Tsuge and Hibino [*Int. Chem. Eng.*, **21,** 66 (1981)], and Takahashi and Miyahara [ibid., p. 224]—have solved the equations resulting from a force balance on the forming bubble, taking into account buoyancy, surface tension, inertia, and viscous-drag forces for both conditions of constant flow through the orifice and constant pressure in the gas chamber. The design method is complex and tedious and involves the solution of algebraic and differential equations. Although Mersmann [*Ger. Chem. Eng.*, **1,** 1 (1978)] claims that the results of Kumar et al. (loc. cit.) well fit experimental data, Lanauze and Harn [*Chem. Eng. Sci.*, **29,** 1663 (1974)] claim differently:

> Further, it has been shown that the mathematical formulation of Kumar's model, including the condition of detachment, could not adequately describe the experimental situation—Kumar's model has several fundamental weaknesses, the computational simplicity being achieved at the expense of physical reality.

In lieu of careful independent checks of predictive accuracy, the results of the comprehensive theoretical work will not be presented here. Simpler, more easily understood predictive methods, for certain important limiting cases, will be presented. As a check on the accuracy of these simpler methods, it will perhaps be prudent to calculate the bubble diameter from the graphical representation by Mersmann (loc. cit.) of the results of Kumar et al. (loc. cit.) and the review by Kulkarni et al. (op. cit.)

For conditions approaching constant flow through the orifice, a relationship derived by equating the buoyant force to the inertia force of the liquid [Davidson et al., *Trans. Instn. Chem. Engrs.*, **38,** 335 (1960)] (dimensionally consistent),

$$d_b = 1.378 \times \frac{6Q^{6/5}}{\pi g^{3/5}} \tag{14-209}$$

fits experimental data reasonably well. Surface tension and liquid viscosity tend to increase the bubble size—at a low Reynolds number. The effect of surface tension is greater for large orifice diameters. The magnitude of the diameter increase due to high liquid viscosity can be obtained from Eq. (14-208).

For conditions approaching constant pressure at the orifice entrance, which probably simulates most industrial applications, there is no independently verified predictive method. For air at near atmospheric pressure sparged into relatively inviscid liquids (11 ~ 100 cP), the correlation of Kumar et al. [*Can. J. Chem. Eng.*, **54,** 503 (1976)] fits experimental data well. Their correlation is presented here as Fig. 14-92.

Wilkinson et al. (op. cit.) make the following observation about the effect of gas density on bubble size: "The fact that the bubble size decreases slightly for higher gas densities can be explained on the basis of a force balance."

Jet Regime With further rate increases, turbulence occurs at the orifice, and the gas stream approaches the appearance of a continuous jet that breaks up 7.6 to 10.2 cm above the orifice. Actually, the stream consists of large, closely spaced, irregular bubbles with a rapid swirling motion. These bubbles disintegrate into a cloud of smaller ones of random size distribution between 0.025 cm or smaller and about 1.25 cm, with a mean size for air and water of about 0.4 cm (Leibson et al., loc. cit.). According to Wilkinson et al. (op. cit.), jetting begins when

$$N_{We,g} = \frac{\rho_g d_o U_o^2}{\sigma} \leq 2 \tag{14-210}$$

There are many contradictory reports about the jet regime, and theory, although helpful (see, for example, Siberman, loc. cit.), is as yet unable to describe the phenomena observed. The correlation of Kumar et al. (Fig. 14-92) is recommended for air-liquid systems.

Formation at Multiple Orifices At high velocities, coalescence of bubbles formed at individual orifices occurs; Helsby and Tuson [*Research* (London), **8,** 270 (1955)], for example, observed the frequent coalescence of bubbles formed in pairs or in quartets at an orifice. Multiple orifices spaced by the order of magnitude of the orifice diameter increase the probability of coalescence, and when the magnitude is small (as in a sintered plate), there is invariably some. The broken lines of Fig. 14-92 presumably represent zones of increased coalescence and relatively less effective dispersion as the gas rate through porous-carbon tubes is increased. Savitskaya [*Kolloidn. Zh.*, **13,** 309 (1951)] found that the average bubble size formed at the surface of a porous plate was such as to maintain constancy of the product of bubble specific surface and interfacial tension as the latter was varied by addition of a surfactant. Konig et al. [*Ger. Chem. Eng.*, **1,** 199 (1978)] produced bubble sizes varying from 0.5 to 4 mm by the use of two porous-plate spargers and one perforated-plate sparger with superficial gas velocities from 1 to 8 cm/s. The small bubble sizes were stabilized by adding up to 0.5 percent of various alcohols to water.

At high-flow rates through perforated plates such as those that occur in distillation columns, Calderbank and Rennie [*Trans. Instn. Chem. Engrs.*, **40,** T3 (1962)]; Porter et al. [ibid., **45,** T265 (1967)]; Rennie and Evans [*Br. Chem. Eng*, **7,** 498 (1962)]; and Valentin (op. cit., Chap. 3) have investigated and discussed the effect of the flow conditions through the multiple orifices on the froths and foams that occur above perforated plates.

Entrainment and Mechanical Disintegration Gas can be entrained into a liquid by a solid or a stream of liquid falling from the gas phase into the liquid, by surface ripples or waves, and by the vertical swirl of a mass of agitated liquid about the axis of a rotating agitator. Small bubbles probably form near the surface of the liquid and are caught into the path of turbulent eddies, whose velocity exceeds the terminal velocity of the bubbles. The disintegration of a submerged mass of gas takes place by the turbulent tearing of smaller bubbles away from the exterior of the larger mass or by the influence of surface tension on the mass when it is attenuated by inertial or shear forces into a cylindrical or disk form. A fluid cylinder that is greater in length than in circumference is unstable and tends to break spontaneously into two or more spheres. These effects account for the action of fluid attrition and of an agitator in the disintegration of suspended gas. Quantitative correlations for gas entrainment by liquid jets and in agitated vessels will be given later.

Foams Two excellent reviews (Shedlovsky, op. cit.; Lemlich, op. cit.; Herzhaft, op. cit.; and Heller et al., op. cit) covering the literature pertinent to foams have been published. A foam is formed when bubbles rise to the surface of a liquid and persist for a while without coalescence with one another or without rupture into the vapor space. The formation of foam, then, consists simply of the formation, rise, and aggregation of bubbles in a liquid in which foam can exist. The life of foams varies over many magnitudes—from seconds to years—but in general is finite. Maintenance of a foam, therefore, is a dynamic phenomenon.

Gravitational force favors the separation of gas from liquid in a disperse system, causing the bubbles to rise to the liquid surface and the liquid contained in the bubble walls to drain downward to the main body of the liquid. Interfacial tension favors the coalescence and ultimate disappearance of bubbles; indeed, it is the cause of bubble destruction upon the rupture of the laminae.

The viscosity of the liquid in a film opposes the drainage of the film and its displacement by the approach of coalescing bubbles. The higher the viscosity, the slower will be the film-thinning process; furthermore, if viscosity increases as the film grows thinner, the process becomes self-retarding. The viscosity of films appears to be greater than that of the main body of the parent liquid in many cases. Sometimes this is a simple temperature effect, the film being cooler because of evaporation; sometimes it is a concentration effect, with dissolved or fine suspended solids migrating to the interface and producing classical or anomalous increases in viscosity; at yet other times, the effect seems to occur without explanation.

If the liquid laminae of a foam system can be converted to impermeable solid membranes, the film viscosity can be regarded as having become infinite, and the resulting solid foam will be permanent. Likewise, if the laminae are composed of a gingham plastic or a thixotrope, the foam will be permanently stable for bubbles whose buoyancy does not permit exceeding the yield stress. For other non-newtonian fluids, however, and for all newtonian ones, no matter how viscous, the viscosity can only delay but never prevent foam disappearance. The popular theory, held since the days of Plateau, that foam life is proportional to surface viscosity and inversely proportional to interfacial tension, is not correct, according to Bikerman (op. cit., p. 161), who points out that it is contradicted by experiment.

The idea that foam films drain to a critical thickness at which they spontaneously burst is also rejected by Bikerman. Foam stability, rather, is keyed to the existence of a surface skin of low interfacial tension immediately overlying a solution bulk of higher tension, latent until it is exposed by rupture of the superficial layer [Maragoni, *Nuovo Cimento,* **2** (5–6), 239 (1871)]. Such a phenomenon of surface elasticity, resulting from concentration differences between bulk and surface of the liquid, accounts for the ability of bubbles to be penetrated by missiles without damage. It is conceivable that films below a certain thickness no longer carry any bulk of solution and hence have no capacity to close surface ruptures, thus becoming vulnerable to mechanical damage that will destroy them. The Maragoni phenomenon is consistent also with the observation that neither pure liquids nor saturated solutions will sustain a foam, since neither extreme will allow the necessary differences in concentration between surface and bulk of solution.

The specific ability of certain finely divided, insoluble solids to stabilize foam has long been known [Berkman and Egloff, op. cit., p. 133; and Bikerman, op. cit., Chap. 11]. Bartsch [*Kolloidchem. Beih,* **20,** 1 (1925)] found that the presence of fine galena greatly extended the life of air foam in aqueous isoamyl alcohol, and the finer the solids, the greater the stability. Particles on the order of 50 μm length extended the life from 17 seconds to several hours. This behavior is consistent with theory, which indicates that a solid particle of medium contact angle with the liquid will prevent the coalescence of two bubbles with which it is in simultaneous contact. Quantitative observations of this phenomenon are scanty.

Berkman and Egloff explain that some additives increase the flexibility or toughness of bubble walls, rather than their viscosity, to render them more durable. They cite as illustrations the addition of small quantities of soap to saponin solutions or of glycerin to soap solution to yield much more stable foam. The increased stability with ionic additives is probably due to electrostatic repulsion between charged, nearly parallel surfaces of the liquid film, which acts to retard draining and hence rupture.

Characteristics of Dispersion

Properties of Component Phases As discussed in the preceding subsection, dispersions of gases in liquids are affected by the viscosity of the liquid, the density of the liquid and of the gas, and the interfacial tension between the two phases. They also may be affected directly by the composition of the liquid phase. Both the formation of bubbles and their behavior during their lifetime are influenced by these quantities as well as by the mechanical aspects of their environment.

Viscosity and density of the component phases can be measured with confidence by conventional methods, as can the interfacial tension between a pure liquid and a gas. The interfacial tension of a system involving a solution or micellar dispersion becomes less satisfactory, because the interfacial free energy depends on the concentration of solute at the interface. Dynamic methods and even some of the so-called static methods involve the creation of new surfaces. Since the establishment of equilibrium between this surface and the solute in the body of the solution requires a finite amount of time, the value measured will be in error if the measurement is made more rapidly than the solute can diffuse to the fresh surface. Eckenfelder and Barnhart (Am. Inst. Chem. Engrs., 42d national meeting, Repr. 30, Atlanta, 1960) found that measurements of the surface tension of sodium lauryl sulfate solutions by maximum bubble pressure were higher than those by DuNuoy tensiometer by 40 to 90 percent, the larger factor corresponding to a concentration of about 100 ppm, and the smaller to a concentration of 2500 ppm of sulfate.

Even if the interfacial tension is measured accurately, there may be doubt about its applicability to the surface of bubbles being rapidly formed in a solution of a surface-active agent, for the bubble surface may not have time to become equilibrated with the solution. Coppock and Meiklejohn [*Trans. Instn. Chem. Engrs.*, **29,** 75 (1951)] reported that bubbles formed in the single-bubble regime at an orifice in a solution of a commercial detergent had a diameter larger than that calculated in terms of the measured surface tension of the solution [Eq. (14-206)]. The disparity is probably a reflection of unequilibrated bubble laminae.

One concerned with the measurement of gas-liquid interfacial tension should consult the useful reviews of methods prepared by Harkins [in Chap. 9 of Weissberger, *Techniques of Organic Chemstry*, 2d ed., vol. 1, part 2, Interscience, New York, 1949), Schwartz and coauthors [*Surface Acttve Agents*, vol. 1, Interscience, New York, 1949, pp. 263–271; *Surface Active Agents and Detergents*, vol. 2, Interscience, New York, 1958, pp. 389–391, 417–418], and by Adamson [*Physical Chemistry of Surfaces*, Interscience, New York, 1960].

Dispersion Characteristics The chief characteristics of gas-in-liquid dispersions, like those of liquid-in-gas suspensions, are heterogeneity and instability. The composition and structure of an unstable dispersion must be observed in the dynamic situation by looking at the mixture, with or without the aid of optical devices, or by photographing it, preferably in nominal steady state; photographs usually are required for quantitative treatment. Stable foams may be examined after the fact of their creation if they are sufficiently robust or if an immobilizing technique such as freezing is employed [Chang et al., *Ind. Eng Chem.*, **48,** 2035 (1956)].

The rate of rise of bubbles has been discussed by Kulkarni, op. cit.; Benfratello, *Energ Elettr.*, **30,** 80 (1953); Haberman and Morton, *Report 802: David W. Taylor Model Basin*, Washington, September 1953; Jackson, loc. cit.; Valentin, op. cit., Chap. 2; Soo, op. cit., Chap. 3; Calderbank, loc. cit., p. CE220; and Levich, op. cit., Chap. 8). A comprehensive and apparently accurate predictive method has been published [Jamialahamadi et al., *Trans ICE*, **72,** part A, 119–122 (1994)]. Small bubbles (below 0.2 mm in diameter) are essentially rigid spheres and rise at terminal velocities that place them clearly in the laminar-flow region; hence their rising velocity may be calculated from Stokes' law. As bubble size increases to about 2 mm, the spherical shape is retained, and the Reynolds number is still sufficiently small (<10) that Stokes' law should be nearly obeyed.

As bubble size increases, two effects set in, however, that alter the velocity. At about $N_{Re} = 100$, a wobble begins that can develop into a helical path if the bubbles are not liberated too closely to one another [Houghton, McLean, and Ritchie, *Chem. Eng. Sci.*, **7,** 40 (1957); and Houghton et al., ibid., p. 111]. Furthermore, for bubbles in the range of 1 mm and larger (until distortion becomes serious) internal circulation can set in [Garner and Hammerton, *Chem. Eng. Sci.*, **3,** (1954); and Haberman and Morton, loc. cit.], and according to theoretical analyses by Hadamard and Rybczynski and given by Levich (op. cit.), the drag coefficient for a low-viscosity dispersed phase and a high-viscosity continuous phase will approach two-thirds of the drag coefficient for rigid spheres, namely $C_D = 16/N_{Re}$. The rise velocity of a circulating bubble or drop will thus be 1.5 times that of a rigid sphere. Redfield and Houghton [*Chem. Eng. Sci.*, **20,** 131 (1965)] have found that CO_2 bubbles rising in pure water agree with the theoretical solution for a circulating drop below $N_{Re} = 1$. Many investigators (see Valentin, op. cit.) have found that extremely small quantities of impurities can retard or stop this internal circulation. In this behavior may lie the explanation of the fact that the addition of long-chain fatty acids to water to produce a concentration of 1.5×10^{-4} molar markedly reduces the rate of rise of bubbles [Stuke, *Naturwissenschaften*, **39,** 325 (1952)].

Above diameters of about 2 mm, bubbles begin to change to ellipsoids, and above 1 cm they become lens-shaped, according to Davies and Taylor [*Proc. Roy. Soc.* (London), **A200,** 379 (1950)]. The rising velocity in thin liquids for the size range 1 mm $< D_B <$ 20 mm has been reported as 20 to 30 cm/s by Haberman and Morton (op. cit.) and Davenport, Richardson, and Bradshaw [*Chem. Eng. Sci.*, **22,** 1221 (1967)]. Schwerdtieger [ibid., **23,** 937 (1968)] even found the same for argon bubbles rising in mercury. Surface-active agents have no effect on the rise velocity of bubbles larger than 4 mm in thin liquids (Davenport et al., loc. cit.).

Above a Reynolds number of the order of magnitude of 1000, bubbles assume a helmet shape, with a flat bottom (Eckenfelder and Barnhart, loc. cit.; and Leibson et al., loc. cit.). After bubbles become large enough to depart from Stokes' law at their terminal velocity, behavior is generally complicated and erratic, and the reported data scatter considerably. The rise can be slowed, furthermore, by a wall effect if the diameter of the container is not greater than 10 times the diameter of the bubbles, as shown by Uno and Kintner [*AIChE J.*, **2,** 420 (1956); and Collins, *J. Fluid Mech.*, **28**(1), 97 (1967)]. Work has

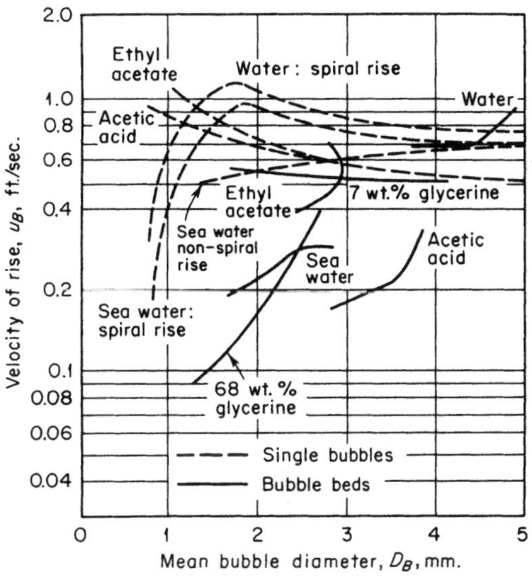

FIG. 14-93 Velocity of rising bubbles, singly and in clouds. To convert feet per second to meters per second, multiply by 0.305. [*From* Chem. Eng. Sci., **7,** *48 (1957).*]

been done to predict the rise velocity of large bubbles [Rippin and Davidson, *Chem. Eng. Sci.*, **22,** 217 (1967); Grace and Harrison, ibid., 1337; Mendelson, *AIChE J.*, **13,** 250 (1967); Cole, ibid.; Lehrer, *J. Chem. Eng. Japan*, **9,** 237 (1976); and Lehrer, *AIChE J.*, **26,** 170 (1980)]. The works of Lehrer present correlations that accurately predict rise velocities for a wide range of system properties. Reviews of the technical literature concerning the rise of single bubbles and drops have been published by Kulkarni et al. (op. cit.) and Clift, Grace, and Weber (*Bubbles, Drops and Particles*, Academic, New York, 1978). Mendelson has used a wave theory to predict the terminal velocity, and Cole has checked the theory with additional data. The other authors listed solved some simplified form of the Navier-Stokes equations. Jamialahmadi et al., loc. cit., have developed a single equation predictive method for bubble rise velocity, which covers the entire range of bubble diameters.

When bubbles are produced in clouds, as by a porous disperser, their behavior during rising is further complicated by interaction among themselves. In addition to the tendency for small bubbles to coalesce and large ones to disintegrate, there are two additional opposing influences on the rate of rise of bubbles of any particular size: (1) A "chimney effect" can develop in which a massive current upward appears at the axis of the bubble stream, leading to increased net bubble velocity; and (2) the proximity of the bubbles to one another can result in a hindered-settling condition, leading to reduced average bubble velocity. Figure 14-93 shows the data of Houghton et al. (op. cit.) for clouds of bubbles compared with their single-bubble data for pure water and seawater and of Peebles and Garber [*Chem. Eng. Progr.*, **49,** 88 (1953)] for acetic acid and ethyl acetate. The bubble clouds were produced with a sintered-glass plate of mean pore size (inferred from air wet-permeability data) of 81 μm.

The difference between the curves for pure water and seawater again illustrates the significance of small concentrations of solute with respect to bubble behavior. In commercial bubble columns and agitated vessels coalescence and breakup are so rapid and violent that the rise velocity of a single bubble is meaningless. The average rise velocity can, however, be readily calculated from holdup correlations that will be given later.

The quantitative examination of bubble systems is aided by the use of proper illumination and photography. The formation of bubbles at single sources often is sufficiently periodic to be stopped by stroboscopic light. Clouds of rising bubbles are more difficult to assess and require careful technique. Satisfactory photographic methods have been developed by Vermenlen, Williams, and Langlois [*Chem. Eng. Progr.*, **51,** 85 (1955)] and by Calderbank [*Trans. Instn. Chem. Engrs.*, **36,** 443 (1958)] and are described by these authors. Calderbank's technique resulted in particularly precise measurements that permitted a good estimation of the surface area of the dispersed bubbles.

Methods of Gas Dispersion The problem of dispersing a gas in a liquid may be attacked in several ways: (1) The gas bubbles of the desired size or which grow to the desired size may be introduced directly into the liquid; (2) a volatile liquid may be vaporized by either decreasing the system pressure or increasing its temperature; (3) a chemical reaction may produce a gas; or (4) a massive bubble or stream of gas is disintegrated by fluid shear and/or turbulence in the liquid.

Spargers: Simple Bubblers The simplest method of dispersing gas in a liquid contained in a tank is to introduce the gas through an open-end standpipe, a horizontal perforated pipe, or a perforated plate at the bottom of the tank. At ordinary gassing rates (corresponding to the jet regime), relatively large bubbles will be produced regardless of the size of the orifices.

Perforated-pipe or -plate spargers usually have orifices 3 to 12 mm in diameter. Effective design methods to minimize maldistribution are presented in the fifth edition of this handbook, p. 5–47, 1973, and by Knaebel [*Chem. Eng.*, 116 (Mar. 9, 1981)]. For turbulent flow conditions into the sparger, the following relationship will allow design of a perforated-pipe sparger for a given degree of maldistribution provided $N_h > 5$ and length/diameter <300.

$$d_P = 0.95(N_h C_v)^{1/2} \times \left(\frac{d_h}{\Delta U_h / U_h} \right)^{1/4} \qquad (14\text{-}211)$$

where d_p = pipe diameter, d_h = sparging hole diameter, N_h = number of holes in sparger, C_v = orifice coefficient for sparger hole (see *Chemical Engineers' Handbook,* 5th ed., pp. 5–13, 5–34), U_h = average velocity through sparger holes, ΔU_h = difference between maximum and minimum velocities through sparger holes, and $\Delta U_h/U_h$ = fractional maldistribution of flow through sparger holes.

Simple spargers are used as agitators for large tanks, principally in the cement and oil industries. Kauffman [*Chem. Metall. Eng.*, **37,** 178–180 (1930)] reported the following air rates for various degrees of agitation in a tank containing 2.7 m (9 ft) of *liquid:*

Degree of agitation	Air rate, $m^3/(m^2$ tank cross section, min)
Moderate	0.0033
Complete	0.0066
Violent	0.016

For a liquid depth of 0.9 m (3 ft), Kauffman recommended that the listed rates be doubled.

An air lift consisting of a sparger jetting into a draft tube with ports discharging at several heights has been recommended by Heiser [*Chem. Eng.*, **55**(1), 135 (1948)] for maintaining agitation in a heavy, coarse slurry, the level of which varies widely. The design is illustrated in Fig. 14-94.

The ability of a sparger to blend miscible liquids might be described in terms of a fictitious diffusivity. Siemes did so, reporting that the agitation produced by a stream of bubbles rising in a tube with a superficial velocity of about 8.2 cm/s corresponded to an apparent diffusion coefficient as large as 75 cm²/s [*Chem. Ing. Tech.*, **29,** 727 (1957)]. The blending rate thus is several orders of magnitude higher than it would be by natural diffusive action. These results are typical of subsequent investigations on back mixing, which will be discussed in more detail later.

Lehrer [*Ind. Eng. Chem. Process Des. Dev.*, **7,** 226 (1968)] conducted liquid-blending tests with air sparging in a 0.61-m-diameter by 0.61-m-tall vessel and found that an air volume equal to about one-half of the vessel volume gave thorough blending of inviscid liquids of equal viscosities. Using an analogy to mechanically agitated vessels in which equal tank turnovers give equal blend times, one would expect this criterion to be applicable to other vessel sizes. Liquids of unequal density would require somewhat more air.

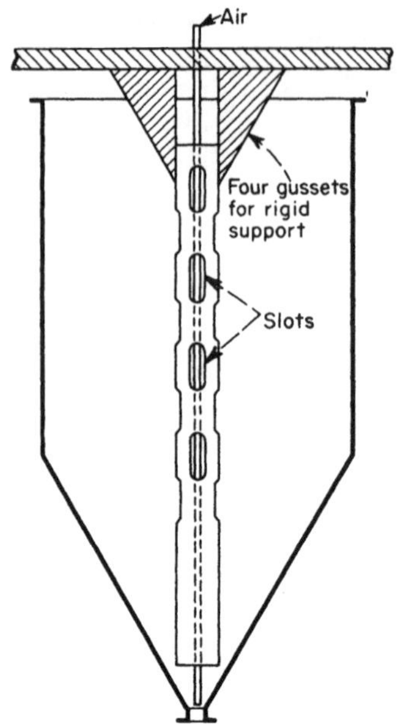

FIG. 14-94 Slotted air lift for agitation of a variable-level charge. [*From Chem. Eng.*, ***55***(*1*), *135* (*1948*).]

Open-end pipes, perforated plates, and ring- or cross-style perforated-pipe spargers are used without mechanical agitation to promote mass transfer, as in chlorinators and biological sewage treatment. In the "quiescent regime" (superficial gas velocity less than 4.57 to 6.1 cm/s [0.15 to 0.2 ft/s]) the previously mentioned spargers are usually operated at orifice Reynolds numbers in excess of 6000 in order to get small bubbles so as to increase the interfacial area and thus increase mass transfer. In the "turbulent regime" (superficial gas velocity greater than 4.57 to 6.1 cm/s), sparger design is not critical because a balance between coalescence and breakup is established very quickly according to Towell et al. [*AIChE Symp. Ser. No. 10,* 97 (1965)]. However, a reasonably uniform orifice distribution over the column cross section is desirable, and according to Fair [*Chem. Eng.*, **74,** 67 (July 3, 1967); 207 (July 17, 1967)] the orifice velocity should be less than 75 to 90 m/s.

Porous Septa In the quiescent regime porous plates, tubes, disks, or other shapes that are made by bonding or sintering together carefully sized particles of carbon, ceramic, polymer, or metal are frequently used for gas dispersion, particularly in foam fractionators. The resulting septa may be used as spargers to produce much smaller bubbles than will result from a simple bubbler. Figure 14-95 shows a comparison of the bubbles emitted by a perforated-pipe sparger [0.16-cm orifices] and a porous carbon septum (120 μm pores). The gas flux through a porous septum is limited on the lower side by the requirement that, for good performance, the whole sparger surface should bubble more or less uniformly, and on the higher side by the onset of serious coalescence at the surface of the septum, resulting in poor dispersion. In the practical range of fluxes, the size of the bubbles produced depends on both the size of pores in the septum and the pressure drop imposed across it, being a direct function of both.

Table 14-23 lists typical grades of porous carbon, silica, alumina, stainless steel (type 316), and polymers commercially available.

Porous media are also manufactured from porcelain, glass, silicon carbide, and a number of metals: Monel, Inconel, nickel, bronze, Hastelloy C, Stellite L-605, gold, platinum, and many types of stainless steel. Two manufacturers of porous septa are Mott (mottcorp.com) and Pall (pall.com). The air permeabilities of Table 14-23 indicate the relative flow resistances of the various grades to homogeneous fluid but may not be used in designing a disperser for submerged

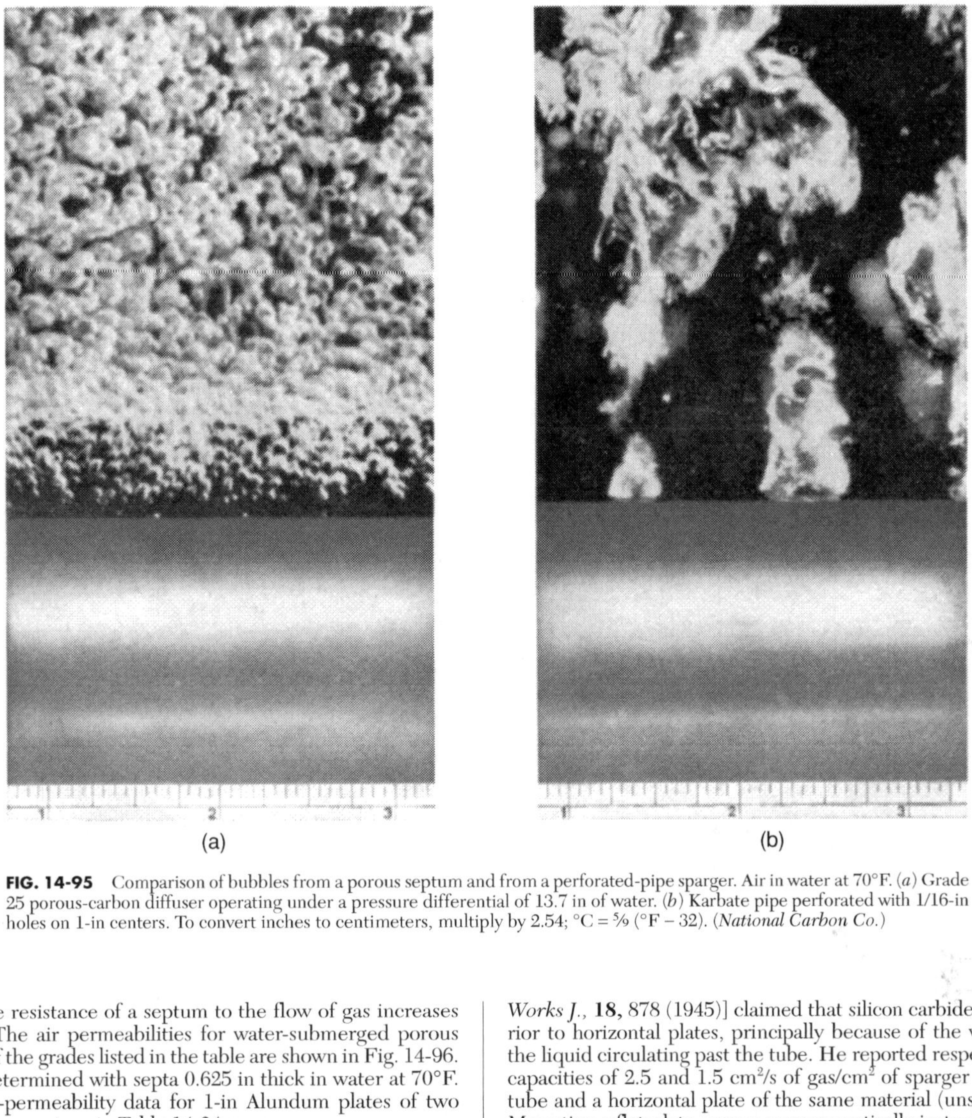

FIG. 14-95 Comparison of bubbles from a porous septum and from a perforated-pipe sparger. Air in water at 70°F. (*a*) Grade 25 porous-carbon diffuser operating under a pressure differential of 13.7 in of water. (*b*) Karbate pipe perforated with 1/16-in holes on 1-in centers. To convert inches to centimeters, multiply by 2.54; °C = 5⁄9 (°F − 32). (*National Carbon Co.*)

operation, for the resistance of a septum to the flow of gas increases when it is wet. The air permeabilities for water-submerged porous carbon of some of the grades listed in the table are shown in Fig. 14-96. The data were determined with septa 0.625 in thick in water at 70°F. Comparable wet-permeability data for 1-in Alundum plates of two grades of fineness are given in Table 14-24.

The gas rate at which coalescence begins to reduce the effectiveness of dispersion appears to depend not only on the pore size and pore structure of the dispersing medium but also on the liquid properties, liquid depth, agitation, and other features of the sparging environment; coalescence is strongly dependent on the concentration of surfactants capable of forming an electrical double layer and thus produce ionic bubbles, long-chain alcohols in water being excellent examples. For porous-carbon media, the manufacturer suggests that the best dispersion performance will result if the broken-line regions of Fig. 14-96 are avoided. For porous stainless-steel spargers, which extend to a lower pore size than carbon, Micro Metallic Division, Pall Corp., recommends (Release 120A, 1959) a working limit of 8 ft/min (0.044 m/s) to avoid serious coalescence. This agrees with the data reported by Konig et al. (loc. cit.), in which 0.08 m/s was used and bubbles as small as 1 mm were produced from a 5-μm porous sparger.

Slabs of porous material are installed by grouting or welding together to form a diaphragm, usually horizontal. Tubes are prone to produce coalesced gas at rates high enough to cause bubbling from their lower faces, but they have the advantage of being demountable for cleaning or replacement (U.S. Patent 2,328,655). Roe [*Sewage Works J.*, **18,** 878 (1945)] claimed that silicon carbide tubes are superior to horizontal plates, principally because of the wiping action of the liquid circulating past the tube. He reported respective maximum capacities of 2.5 and 1.5 cm^3/s of gas/cm^2 of sparger for a horizontal tube and a horizontal plate of the same material (unspecified grade). Mounting a flat-plate porous sparger vertically instead of horizontally seriously reduces the effectiveness of the sparger for three reasons: (1) The gas is distributed over a reduced cross section; (2) at normal rates, the lower portion of the sparger may not operate because of difference in hydrostatic head; and (3) there is a marked tendency for bubbles to coalesce along the sparger surface. Bone (M.S. thesis in chemical engineering, University of Kansas, 1948) found that the oxygen sulfite solution coefficient for a 3.2- by 10-cm rectangular porous carbon sparger was 26 to 41 percent lower for vertical than for horizontal operation of the sparger, the greatest reduction occurring when the long dimension was vertical.

Precipitation and Generation Methods For a thorough understanding of the phenomena involved, bubble nucleation should be considered. A discussion of nucleation phenomena is beyond the scope of this handbook; however, a starting point with recent references is Deng, Lee, and Cheng, *J. Micromech. Microeng.*, **15,** 564 (2005), and Jones, Evans, and Galvin, *Adv. Colloid and Interface Sci.* **80,** 27 (1999).

Gas Dispersion—Vessel Headspace Boerma and Lankester have measured the surface aeration of a nine-bladed disk-type turbine (NOTE: A well-designed pitched-blade turbine will give equal or better performance). In a fully baffled vessel, the optimum depth to obtain maximum gas dispersion was 15 to 50 percent of the impeller diameter.

TABLE 14-23 Characteristics of Porous Septa

			Air-permeability data		
Grade	Avg. % porosity	Avg. pore diam.	Diaphragm thickness, in	Pressure differential, in water	Air flow, cu ft/(sq ft)(min)
		Alundum porous alumina*			
P2220		25	1	2	0.35
P2120	36	60	1	2	2
P260	35	164	1	2	15
P236	34	240	1	2	40
P216		720	1	2	110
		National porous carbon†			
60	48	33	1	2	
45	48	58	1	2	2
25	48	120	1	2	13
		Filtros porous silica‡			
Extra fine	26.0	55	1.5	2	1–3
Fine	28.8	110	1.5	2	4–8
Medium fine	31.1	130	1.5	2	9–12
Medium	33.7	150	1.5	2	13–20
Medium coarse	33.8	200	1.5	2	21–30
Coarse	34.5	250	1.5	2	31–59
Extra coarse	36.5	300	1.5	2	60–100
		Porous plastic§			
Teflon		9	0.125	1.38	5
Kel-F		15	0.125	1.38	13
		Micro Metallic porous stainless steel§,¶			
H	45	5	0.125	1.38	1.8
G	50	10	0.125	1.38	3
F	50	20	0.125	1.38	5
E	50	35	0.125	1.38	18
D	50	65	0.125	1.38	60
C	55	165	0.125	27.7	990

*Data by courtesy of Norton Co., Worcester, Mass. A number of other grades between the extremes listed are available.
†Data by courtesy of National Carbon Co., Cleveland, Ohio.
‡Data by courtesy of Filtros Inc., East Rochester, N.Y.
§Data by courtesy of Pall Corp., Glen Cove, N.Y.
¶Similar septa made from other metals are available.

In a vessel with baffles extending only halfway to the liquid surface the optimum impeller submergence increased with agitator speed because of the vortex formed. At optimum depth, the following correlation is recommended for larger vessels:

$$Q = 0.00015(N/10)^{2.5}(D/0.1)^{4.5} \quad (14\text{-}214)$$

with Q in m³/s, N in rps, and D in m. Thus, a typical 3-m-diameter plant-size vessel, with four-blade pitched impeller ($D = 1$ m) operating at 2 rps will give gas dispersion from the headspace into the batch of $Q = 0.00015(2/10)^{2.5}(2/0.1)^{4.5} = 1.9$ m³/s (4000 ft³/min).

Borema et al. (op. cit.) recommend headspace gas dispersion with partial baffling for fatty oils hydrogenation in stirred reactors: "The hydrogen in the head-space of the closed reactor can again be brought into contact with the liquid by a stirrer under the liquid level, . . ." And Penney in Paul et al. (loc. cit.) says, "partial baffling can be very effective to produce a vortex, which can effectively drawdown gas from the headspace. . . ." However, Middleton et al. in Paul et al. (op. cit) say, "The simplest self-inducer for an agitated vessel is an impeller located near the surface, sometimes with the upper part of the baffles removed so as to encourage the formation of a vortex. This is, however, a sensitive and unstable arrangement. It is better, although probably more expensive, to use a self-inducing impeller system in which gas is drawn down a hollow shaft to the low-pressure region behind the blades of a suitable, often shrouded impeller. . . . Various proprietary designs are available, such as the Ekato gasjet, Prasair AGR and the Frings Fribo-rator. . . ." Figure 14-97 illustrates a gas-inducing hollow shaft/hollow impeller agitator. In many hydrogenation reactors, the impeller just underneath the free surface has, without any doubt, performed admirably; consequently, one must consider this fact very carefully before using a self-inducing impeller.

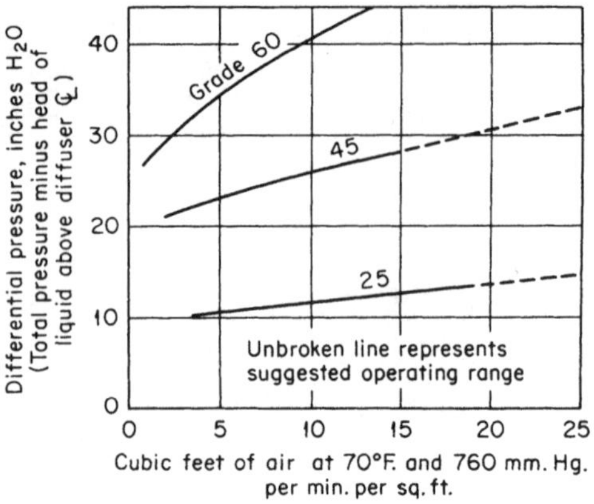

FIG. 14-96 Pressure drop across porous-carbon diffusers submerged in water at 70°F. To convert feet per minute to meters per second, multiply by 0.0051; to convert inches to millimeters, multiply by 25.4; °C = 5⁄9 (°F − 32). (*National Carbon Co.*)

Gas dispersion through the free surface by mechanical aerators is commonplace in aerobic waste-treatment lagoons. Surface aerators are generally of three types: (1) large-diameter flow-speed turbines operating just below the free surface of the liquid, often pontoon-mounted; (2) small-diameter high-speed (normally motor-speed) propellers operating in draft tubes, the units of which are always pontoon-mounted; and (3) hollow-tube turbines (Fig. 14-97). An example of the turbine type is illustrated in Fig. 14-98 and the propeller type is illustrated in Fig. 14-99. There are several other styles of the turbine type; for instance, Mixing Equipment Co., Inc., (www.lightninmixers.com), uses an unshrouded 45° axial-flow turbine [see Dykman and Michel, *Chem. Eng.*, **117** (Mar. 10, 1969)], and Infilco (www.infilcodegremont.com) makes a unit that has a large-diameter vaned disk operating just below the free surface with a smaller-diameter submerged-disk turbine for additional solids suspension.

Aeration injectors, like the one shown in Fig. 14-100 by Penberthy [a division of Houdialle Industries (penberthy-online.com/jet1.asp)], are used to provide mass transfer in gas-liquid applications, and simple impingement aerators (Fig. 14-101) are sometimes used for mass-transfer applications.

Equipment Selection Ideally, selection of equipment to produce a gas-in-liquid dispersion should be made on the basis of a complete economic analysis. The design engineer and especially the pilot-plant engineer seldom have sufficient information or time to do a complete economic analysis. In the following discussion, some

TABLE 14-24 Wet Permeability of Alundum Porous Plates 1 in Thick*

Dry permeability at 2 in of water differential, cu ft/(min)(sq ft)	Pressure differential across wet plate, in of water	Air flow through wet plate, cu ft/(min)(sq ft)
4.3	20.67	2.0
	21.77	3.0
	22.86	4.0
	23.90	5.0
55.0	4.02	1.0
	4.14	2.0
	4.22	3.0
	4.27	4.0
	4.30	5.0

*Data by courtesy of Norton Company, Worcester, Mass. To convert inches to centimeters, multiply by 2.54; to convert feet per minute to meters per second, multiply by 0.0051.

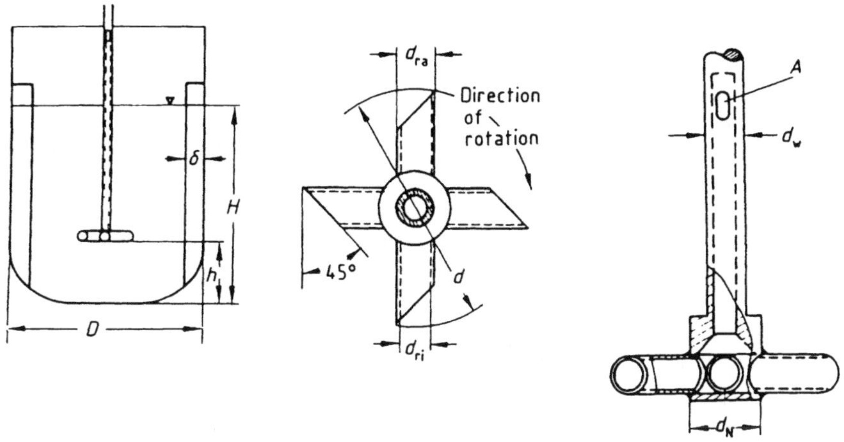

FIG. 14-97 Installation and dimensions of a tube stirrer: $h/d = 1$; $H/D \approx 1$; $D/\delta = 10$; $A = 1.5\ d_w^2$; $D/d_N = 10$; $d/d_N = 3$; $d/d_{ri} = 7.5$; $d/d_{ra} = 6$. [*Zlokarnik,* Ullman's Encyclopedia of Industrial Chemistry, *Sec. 25, VCH, Weinheim, Germany, 1988.*]

guidelines are given as to what equipment might be feasible and what equipment might prove most economical.

For producing foam for foam-separation processes, perforated-plate or porous-plate spargers are normally used. Mechanical agitators are often not effective in the light foams needed in foam fractionation. Dissolved-air flotation, based on the release of a pressurized flow in which oxygen was dissolved, has been shown to be effective for particulate removal when sparged air failed because the bubbles formed upon precipitation are smaller—down to 80 μm—than bubbles possible with sparging, typically 1000 μm [Grieves and Ettelt, *AIChE J.*, **13,** 1167 (1967)]. Mechanically agitated surface aerators such as the Wemco-Fagergren flotation unit (Fig. 14-102) are used extensively for ore flotation.

To produce foam in batch processes, mechanical agitators are used almost exclusively. The gas can either be introduced through the free surface by the entraining action of the impeller or alternatively sparged beneath the impeller. In such batch operation, the liquid level gradually rises as the foam is generated; thus, squatly impellers such as turbines are rapidly covered with foam and must almost always be sparged from below. Tall impellers such as wire whips (Fig. 14-103) are especially well suited to entrain gas from the vapor space. Intermeshing wire whips are standard kitchen utensils for producing foamed meringues, consisting of air, vegetable oil, and egg whites. For a new application, generally some experimentation with different impellers is necessary in order to get the desired fine final bubble size without getting frothing over initially. For producing foams continually, an aspirating venturi nozzle and restrictions in pipes such as baffles and metal gauzes are generally most economical.

For gas absorption, the equipment possibilities are generally packed columns; plate distillation towers, possibly with mechanical agitation on every plate; deep-bed contactors (bubble columns or sparged lagoons); and mechanically agitated vessels or lagoons. Packed towers and plate distillation columns are discussed elsewhere. Generally these devices are used when a relatively large number of stages (more than two or three) is required to achieve the desired result practically.

The volumetric mass-transfer coefficients and heights of transfer units for bubble columns and packed towers have been compared for absorption of CO_2 into water by Houghton et al. [*Chem. Eng. Sci.*, **7,** 26 (1957)]. The bubble column will tolerate much higher vapor velocities, and in the overlapping region (superficial gas velocities of 0.9 to 1.8 cm/s), the bubble column has about three times higher mass-transfer coefficient and about 3 times greater height of transfer unit. The liquid in a bubble column is, for practical purposes, quite well mixed; thus, chemical reactions and component separations requiring significant plug flow of the liquid cannot be effected with bubble columns. Bubble columns and agitated vessels are the ideal equipment for processes in which the fraction of gas absorbed need not be great, possibly the gas can be recycled, and the liquid can or should be well mixed. The gas phase in bubble columns is not nearly so well back-mixed as the liquid, and often plug flow of the gas is a logical assumption, but in agitated vessels the gas phase is also well mixed.

The choice of a bubble column or an agitated vessel depends primarily on the solubility of the gas in the liquid, the corrosiveness of the liquid (often a gas compressor can be made of inexpensive material, whereas a mechanical agitator may have to be made of exotic, expensive materials), and the rate of chemical reaction as compared with the mass-transfer rate. Bubble columns and agitated vessels are seldom used for gas absorption except in chemical reactors. As a general rule, if the overall reaction rate is five times greater than the mass-transfer rate in a simple bubble column, a mechanical agitator will be most economical unless the mechanical agitator would have to be made from considerably more expensive material than the gas compressor.

In bubble columns and simply sparged lagoons, selecting the sparger is a very important consideration. In the turbulent regime (superficial gas velocity greater than 4.6 to 6 cm/s), inexpensive

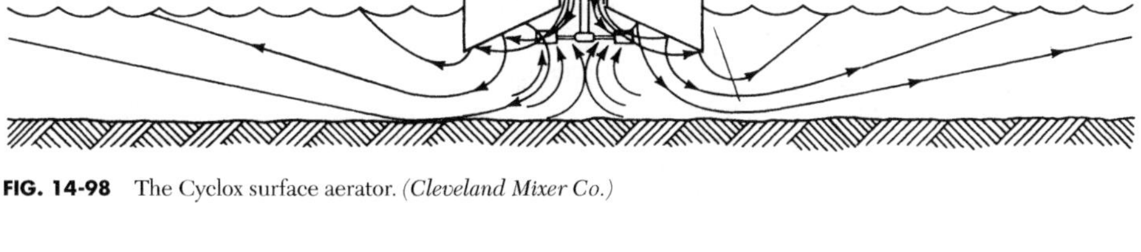

FIG. 14-98 The Cyclox surface aerator. *(Cleveland Mixer Co.)*

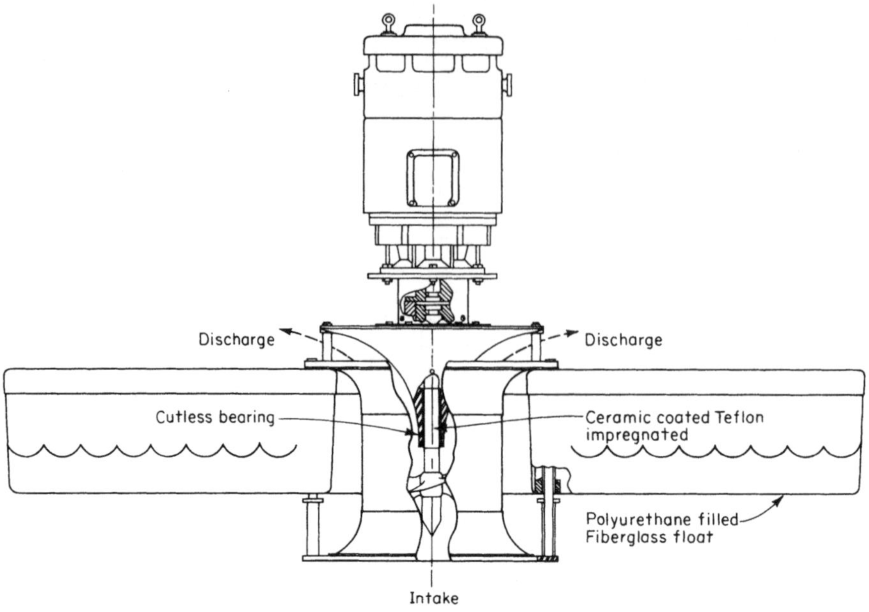

FIG. 14-99 Propeller-type surface aerator. *(Ashbrook-Simon-Hartley Corp.)*

perforated-pipe spargers should be used. Often the holes must be placed on the pipe bottom in order to make the sparger free-draining during operation. In the quiescent regime, porous septa will often give considerably higher overall mass-transfer coefficients than perforated plates or pipes because of the formation of tiny bubbles that do not coalesce. Chain and coworkers (*First International Symposium on Chemical Microbiology,* World Health Organization, Monograph Ser. 10, Geneva, 1952) claimed that porous disks are about twice as effective as open-pipe and ring spargers for the air oxidation of sodium sulfite. Eckenfelder [*Chem. Eng. Progr.,* **52**(7), 290 (1956)] has compared the oxygen-transfer capabilities of various devices on the basis of the operating power required to absorb a given quantity of O_2. The installed cost of the various pieces of equipment probably would not vary sufficiently to warrant being including in an economic analysis. Surface mechanical aerators are not included in this comparison. Of the units compared, it appears that porous tubes give the most efficient power usage. Kalinske (*Adv. Biol. Waste Treatment,* 1963, p. 157) has compared submerged sparged aerators with mechanical surface aerators. He has summarized this comparison in *Water Sewage Works,* **33** (January 1968). He indicates that surface aerators are significantly more efficient than subsurface aeration, both for oxygen absorption and for gas-stripping operations.

Zlokarnik and Mann (paper at Mixing Conf., Rindge, New Hampshire, August 1975) have found the opposite of Kalinske, i.e., subsurface diffusers, subsurface sparged turbines, and surface aerators compare approximately as 4:2:1 respectively in terms of O_2 transfer efficiency; however, Zlokarnik [*Adv. Biochem. Eng.,* **11,** 157 (1979)] later indicates that the scale-up correlation used earlier might be somewhat inaccurate. When all available information is considered, it appears that with near-optimum design any of the aeration systems (diffusers, submerged turbines, or surface impellers) should give a transfer efficiency of at least 2.25 kg O_2/kWh. Thus, the final selection should probably be made primarily on the basis of operational reliability, maintenance, and capital costs.

Mass Transfer Mass transfer in plate and packed gas-liquid contactors has been covered earlier in this subsection. Attention here will be limited to deep-bed contactors (bubble columns and agitated vessels). Theory underlying mass transfer between phases is discussed in Sec. 5 of this handbook.

To design deep-bed contactors for mass-transfer operations, one must have, in general, predictive methods for the following design parameters:

- Flooding (for both columns and agitator impellers)
- Agitator power requirements

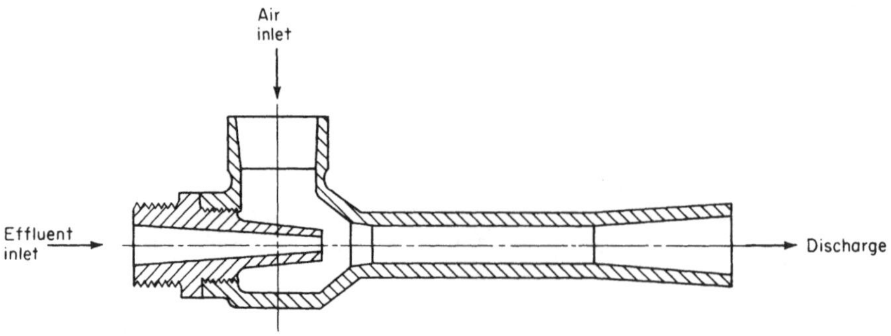

FIG. 14-100 Aeration ejector. *(Penberthy, a division of Houdaille Industries, Inc.)*

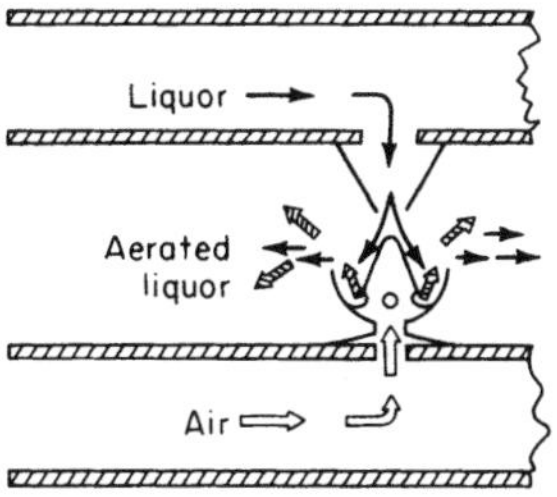

FIG. 14-101 Impingement aerator.

- Gas-phase and liquid-phase mass-transfer coefficients
- Interfacial area
- Interface resistance
- Mean concentration driving force for mass transfer

In most cases, available methods are incomplete or unreliable, and some supporting experimental work is necessary, followed by scale-up. The methods given here should allow theoretical feasibility studies, help minimize experimentation, and permit a measure of optimization in final design.

Flooding of Agitator Impellers Impeller flooding correlations for six-blade disk (6BD) Rusthon turbines and six-blade disk Smith turbines (Chemineer designation: CD-6) are presented by Bakker, Myers, and Smith [*Chem. Eng.*, **101,** 98 (Dec. 1994)] and a review of impeller flooding. The Bakker et al. (loc. cit.) correlation is

$$Q/ND^3 = C_{FL} N_{Fr} (D/T)^{3.5} \qquad (14\text{-}215)$$

where $C_{FL} = 30$ for a 6BD impeller and $C_{FL} = 70$ for a concave blade CD-6 impeller and N_{Fr} = Froude number = N^2D/g; Q = gas flow rate at flooding, m^3/s; N = impeller speed, rps; D = impeller diameter, m; and T = tank diameter, m. Note that the CD-6 impeller will handle 70/30 = 2.33 times the gas a 6BD will handle, without flooding, at the same N and D; this is the great advantage of the CD-6 along with lower power decrease as the gas flow rate increases.

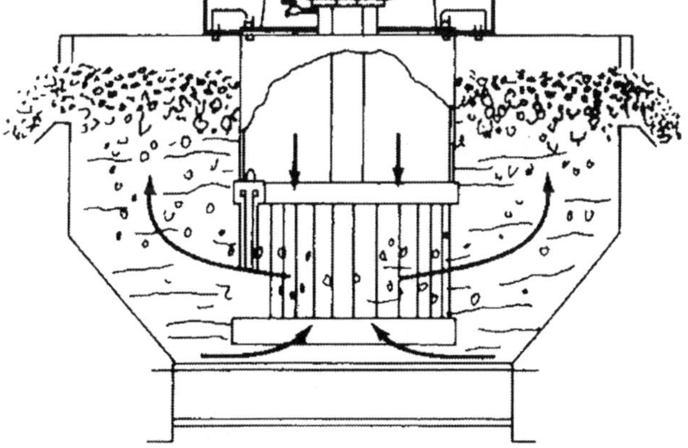

FIG. 14-102 The Wemco-Fagergren flotation machine. [*From www.tucottbus.de/BTU/Fak4/Aufbtech/pages/pbrr_N 7 Sep2-(elstat-flotat).pdf*].

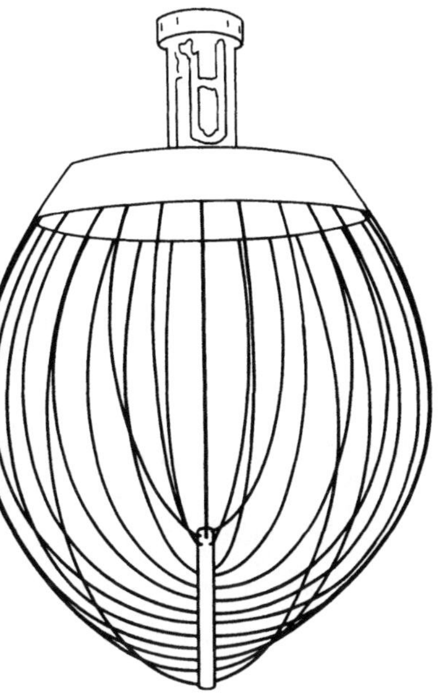

FIG. 14-103 Wire whip.

Gassed Impeller Power Bakker et al. (op. cit.) have given a gassed power correlation for the 6BD and CD-6 impellers.

$$P_g/P_u = [1 - (b - a\mu)]N_{Fr}^d \tanh (cN_A) \qquad (14\text{-}216a)$$

where P_g = gassed power, W; P_u = ungassed power, W; $N_A = Q/ND^3$; and the constants of Eq. (14-216*a*) are given in Table 14-25.

As mentioned previously, the CD-6 suffers much less power decrease with increased gassing compared to the 6BD. For example, at $N_A = 0.15$; $P_g/P_u = 0.7$ for the CD-6 and 0.5 for the CD-6.

The ungassed power is calculated by

$$P_u = N_p \rho / N^3 D^5 \qquad (14\text{-}216b)$$

where the impeller power numbers N_p are given Table 14-25.

Bakker et al. (op. cit.) and Sensel et al. (op. cit.) have given correlations for gas holdup in agitated vessels. The Bakker et al. correlation is

$$\varepsilon = C_\varepsilon = (P_g/V)^A v_{sg}^B \qquad (14\text{-}217)$$

where $C_\varepsilon = 0.16$, $A = 0.33$, $B = 0.67$; V = batch volume, m^3; v_{sg} = superficial gas velocity = $Q/[(\pi/4)T^2]$; T = tank diameter, m. Equation (14-217) applies for both 6BD and CD-6.

Interfacial Area This consideration in agitated vessels has been reviewed and summarized by Tatterson (op. cit.). Predictive methods for interfacial area are not presented here because correlations are given for the overall volumetric mass transfer coefficient liquid phase controlling mass transfer.

TABLE 14-25 Constants in Eq. (14-216) and Impeller Power Numbers

Impeller type	a	b	c	d	N_p
6BD	0.72	0.72	24	0.25	3.2
CD-6	0.12	0.44	12	0.37	5.5

Overall Mass-Transfer Coefficient In systems with relatively sparing soluble gases, where the gas-phase resistance is negligible, the mass-transfer rate can be determined by using the concept of an overall volumetric mass-transfer coefficient k_La as follows:

$$M_s = k_La(C_s^\circ - C_{s,b}) \tag{14-218}$$

where M_s = solute molar mass-transfer rate, kg·mol/s; k_La = overall mass-transfer coefficient, 1/s; C_s° = solute concentration in equilibrium with the liquid phase, kg·mol/s; and $C_{s,b}$ = solute concentration in bulk of liquid.

Bakker et al. (op. cit.) have given a correlation for k_La for aqueous systems in the absence of significant surface active agents.

$$k_La = C_{kLa}(P_g/V)^a v_{sg}^b \tag{14-219}$$

where C_{kLa} = 0.015, 1/s; Eq. (14-219) applies for both 6BD and CD-6.

Interfacial Phenomena These can significantly affect overall mass transfer. Deckwer, *Bubble Column Reactors*, Wiley, Hoboken, N.J., 1992, has covered the effect of surfactants on mass transfer in bubble columns. In fermentation reactors, small quantities of surface-active agents (especially antifoaming agents) can drastically reduce overall oxygen transfer (Aiba et al., op. cit., pp. 153, 154), and in aerobic mechanically aerated waste-treatment lagoons, overall oxygen transfer has been found to be from 0.5 to 3 times that for pure water from tests with typical sewage streams (Eckenfelder et al., op. cit., p. 105).

One cannot quantitatively predict the effect of the various interfacial phenomena; thus, these phenomena will not be covered in detail here. The following literature gives a good general review of the effects of interfacial phenomena on mass transfer: Goodridge and Robb, *Ind. Eng. Chem. Fund.*, **4,** 49 (1965); Calderbank, *Chem. Eng. (London),* CE 205 (1967); Gal-Or et al., *Ind. Eng. Chem.*, **61**(2), 22 (1969); Kintner, *Adv. Chem. Eng.*, **4** (1963); Resnick and Gal-Or, op. cit., p. 295; Valentin, loc. cit.; and Elenkov, loc. cit., and *Ind. Eng. Chem. Ann. Rev. Mass Transfer,* **60**(1), 67 (1968); **60**(12), 53 (1968); **62**(2), 41 (1970). In the following outline, the effects of the various interfacial phenomena on the factors that influence overall mass transfer are given. Possible effects of interfacial phenomena are tabulated below:

1. Effect on continuous-phase mass-transfer coefficient
   a. Impurities concentrate at interface. Bubble motion produces circumferential surface-tension gradients that act to retard circulation and vibration, thereby decreasing the mass-transfer coefficient.
   b. Large concentration gradients and large heat effects (very soluble gases) can cause interfacial turbulence (the Marangoni effect), which increases the mass-transfer coefficient.
2. Effect on interfacial area
   a. Impurities will lower static surface tension and give smaller bubbles.
   b. Surfactants can electrically charge the bubble surface (produce ionic bubbles) and retard coalescence (soap stabilization of an oil-water emulsion is an excellent example of this phenomenon), thereby increasing the interfacial area.
   c. Large concentration gradients and large heat effects can cause bubble breakup.
3. Effect on mean mass-transfer driving force
   a. Relatively insoluble impurities concentrate at the interface, giving an interfacial resistance. This phenomenon has been used in retarding evaporation from water reservoirs.
   b. The axial concentration variation can be changed by changes in coalescence. The mean driving force for mass transfer is therefore changed.

Gas Holdup (ε) in Bubble Columns With coalescing systems, holdup may be estimated from a correlation by Hughmark [*Ind. Eng Chem. Process Des. Dev.*, **6,** 218–220 (1967)] reproduced here as Fig. 14-104. For noncoalescing systems, with considerably smaller bubbles, ε can be as great as 0.6 at U_{sg} = 0.05 m/s, according to Mersmann [*Ger. Chem. Eng.*, **1,** 1 (1978)].

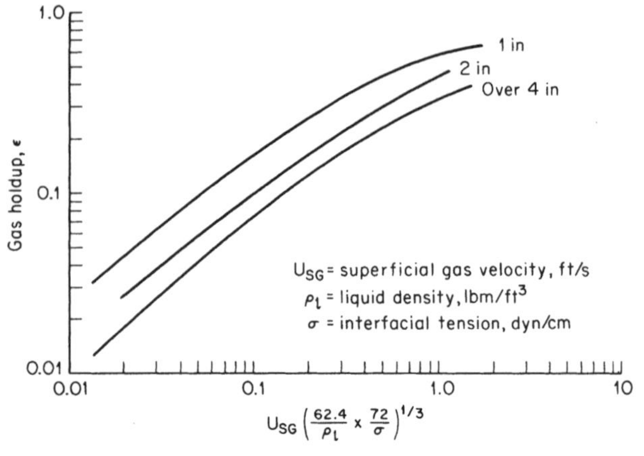

FIG. 14-104 Gas holdup correlation. [Ind. Eng. Chem. Process Des. Dev., **6,** 218 (*1967*).]

It is often helpful to use the relationship between ε and superficial gas velocity (U_{sg}) and the rise velocity of a gas bubble relative to the liquid velocity ($U_r + U_L$, with U_L defined as positive upward):

$$\varepsilon = \frac{U_{sg}}{U_r + U_L} \tag{14-220}$$

Rise velocities of bubbles through liquids have been discussed previously.

For a better understanding of the interactions between parameters, it is often helpful to calculate the effective bubble rise velocity U_r from measured valves of ε; for example, the data of Mersmann (loc. cit.) indicated ε = 0.6 for U_{sg} = 0.05 m/s, giving U_r = 0.083 m/s, which agrees with the data reported in Fig. 14-43 for the rise velocity of bubble clouds. The rise velocity of single bubbles, for d_b ~ 2 mm, is about 0.3 m/s, for liquids with viscosities not too different from water. Using this value in Eq. (14-220) and comparing with Fig. 14-104, one finds that at low values of U_{sg}, the rise velocity of the bubbles is less than the rise velocity of a single bubble, especially for small-diameter tubes, but that the opposite occurs for large values of U_{sg}.

More recent literature regarding generalized correlational efforts for gas holdup is adequately reviewed by Tsuchiya and Nakanishi [*Chem. Eng Sci.*, **47**(13/14), 3347 (1992)] and Sotelo et al. [*Int. Chem. Eng.*, **34**(1), 82–90 (1994)]. Sotelo et al. (op. cit.) have developed a dimensionless correlation for gas holdup that includes the effect of gas and liquid viscosity and density, interfacial tension, and diffuser pore diameter. For systems that deviate significantly from the waterlike liquids for which Fig. 14-104 is applicable, their correlation (the fourth numbered equation in the paper) should be used to obtain a more accurate estimate of gas holdup. Mersmann (op. cit.) and Deckwer et al. (op. cit.) should also be consulted.

Liquid-phase mass-transfer coefficients in bubble columns have been reviewed by Calderbank ("Mixing," loc. cit.), Fair (*Chem. Eng.*, loc. cit.), Mersmann [*Ger. Chem. Eng.* **1,** 1 (1978), *Int. Chem. Eng.*, **32**(3) 397–405 (1991)], Deckwer et al. [*Can. J. Chem. Eng*, **58,** 190 (1980)], Hikita et al. [*Chem. Eng. J.*, **22,** 61 (1981)] and Deckwer and Schumpe [*Chem. Eng. Sci.*, **48**(5), 889–911 (1993)]. The correlation of Ozturk, Schumpe, and Deckwer [*AIChE J.*, **33,** 1473–1480 (1987)] is recommended. Deckwer et al. (op. cit.) have documented the case for using the correlation:

Ozturk et al. (1987) developed a new correlation on the basis of a modification of the Akita-Yoshida correlation suggested by Nakanoh and Yoshida (1980). In addition, the bubble diameter d_b rather than the column diameter was used as the characteristic length as the column diameter has little influence on k_La. The value of d_b was assumed to be approximately constant (d_b = 0.003 m). The correlation was obtained by nonlinear regression is as follows:

$$\left(\frac{k_L a\, d_B}{D_L}\right) = 0.62\left[\frac{\mu_L}{(\rho_L D_L)}\right]^{0.5}\left(\frac{g\rho_L d_B^2}{\sigma}\right)^{0.33}\left(\frac{g\rho_L^2 d_B}{\mu_L^2}\right)^{0.29} \times \left[\frac{U_G}{(g d_B)^{0.5}}\right]^{0.68}\left(\frac{\rho_G}{\rho_L}\right)^{0.04} \quad (14\text{-}221)$$

where $k_L a$ = overall mass-transfer coefficient, d_B = bubble diameter = 0.003 m, D_L = diffusivity of gas in liquid, ρ = density, μ = viscosity, σ = interfacial tension, g = gravitational acceleration.

As mentioned earlier, surfactants and ionic solutions significantly affect mass transfer. Normally, surface affects act to retard coalescence and thus increase the mass transfer. For example, Hikata *et al.* [*Chem. Eng. J.*, **22,** 61–69 (1981)] have studied the effect of KCl on mass transfer in water. As KCI concentration increased, the mass transfer increased up to about 35 percent at an ionic strength of 6 gm/l. Other investigators have found similar increases for liquid mixtures.

Axial Dispersion Backmixing in bubble columns has been extensively studied. Wiemann and Mewes [*Ind. Eng. Chem. Res.*, **44,** 4959 (2005)] and Wild et al. [*Int. J. Chemical Reactor Eng.*, **1,** R7 (2003)] give a long list of references pertaining to backmixing in bubble columns. An excellent review article by Shah et al. [*AIChE J.*, **24,** 369 (1978)] has summarized the literature prior to 1978. Works by Konig et al. [*Ger. Chem. Eng.*, **1,** 199 (1978)], Lucke et al. [*Trans. Inst. Chem. Eng.*, **58,** 228 (1980)], Riquarts [*Ger. Chem. Eng.*, **4,** 18 (1981)], Mersmann (op. cit.), Deckwer (op. cit.), Yang et al. [*Chem. Eng. Sci.*, **47**(9–11), 2859 (1992)], and Garcia-Calvo and Leton [*Chem. Eng. Sci.*, **49**(21), 3643 (1994)] are particularly useful references.

Axial dispersion occurs in both the liquid and the gas phases. The degree of axial dispersion is affected by vessel diameter, vessel internals, gas superficial velocity, and surface-active agents that retard coalescence. For systems with coalescence-retarding surfactants the initial bubble size produced by the gas sparger is also significant. The gas and liquid physical properties have only a slight effect on the degree of axial dispersion, except that liquid viscosity becomes important as the flow regime becomes laminar. With pure liquids, in the absence of coalescence-inhibiting, surface-active agents, the nature of the sparger has little effect on the axial dispersion, and experimental results are reasonably well correlated by the dispersion model. For the liquid phase the correlation recommended by Deckwer et al. (op. cit.), after the original work by Baird and Rice [*Chem. Eng. J.*, **9,** 171(1975)] is as follows:

$$\frac{E_L}{(DU_G)} = 0.35\left(\frac{gD}{U_G^2}\right)^{1/3} \quad (14\text{-}222)$$

where E_L = liquid-phase axial dispersion coefficient, U_G = superficial velocity of the gas phase, D = vessel diameter, and g = gravitational acceleration.

The recommended correlation for the gas-phase axial-dispersion coefficient is given by Field and Davidson (loc. cit.):

$$E_G = 56.4\, D^{1.33}\left(\frac{U_G}{\varepsilon}\right)^{3.56} \quad (14\text{-}223)$$

where E_G = gas-phase axial-dispersion coefficient, m²/s; D = vessel diameter, m; U_G = superficial gas velocity, m/s; and ε = fractional gas holdup, volume fraction.

The correlations given in the preceding paragraphs are applicable to vertical cylindrical vessels with pure liquids without coalescence inhibitors. For other vessel geometries such as columns of rectangular cross section, packed columns, and coiled tubes, the work of Shah et al. (loc. cit.) should be consulted. For systems containing coalescence-inhibiting surfactants, axial dispersion can be vastly different from that in systems in which coalescence is negligible. Konig et al. (loc. cit.) have well demonstrated the effects of surfactants and sparger type by conducting tests with weak alcohol solutions using three different porous spargers. With pure water, the sparger—and, consequently, initial bubble size—had little effect on back mixing because coalescence produced a dynamic-equilibrium bubble size not far above the sparger. With surfactants, the average bubble size was smaller than the dynamic-equilibrium bubble size. Small bubbles produced minimal back mixing up to $\varepsilon \approx 0.40$; however, above $\varepsilon \approx 0.40$ backmixing increased very rapidly as U_G increased The rapid increase in back mixing as ε exceeds 0.40 was postulated to occur indirectly because a bubble carries upward with it a volume of liquid equal to about 70 percent of the bubble volume, and, for $\varepsilon \approx 0.40$, the bubbles carry so much liquid upward that steady, uniform bubble rise can no longer be maintained and an oscillating, slugging flow develops, which produces fluctuating pressure at the gas distributor and the formation of large eddies. The large eddies greatly increase backmixing. For the air alcohol-water system, the minimum bubble size to prevent unsteady conditions was about 1, 1.5, and 2 mm for U_G = 1, 3, and 5 cm/s, respectively. Any smaller bubble size produced increased backmixing. The results of Konig et al. (loc. cit.) clearly indicate that the interaction of surfactants and sparger can be very complex; thus, one should proceed very cautiously in designing systems for which surfactants significantly retard coalescence. Caution is particularly important because surfactants can produce either much more or much less backmixing than surfactant-free systems, depending on the bubble size, which, in turn, depends on the sparger utilized.

PHASE SEPARATION

Gases and liquids may be intentionally contacted as in absorption and distillation, or a mixture of phases may occur unintentionally as in vapor condensation from inadvertent cooling or liquid entrainment from a film. Regardless of the origin, it is usually desirable or necessary ultimately to separate gas-liquid dispersions. While separation will usually occur naturally, the rate is often economically intolerable and separation processes are employed to accelerate the step.

GAS-PHASE CONTINUOUS SYSTEMS

Practical separation techniques for liquid particles in gases are discussed. Since gas-borne particulates include both liquid and solid particles, many devices used for dry-dust collection (discussed in Sec. 17 under "Gas-Solids Separation") can be adapted to liquid-particle separation. Also, the basic subject of particle mechanics is covered in Sec. 6. Separation of liquid particulates is frequently desirable in chemical processes such as in countercurrent-stage contacting because liquid entrainment with the gas partially reduces true countercurrency. Separation before entering another process step may be needed to prevent corrosion, to prevent yield loss, or to prevent equipment damage or malfunction. Separation before the atmospheric release of gases may be necessary to prevent environmental problems and for regulatory compliance.

GENERAL REFERENCES

G-1. Buonicore and Davis, eds., *Air Pollution Engineering Manual,* Van Nostrand Reinhold, New York, 1992.
G-2. Calvert and Englund, eds., *Handbook of Air Pollution Technology,* Wiley, New York, 1984.
G-3. Cheremisinoff, ed., *Encyclopedia of Environmental Control Technology,* vol. 2, Gulf Pub., Houston, 1989.
G-4. McKetta, *Unit Operations Handbook,* vol. 1–2, Dekker, New York, 1992.
G-5. Wark and Warner, *Air Pollution: Its Origin and Control,* 2d ed., Harper & Row, New York, 1981.
G-6. Hesketh, *Air Pollution Control,* 1979; *Fine Particles in Gaseous Media,* Ann Arbor Science Pubs., Ann Arbor, MI, 1977.
G-7. Stern, *Air Pollution,* 3d ed., vols. 3–5, Academic, Orlando, FL, 1976–77.

G-8. Strauss, *Industrial Gas Cleaning,* 2d ed., Pergamon, New York, 1975.
G-9. Theodore and Buonicore, *Air Pollution Control Equipment; Selection, Design, Operation and Maintenance,* Prentice Hall, Englewood Cliffs, NJ, 1982.
G-10. Wang and Pereira, eds., *Handbook of Environmental Engineering,* vol. 1, Humana, Clifton, NJ 1979.
G-11. Cheremisinoff and Young, *Air Pollution Control and Design Handbook,* parts 1–2, Dekker, New York, 1977.
G-12. Nonhebel, *Gas Purification Processes for Air Pollution Control,* Newnes-Butterworth, London, 1972.

Sampling
R-1. *Code of Federal Regulations,* 40 (CFR 40), subchapter C—Air Programs, parts 50–99, Office of the Federal Register, Washington.
R-2. Ref. G-11, part 1, pp. 65–121.
R-3. Cooper and Rossano, *Source Testing for Air Pollution Control,* Environmental Science Services, Wilton, Connecticut, 1970.
R-4. Ref. G-7, vol. 3, pp. 525–587.
R-5. *Methods of Air Sampling and Analysis,* 2d Ed., American Public Health Assoc., Washington, 1977.
R-6. Stockham and Fochtman, *Particle Size Analysis,* Ann Arbor Science Pubs., Ann Arbor, Michigan, 1977.
R-7. Ref. G-2, Ch. 31, pp. 785–832.
R-8. Ref. G-8, Ch. 2, pp. 39–79.

Specific
R-9. Calvert, Goldchmid, Leith, and Mehta, NTIS Publ. PB-213016, 213017, 1972.
R-10. Calvert, *J. Air Pollut. Control Assoc.* **24,** 929 (1974).
R-11. Calvert, *Chem. Eng.,* **84**(18), 54 (1977).
R-12. Calvert, Yung, and Leung, NTIS Publ. PB-248050, 1975.
R-13. Calvert and Lundgren, *J. Air Pollut. Control Assoc.,* **18,** 677 (1968).
R-14. Calvert, Lundgren, and Mehta, *J. Air Pollut. Control Assoc.,* **22,** 529 (1972).
R-15. Yung, Barbarika, and Calvert, *J. Air Pollut. Control Assoc.,* **27,** 348, (1977).
R-16. Katz, M.S. thesis, Pennsylvania State University, 1958.
R-17. York and Poppele, *Chem. Eng. Prog.,* **59**(6), 45 (1963).
R-18. York, *Chem. Eng. Prog.,* **50,** 421 (1954).
R-19. Ref. G-2, Ch. 10, pp. 215–248.
References with the notation (R-) are cited in the text.

Definitions: Mist and Spray Little standardization has been adopted in defining gas-borne liquid particles, and this frequently leads to confusion in the selection, design, and operation of collection equipment. Aerosol applies to suspended particulate, either solid or liquid, which is slow to settle by gravity and to particles from the submicrometer range up to 10 to 20 μm. Mists are fine suspended liquid dispersions usually resulting from condensation and ranging upward in particle size from around 0.1 μm. Spray refers to entrained liquid droplets. The droplets may be entrained from atomizing processes previously discussed under "Liquid-in-Gas Dispersions" in this section. In such instances, size will range from the finest particles produced up to a particle whose terminal settling velocity is equal to the entraining gas velocity if some settling volume is provided. Process spray is often created unintentionally, such as by the condensation of vapors on cold duct walls and its subsequent reentrainment, or from two-phase flow in pipes, gas bubbling through liquids, and entrainment from boiling liquids. Entrainment size distribution from sieve trays has been given by Cheng and Teller [*Am. Inst. Chem. Eng. J.,* **7**(2), 282 (1961)] and evaporator spray by Garner et al. [*Trans. Inst. Chem. Eng.,* **32,** 222 (1954)]. In general, spray can range downward in particle size from 5000 μm. There can be overlapping in size between the coarsest mist particles and the finest spray particles, but some authorities have found it convenient arbitrarily to set a boundary of 10 μm between the two. Actually, considerable overlap exists in the region of 5 to 40 μm. Table 14-26 lists typical ranges of particle size created by different mechanisms. The sizes actually entrained can be influenced by the local gas velocity. Figure 14-105 compares the approximate size range of liquid particles with other particulate material and the approximate applicable size range of collection devices. Figure 17-34 gives an expanded chart by Lapple for solid particles. Mist and fog formation has been discussed previously.

Gas Sampling The sampling of gases containing mists and sprays may be necessary to obtain data for collection-device design, in which case particle-size distribution, total mass loading, and gas volume, temperature, pressure, and composition may all be needed. Other reasons for sampling may be to determine equipment performance, measure yield loss, or determine compliance with regulations.

Location of a sample probe in the process stream is critical especially when larger particles must be sampled. Mass loading in one portion of a duct may be severalfold greater than in another portion as affected by flow patterns. Therefore, the stream should be sampled at a number of points. The U.S. Environmental Protection Agency (R-1) has specified 8 points for ducts between 0.3 and 0.6 m (12 and 24 in) and 12 points for larger ducts, provided there are no flow disturbances for eight pipe diameters upstream and two downstream from the sampling point. When only particles smaller than 3 μm are to be sampled, location and number of sample points are less critical since such particles remain reasonably well dispersed by brownian motion. However, some gravity settling of such particles and even gases of high density have been observed in long horizontal breeching. Isokinetic sampling (velocity at the probe inlet is equal to local duct velocity) is required to get a representative sample of particles larger than 3 μm (error is small for 4- to 5-μm particles). Sampling methods and procedures for mass loading have been developed (R-1 through R-8).

Particle Size Analysis Many particle-size-analysis methods suitable for dry-dust measurement are unsuitable for liquids because of coalescence and drainage after collection. Measurement of particle sizes in the flowing aerosol stream by using a cascade impactor is one

TABLE 14-26 Particle Sizes Produced by Various Mechanisms

Mechanism or process	Particle-size range, μm
Liquid pressure spray nozzle	100–5000
Gas-atomizing spray nozzle	1–100
Gas bubbling through liquid or boiling liquid	20–1000
Condensation processes with fogging	0.1–30
Annular two-phase flow in pipe or duct	10–2000

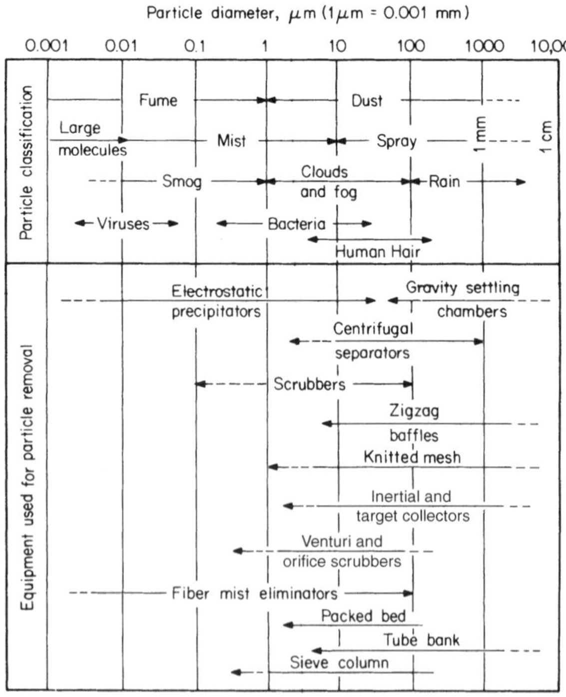

FIG. 14-105 Particle classification and useful collection equipment versus particle size.

of the better means. The impacting principle has been described by Ranz and Wong [*Ind. Eng. Chem.*, **44,** 1371 (1952)] and Gillespie and Johnstone [*Chem. Eng. Prog.*, **51,** 75F (1955)]. The Andersen, Sierra, and University of Washington impactors may be used if the sampling period is kept short so as not to saturate the collection substrate. An impactor designed specifically for collecting liquids has been described by Brink, Kennedy, and Yu [*Am. Inst. Chem. Eng. Symp. Ser.*, **70**(137), 333 (1974)].

Collection Mechanisms Mechanisms which may be used for separating liquid particles from gases are (1) gravity settling, (2) inertial (including centrifugal) impaction, (3) flow-line interception, (4) diffusional (brownian) deposition, (5) electrostatic attraction, (6) thermal precipitation, (7) flux forces (thermophoresis, diffusiophoresis, Stefan flow), and (8) particle agglomeration (nucleation) techniques. Equations and parameters for these mechanisms are given in Table 17-2. Most collection devices rarely operate solely with a single mechanism, although one mechanism may so predominate that it may be referred to, for instance, as an inertial-impaction device.

After collection, liquid particles coalesce and must be drained from the unit, preferably without reentrainment. Calvert (R-12) has studied the mechanism of reentrainment in a number of liquid-particle collectors. Four types of reentrainment were typically observed: (1) transition from separated flow of gas and liquid to a two-phase region of separated-entrained flow, (2) rupture of bubbles, (3) liquid creep on the separator surface, and (4) shattering of liquid droplets and splashing. Generally, reentrainment increased with increasing gas velocity. Unfortunately, in devices collecting primarily by centrifugal and inertial impaction, primary collection efficiency increases with gas velocity; thus overall efficiency may go through a maximum as reentrainment overtakes the incremental increase in efficiency. Prediction of collection efficiency must consider both primary collection and reentrainment.

Procedures for Design and Selection of Collection Devices Calvert and coworkers (R-9 to R-12 and R-19) have suggested useful design and selection procedures for particulate-collection devices in which direct impingement and inertial impaction are the most significant mechanisms. The concept is based on the premise that the mass median aerodynamic particle diameter d_{p50} is a significant measure of the difficulty of collection of the liquid particles and that the collection device cut size d_{pc} (defined as the aerodynamic particle diameter collected with 50 percent efficiency) is a significant measure of the capability of the collection device. The aerodynamic diameter for a particle is the diameter of a spherical particle (with an arbitrarily assigned density of 1 g/cm^3) which behaves in an air stream in the same fashion as the actual particle. For real spherical particles of diameter d_p, the equivalent aerodynamic diameter d_{pa} can be obtained from the equation $d_{pa} = dp(\rho_p C')^{1/2}$, where ρ_p is the apparent particle density (mass/volume) and C' is the Stokes-Cunningham correction factor for the particle size, all in consistent units. If particle diameters are expressed in micrometers, ρ_p can be in grams per cubic centimeter and C' can be approximated by $C' = 1 + A_c(2\lambda/D_p)$, where A_c is a constant dependent upon gas composition, temperature, and pressure ($A_c = 0.88$ for atmospheric air at 20°C) and λ is the mean free path of the gas molecules ($\lambda = 0.10$ μm for 20°C atmospheric air). For other temperatures or pressures, or gases other than air, calculations using these more precise equations may be made: $A_c = 1.257 + 0.4 \exp [-1.1 (d_p/2\lambda)]$ and $\lambda = \mu_g/0.499\rho_g \times \mu_m$ (where μ_g is the gas viscosity, kg/m·h; p_g is gas density, g/cm^3; and μ_m is the mean molecular speed, m/s. $u_m = [8R_uT/\pi M]^{0.5}$, where R_u is the universal gas constant, 8.315 kJ/kg·mol·K; T is the gas absolute temperature, K; and M is the molar mass or equivalent molecular weight of the gas. (π is the usual geometric constant.) For test purposes (air at 25°C and 1 atm), $p_g = 1.183$ kg/m, $\mu_g = 0.0666$ kg/m·h, $\lambda = 0.067$ μm, and $u_m = 467$ m/s. For airborne liquid particles, the assumption of spherical shape is reasonably accurate, and ρ_p is approximately unity for dilute aqueous particles at ambient temperatures. C' is approximately unity at ambient conditions for such particles larger than 1 to 5 μm, so that often the actual liquid particle diameter and the equivalent aerodynamic diameter are identical.

When a distribution of particle sizes which must be collected is present, the actual size distribution must be converted to a mass distribution by aerodynamic size. Frequently the distribution can be represented or approximated by a log-normal distribution (a straight line on a log-log plot of cumulative mass percent of particles versus diameter) which can be characterized by the mass median particle diameter d_{p50} and the standard statistical deviation of particles from the median σ_g. σ_g can be obtained from the log-log plot by $\sigma_g = D_{pa50}/D_{pe}$ at 15.87 percent $= D_{pe}$ at 84.13 percent$/D_{pa50}$.

The grade efficiency η of most collectors can be expressed as a function of the aerodynamic particle size in the form of an exponential equation. It is simpler to write the equation in terms of the particle penetration P_t (those particles not collected), where the fractional penetration $P_t = 1 - \eta$, when η is the fractional efficiency. The typical collection equation is

$$P_t = e^{(-A_a D_{pa} B)} \tag{14-224}$$

where A_a and B are functions of the collection device. Calvert (R-12) has determined that for many devices in which the primary collection mechanism is direct interception and inertial impaction, such as packed beds, knitted-mesh collectors, zigzag baffles, target collectors such as tube banks, sieve-plate columns, and venturi scrubbers, the value of B is approximately 2.0. For cyclonic collectors, the value of B is approximately 0.67. The overall integrated penetration $\bar{P}_t$ for a device handling a distribution of particle sizes can be obtained by

$$\bar{P}_t = \int_0^W \left(\frac{dW}{W}\right) P_t \tag{14-225}$$

where (dW/W) is the mass of particles in a given narrow size distribution and P_t is the average penetration for that size range. When the particles to be collected are log-normally distributed and the collection device efficiency can be expressed by Eq. (14-224), the required overall integrated collection efficiency $\bar{P}_t$ can be related to the ratio of the device aerodynamic cut size D_{pc} to the mass median aerodynamic particle size D_{pa50}. This required ratio for a given distribution and collection is designated R_{rL} and these relationships are illustrated graphically in Fig. 14-106. For the many devices for which B is approximately 2.0, a simplified plot (Fig. 14-107) is obtained. From these figures, by knowing the desired overall collection efficiency and particle distribution, the value of R_{rL} can be read. Substituting the mass median particle diameter gives the aerodynamic cut size

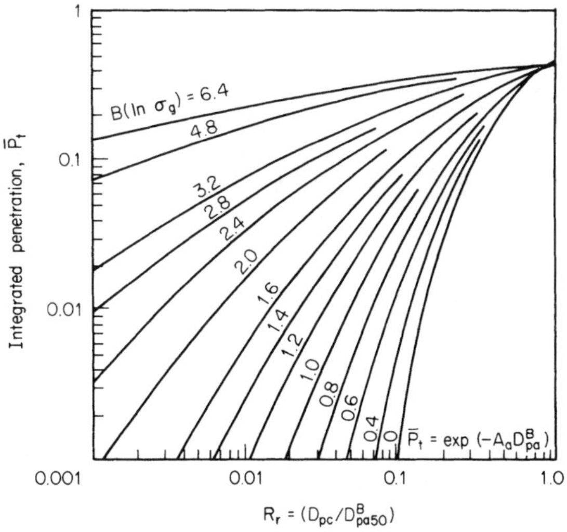

FIG. 14-106 Overall integrated penetration as a function of particle-size distribution and collector parameters. (*Calvert, Yung, and Leung, NTIS Publ. PB-248050, 1975.*)

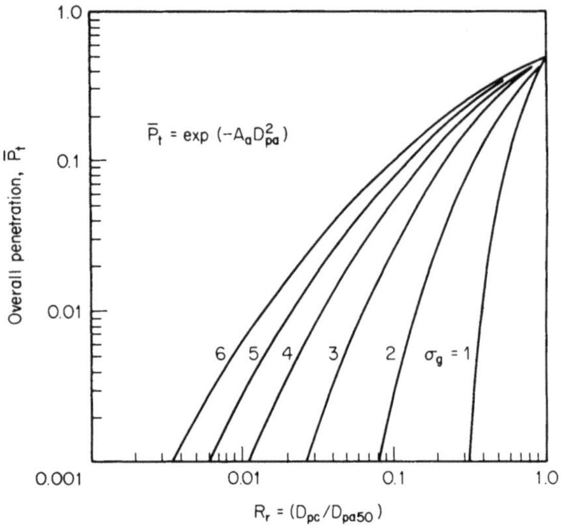

FIG. 14-107 Overall integrated penetration as a function of particle-size distribution and collector cut diameter when $B = 2$ in Eq. (14-224). (*Calvert, Goldshmid, Leith, and Mehta, NTIS Publ. PB-213016, 213017, 1972.*)

required from the collection device being considered. Therefore, an experimental plot of aerodynamic cut size for each collection device versus operating parameters can be used to determine the device suitability.

Collection Equipment

Gravity Settlers Gravity can act to remove larger droplets. Settling or disengaging space above aerated or boiling liquids in a tank or spray zone in a tower can be very useful. If gas velocity is kept low, all particles with terminal settling velocities (see Sec. 6) above the gas velocity will eventually settle. Increasing vessel cross section in the settling zone is helpful. Terminal velocities for particles smaller than 50 μm are very low and generally not attractive for particle removal. Laminar flow of gas in long horizontal paths between trays or shelves on which the droplets settle is another effective means of employing gravity. Design equations are given in Sec. 17 under "Gas-Solids Separations." Settler pressure drop is very low, usually being limited to entrance and exit losses.

Centrifugal Separation Centrifugal force can be utilized to enhance particle collection to several hundredfold that of gravity. The design of cyclone separators for dust removal is treated in detail in Sec. 17 under "Gas-Solids Separations," and typical cyclone designs are shown in Fig. 17-43. Dimension ratios for one family of cyclones are given in Fig. 17-36. Cyclones, if carefully designed, can be more efficient on liquids than on solids since liquids coalesce on capture and are easy to drain from the unit. However, some precautions not needed for solid cyclones are necessary to prevent reentrainment.

Tests by Calvert (R-12) show high primary collection efficiency on droplets down to 10 μm and in accordance with the efficiency equations of Leith and Licht [*Am. Inst. Chem. Eng. Symp. Ser.*, 68(126), 196–206 (1972)] for the specific cyclone geometry tested if entrainment is avoided. Typical entrainment points are (1) creep along the gas outlet pipe, (2) entrainment by shearing of the liquid film from the walls, and (3) vortex pickup from accumulated liquid in the bottom (Fig. 14-108*a*). Reentrainment from creep of liquid along the top of the cyclone and down the outlet pipe can be prevented by providing the outlet pipe with a flared conical skirt (Fig. 14-108*b*), which provides a point from which the liquid can drip without being caught in the outlet gas. The skirt should be slightly shorter than the gas outlet pipe but extend below the bottom of the gas inlet. The cyclone inlet gas should not impinge on this skirt. Often the bottom edge of the skirt is V-notched or serrated.

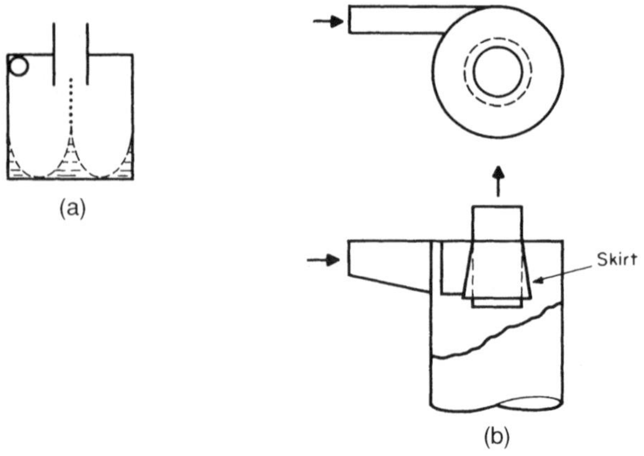

FIG. 14-108 (*a*) Liquid entrainment from the bottom of a vessel by centrifugal flow. (*Rietema and Verver,* Cyclones in Industry, *Elsevier, Amsterdam, 1961.*) (*b*) Gas-outlet skirt for liquid cyclones. (*Stern et al.,* Cyclone Dust Collectors, *American Petroleum Institute, New York, 1955.*)

Reentrainment is generally reduced by lower inlet gas velocities. Calvert (R-12) reviewed the literature on predicting the onset of entrainment and found that of Chien and Ibele (ASME Pap. 62-WA170) to be the most reliable. Calvert applies their correlation to a liquid Reynolds number on the wall of the cyclone, $N_{Re,L} = 4Q_L/h_i\upsilon_L$, where Q_L is the volumetric liquid flow rate, cm³/s; h_i is the cyclone inlet height, cm; and υ_L is the kinematic liquid viscosity, cm²/s. He finds that the onset of entrainment occurs at a cyclone inlet gas velocity V_{ci}, m/s, in accordance with the relationship in $V_{ci} = 6.516 - 0.2865 \ln N_{Re,L}$.

Reentrainment from the bottom of the cyclone can be prevented in several ways. If a typical long-cone dry cyclone is used and liquid is kept continually drained, vortex entrainment is unlikely. However, a vortex breaker baffle in the outlet is desirable, and perhaps a flat disk on top extending to within 2 to 5 cm (0.8 to 2 in) of the walls may be beneficial. Often liquid cyclones are built without cones and have dished bottoms. The modifications described earlier are definitely needed in such situations. Stern, Caplan, and Bush (*Cyclone Dust Collectors,* American Petroleum Institute, New York, 1955) and Rietema and Verver (in Tengbergen, *Cyclones in Industry,* Elsevier, Amsterdam, 1961, chap. 7) have discussed liquid-collecting cyclones.

As with dust cyclones, no reliable pressure-drop equations exist (see Sec. 17), although many have been published. A part of the problem is that there is no standard cyclone geometry. Calvert (R-12) experimentally obtained $\Delta P = 0.000513\ \rho_g(Q_g/h_iW_i)^2(2.8h_iw_i/d_o^2)$, where ΔP is in cm of water; ρ_g is the gas density, g/cm³; Q_g is the gas volumetric flow rate, cm³/s; h_i and w_i are cyclone inlet height and width respectively, cm; and d_o is the gas outlet diameter, cm. This equation is in the same form as that proposed by Shepherd and Lapple [*Ind. Eng. Chem.*, **31,** 1246 (1940)] but gives only 37 percent as much pressure drop.

Liquid cyclone efficiency can be improved somewhat by introducing a coarse spray of liquid in the cyclone inlet. Large droplets which are easily collected collide with finer particles as they sweep the gas stream in their travel to the wall. (See subsection "Wet Scrubbers" regarding optimum spray size.) Cyclones may also be operated wet to improve their operation on dry dust. Efficiency can be improved through reduction in entrainment losses since the dust particles become trapped in the water film. Collision between droplets and dust particles aids collection, and adequate irrigation can eliminate problems of wall buildup and fouling. The most effective operation is obtained by spraying countercurrently to the gas flow in the cyclone inlet duct at liquid rates of 0.7 to 2.0 L/m³ of gas. There are also many proprietary designs of liquid separators using centrifugal force, some of which are illustrated in Fig. 14-109. Many of these were originally

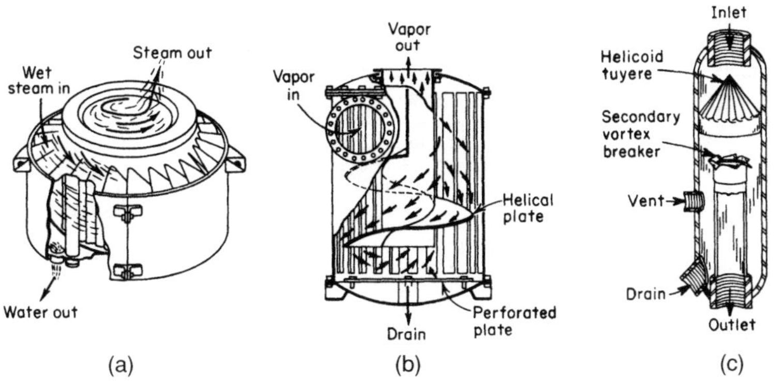

FIG. 14-109 Typical separators using impingement in addition to centrifugal force. (*a*) Hi-eF purifier. (*V. D. Anderson Co.*) (*b*) Flick separator. (*Wurster & Sanger, Inc.*) (*c*) Type RA line separator. (*Centrifix Corp., Bull. 220.*)

developed as steam separators to remove entrained condensate. In some designs, impingement on swirl baffles aids separation.

Impingement Separation Impingement separation employs direct impact and inertial forces between particles, the gas streamlines, and target bodies to provide capture. The mechanism is discussed in Sec. 17 under "Gas-Solids Separations." With liquids, droplet coalescence occurs on the target surface, and provision must be made for drainage without reentrainment. Calvert (R-12) has studied droplet collection by impingement on targets consisting of banks of tubes, zigzag baffles, and packed and mesh beds. Figure 14-110 illustrates some other types of impingement-separator designs.

In its simplest form, an impingement separator may be nothing more than a target placed in front of a flow channel such as a disk at the end of a tube. To improve collection efficiency, the gas velocity may be increased by forming the end into a nozzle (Fig. 14-110*a*). Particle collection as a function of size may be estimated by using the target-efficiency correlation in Fig. 17-39. Since target efficiency will be low for systems with separation numbers below 5 to 10 (small particles, low gas velocities), the mist will frequently be subjected to a number of targets in series as in Fig. 14-110*c, d,* and *g*.

The overall droplet penetration is the product of penetration for each set of targets in series. Obviously, for a distribution of particle sizes, an integration procedure is required to give overall collection efficiency. This target-efficiency method is suitable for predicting efficiency when the design effectively prevents the bypassing or short-circuiting of targets by the gas stream and provides adequate time to accelerate the liquid droplets to gas velocity. Katz (R-16) investigated a jet and target-plate entrainment separator design and found the pressure drop less than would be expected to supply the kinetic energy both for droplet acceleration and gas friction. An estimate based on his results indicates that the liquid particles on the average were being accelerated to only about 60 percent of the gas velocity. The largest droplets, which are the easiest to collect, will be accelerated less than the smaller particles. This factor has a leveling effect on collection efficiency as a function of particle size so that experimental results on such devices may not show as sharp a decrease in efficiency with particle size as predicted by calculation. Such results indicate that in many cases our lack of predicting ability results, not from imperfections in the theoretical treatment, but from our lack of knowledge of velocity distributions within the system.

Katz (R-16) also studied *wave-plate impingement separators* (Fig. 14-110*b*) made up of 90° formed arcs with an 11.1-mm (0.44-in) radius and a 3.8-mm (0.15-in) clearance between sheets. The pressure drop is a function of system geometry. The pressure drop for Katz's system and collection efficiency for seven waves are shown in Fig. 14-111. Katz used the Souders-Brown expression to define a design velocity for the gas between the waves:

$$U = K\sqrt{(\rho_l - \rho_g)/\rho_g} \tag{14-226}$$

K is 0.12 to give U in ms^{-1} (0.4 for ft/s), and ρ_l and ρ_g are liquid and gas densities in any consistent set of units. Katz found no change in efficiency at gas velocities from one-half to 3 times that given by the equation.

Calvert (R-12) investigated *zigzag baffles* of a design more like Fig. 14-110*e*. The baffles may have spaces between the changes in direction or be connected as shown. He found close to 100 per collection for water droplets of 10 μm and larger. Some designs had high efficiencies down to 5 or 8 μm. Desirable gas velocities were 2 to 3.5 m/s (6.6 to 11.5 ft/s), with a pressure drop for a six-pass baffle of 2 to 2.5 cm (0.8 to 1.0 in) of water. On the basis of turbulent mixing, an equation was developed for predicting primary collection efficiency as a function of particle size and collector geometry:

$$\eta = 1 - \exp\left[-\frac{u_{te} n W \theta}{57.3 U_g b \tan\theta}\right] \tag{14-227}$$

where η is the fractional primary collection efficiency; u_{te} is the drop terminal centrifugal velocity in the normal direction, cm/s; U_g is the superficial gas velocity, cm/s; n is the number of rows of baffles or bends; θ is the angle of inclination of the baffle to the flow path, °; W is the width of the baffle, cm; and b is the spacing between baffles in the same row, cm. For conditions of low Reynolds number ($N_{Re,D} <$ 0.1) where Stokes' law applies, Calvert obtains the value for drop terminal centrifugal velocity of $u_{te} = d_p^2\,\rho_p a/18\,\mu_g$, where d_p and ρ_p are the drop particle diameter, cm, and particle density, g/cm^3, respectively; μ_g is the gas viscosity, P; and a is the acceleration due to centrifugal force. It is defined by the equation $a = 2U_g^2 \sin\theta / W \cos^3\theta$. For situations in which Stokes' law does not apply, Calvert recommends substitution in the derivation of Eq. (14-227) for u of drag coefficients from drag-coefficient data of Foust et al. (*Principles of Unit Operations*, Toppan Co., Tokyo, 1959).

Calvert found that reentrainment from the baffles was affected by the gas velocity, the liquid-to-gas ratio, and the orientation of the baffles. Horizontal gas flow past vertical baffles provided the best drainage and lowest reentrainment. Safe operating regions with vertical baffles are shown in Fig. 14-112. Horizontal baffles gave the poorest drainage and the highest reentrainment, with inclined baffles intermediate in performance. Equation (14-228), developed by Calvert, predicts pressure drop across zigzag baffles. The indicated summation must be made over the number of rows of baffles present.

$$\Delta P = \sum_{i=1}^{i=n} 1.02 \times 10^{-3} f_D \rho_g \frac{U_g'^2 A_p}{2A_t} \tag{14-228}$$

ΔP is the pressure drop, cm of water; ρ_g is the gas density, g/cm^3; A_p is the total projected area of an entire row of baffles in the direction of inlet gas flow, cm^2; and A_t is the duct cross-sectional area, cm^2. The

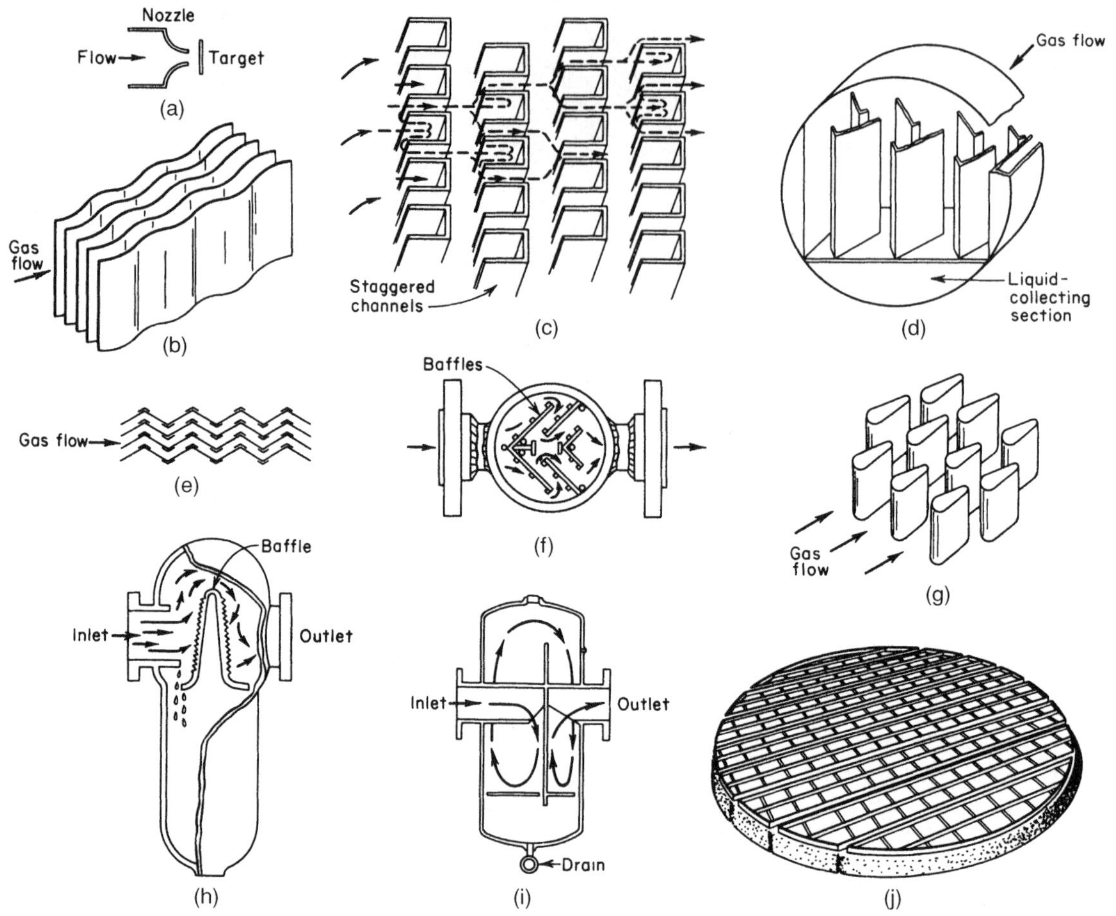

FIG. 14-110 Typical impingement separators. (*a*) Jet impactor. (*b*) Wave plate. (*c*) Staggered channels. (*Blaw-Knox Food & Chemical Equipment, Inc.*) (*d*) Vane-type mist extractor. (*Maloney-Crawford Tank and Mfg. Co.*) (*e*) Peerless line separator. (*Peerless Mfg. Co.*) (*f*) Strong separator. (*Strong Carlisle and Hammond.*) (*g*) Karbate line separator. (*Union Carbide Corporation*) (*h*) Type E horizontal separator. (*Wright-Austin Co.*) (*i*) PL separator. (*Ingersoll Rand.*) (*j*) Wire-mesh demister. (*Otto H. York Co.*)

value f_D is a drag coefficient for gas flow past inclined flat plates taken from Fig. 14-113, while U_g' is the actual gas velocity, cm/s, which is related to the superficial gas velocity U_g by $U_g' = U_g/\cos\theta$. It must be noted that the angle of incidence θ for the second and successive rows of baffles is twice the angle of incidence for the first row. Most of Calvert's work was with 30° baffles, but the method correlates well with other data on 45° baffles.

The Karbate line separator (Fig. 14-110*g*) is composed of several layers of teardrop-shaped target rods of Karbate. A design flow constant K in Eq. (14-226) of 0.035 m/s (1.0 ft/s) is recommended by the

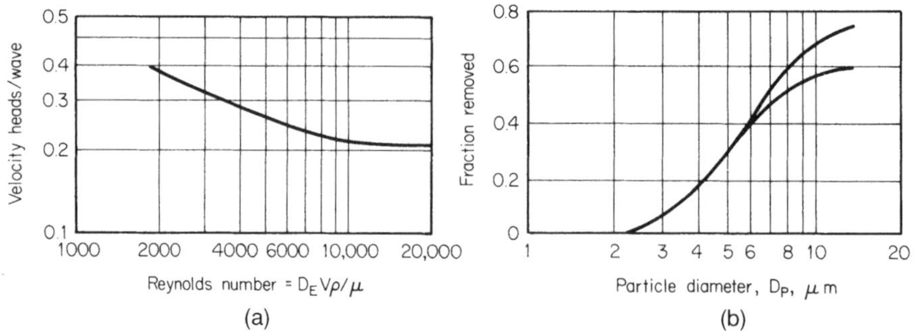

FIG. 14-111 Pressure drop and collection efficiency of a wave-plate separator. (*a*) Pressure drop. (*b*) Efficiency D_E = clearance between sheets. (*Katz, M.S. thesis, Pennsylvania State University, 1958.*)

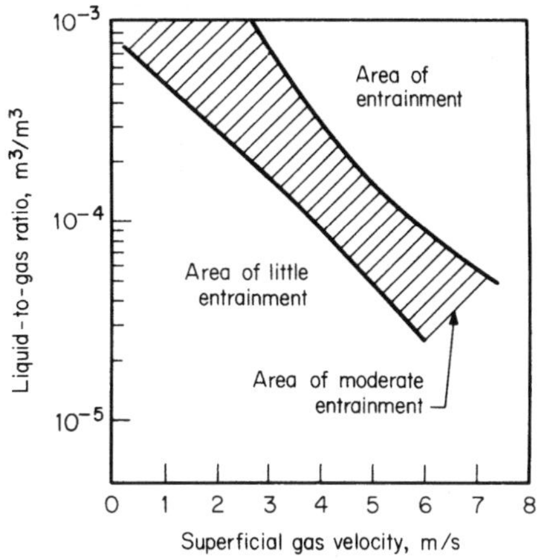

FIG. 14-112 Safe operating region to prevent reentrainment from vertical zigzag baffles with horizontal gas flow. (*Calvert, Yung, and Leung, NTIS Publ. PB-248050, 1975.*)

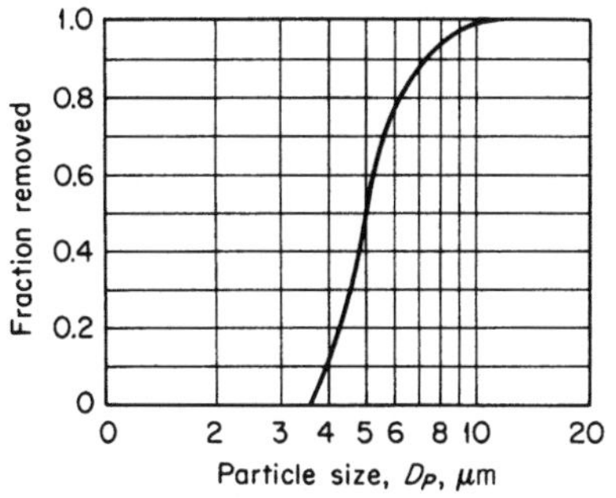

FIG. 14-114 Collection efficiency of Karbate line separator, based on particles with a specific gravity of 1.0 suspended in atmospheric air with a pressure drop of 2.5 cm water gauge. (*Union Carbide Corporation Cat. Sec. S-6900, 1960.*)

manufacturer. Pressure drop is said to be 5½ velocity heads on the basis of the superficial gas velocity. This value would probably increase at high liquid loads. Figure 14-114 gives the manufacturer's reported grade efficiency curve at the design air velocity.

The use of multiple tube banks as a droplet collector has also been studied by Calvert (R-12). He reports that collection efficiency for closely packed tubes follows equations for rectangular jet impaction which can be obtained graphically from Fig. 14-115 by using a dimensional parameter β which is based on the tube geometry; $\beta = 2l_i/b$, where b is the open distance between adjacent tubes in the row (orifice width) and l_i is the impaction length (distance between orifice and impingement plane), or approximately the distance between centerlines of successive tube rows. Note that the impaction parameter K_p is plotted to the one-half power in Fig. 14-115 and that the radius of the droplet is used rather than the diameter. Collection efficiency overall for a given size of particle is predicted for the entire tube bank by

$$\eta = 1 - (1 - \eta_b)^N \qquad (14\text{-}229)$$

where η_b is the collection efficiency for a given size of particle in one stage of a rectangular jet impactor (Fig. 14-115) and N is the number of stages in the tube bank (equal to one less than the number of rows).

For widely spaced tubes, the target efficiency η_g can be calculated from Fig. 17-39 or from the impaction data of Golovin and Putnam [*Ind. Eng. Chem. Fundam.*, **1**, 264 (1962)]. The efficiency of the overall tube banks for a specific particle size can then be calculated from the equation $\eta = 1 - (1 - \eta_t a'/A)^n$, where a' is the cross-sectional area of all tubes in one row, A is the total flow area, and n is the number of rows of tubes.

Calvert reports pressure drop through tube banks to be largely unaffected by liquid loading and indicates that Grimison's correlations in Sec. 6 ("Tube Banks") for gas flow normal to tube banks or data for gas flow through heat-exchanger bundles can be used. However, the following equation is suggested:

$$\Delta P = 8.48 \times 10^{-3} n \rho_g U_g'^2 \qquad (14\text{-}230)$$

where ΔP is cm of water; n is the number of rows of tubes; ρ_g is the gas density, g/cm³; and U_g' is the actual gas velocity between tubes in a row, cm/s. Calvert did find an increase in pressure drop of about 80 to 85 percent above that predicted by Eq. (14-230) in vertical upflow of gas through tube banks due to liquid holdup at gas velocities above 4 m/s.

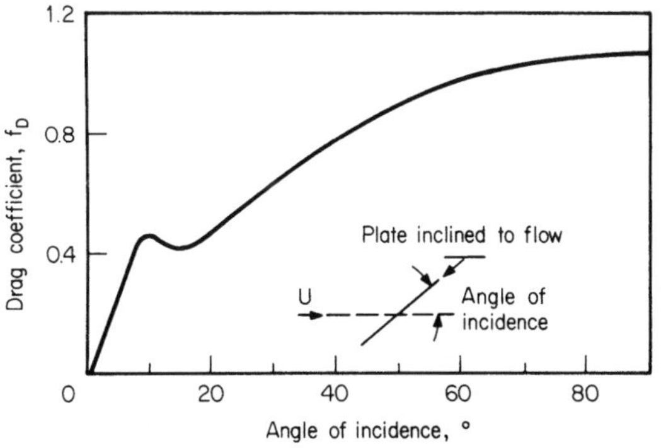

FIG. 14-113 Drag coefficient for flow past inclined flat plates for use in Eq. (14-228). [*Calvert, Yung, and Leung, NTIS Publ. PB-248050; based on Fage and Johansen,* Proc. R. Soc. (London), *116A, 170 (1927).*]

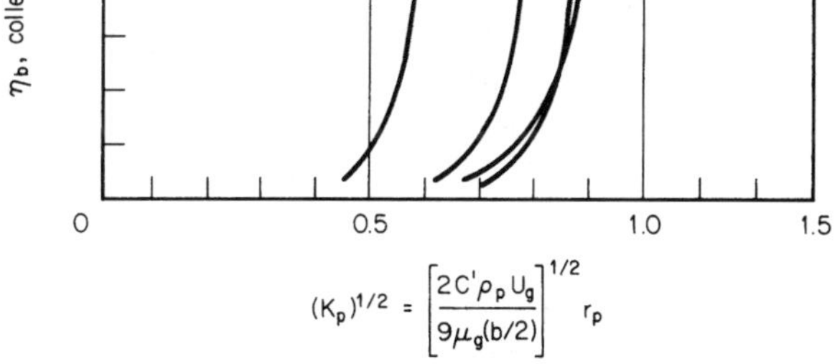

FIG. 14-115 Experimental collection efficiencies of rectangular impactors. C' is the Stokes-Cunningham correction factor; ρ_p, particle density, g/cm³; U_g, superficial gas velocity, approaching the impactor openings, cm/s; and μ_g, gas viscosity, P. [*Calvert, Yung, and Leung, NTIS Publ. PB-248050; based on Mercer and Chow,* J. Coll. Interface Sci., **27,** 75 (*1968*).]

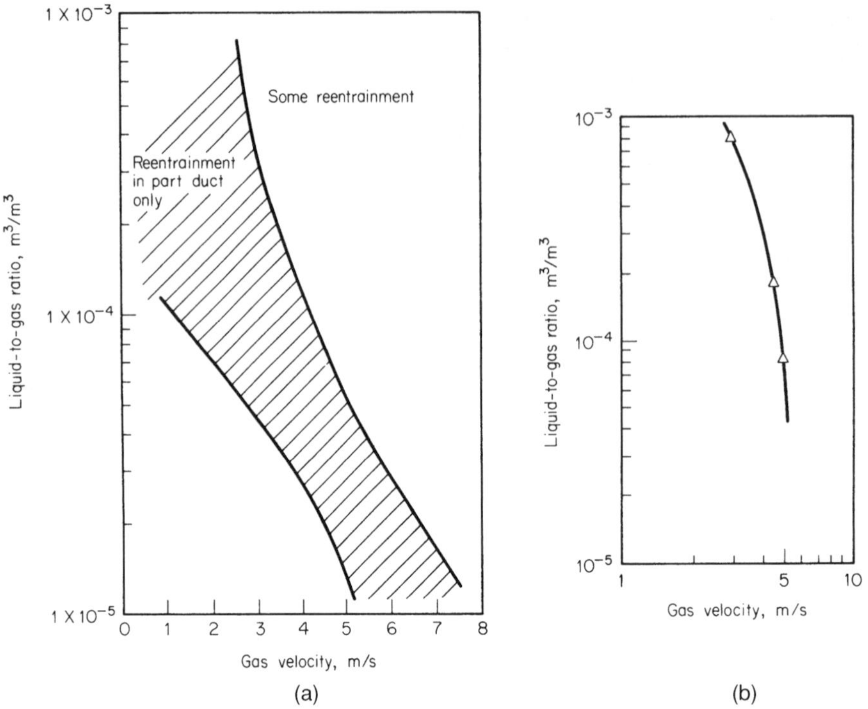

FIG. 14-116 Experimental results showing effect of gas velocity and liquid load on entrainment from (*a*) vertical tube banks with horizontal gas flow and (*b*) horizontal tube banks with upflow. To convert meters per second to feet per second, multiply by 3.281. (*Calvert, Yung, and Leung, NTIS Publ. PB-248050.*)

The onset of liquid reentrainment from tube banks can be predicted from Fig. 14-116. Reentrainment occurred at much lower velocities in vertical upflow than in horizontal gas flow through vertical tube banks. While the top of the cross-hatched line of Fig. 14-116*a* predicts reentrainment above gas velocities of 3 m/s (9.8 ft/s) at high liquid loading, most of the entrainment settled to the bottom of the duct in 1 to 2 m (3.3 to 6.6 ft), and entrainment did not carry significant distances until the gas velocity exceeded 7 m/s (23 ft/s).

Packed-Bed Collectors Many different materials, including coal, coke, broken solids of various types such as brick, tile, rock, and stone, as well as normal types of tower-packing rings, saddles, and special plastic shapes, have been used over the years in packed beds to remove entrained liquids through impaction and filtration. Separators using natural materials are not available as standard commercial units but are designed for specific applications. Coke boxes were used extensively in the years 1920 to 1940 as sulfuric acid entrainment separators (see *Chemical Engineers' Handbook*, 5th ed., p. 18–87) but have now been largely superseded by more sophisticated and efficient devices.

Jackson and Calvert [*Am. Inst. Chem. Eng. J.*, **12**, 1075 (1966)] studied the collection of fine fuel-oil-mist particles in beds of ½-in glass spheres, Raschig rings, and Berl and Intalox saddles. The mist had a mass median particle diameter of 6 μm and a standard deviation of 2.0. The collection efficiency as a function of particle size and gas velocity in a 355-mm- (14-in-) diameter by 152-mm- (6-in-) thick bed of Intalox saddles is given in Fig. 14-117. This and additional work have been generalized by Calvert (R-12) to predict collection efficiencies of liquid particles in any packed bed. Assumptions in the theoretical development are that the drag force on the drop is given by Stokes' law and that the number of semicircular bends to which the gas is subjected, η_1, is related to the length of the bed, Z (cm), in the direction of gas flow, the packing diameter, d_c (cm), and the gas-flow channel width, b (cm), such that $\eta_1 = Z/(d_c + b)$. The gas velocity through the channels, U_{gb} (cm/s), is inversely proportional to the bed free volume for gas flow such that $U_{gb} = U_g[1/(\varepsilon - h_b)]$, where U_g is the gas superficial velocity, cm/s, approaching the bed, ε is the bed void fraction, and h_b is the fraction of the total bed volume taken up with liquid which can be obtained from data on liquid holdup in packed beds. The width of the semicircular channels b can be expressed as a fraction j of the diameter of the packing elements, such that $b = jd_c$. These assumptions (as modified by G. E. Goltz, personal communica-

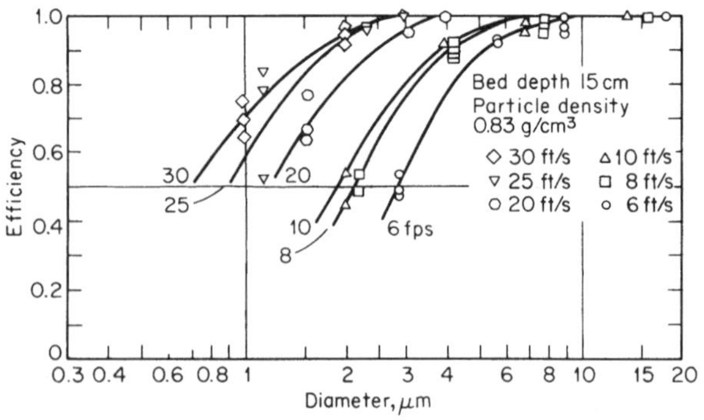

FIG. 14-117 Experimental collection efficiency. ½-in Intalox saddles. To convert feet per second to meters per second, multiply by 0.3048; to convert centimeters to inches, multiply by 0.394; and to convert grams per cubic centimeter to pounds per cubic foot, multiply by 62.43. [*Jackson and Calvert*, Am. Inst. Chem. Eng. J., ***12**, 1975 (1968)*.]

TABLE 14-27 Experimental Values for *j*, Channel Width in Packing as a Fraction of Packing Diameter

Packing size			
cm	in	Type of packing	*j*
1.27	0.5	Berl and Intalox saddles, marbles, Raschig rings	0.192
2.54	1.0	Berl and Intalox saddles, pall rings	0.190
3.8	1.5	Berl and Intalox saddles, pall rings	0.165
7.6–12.7	3–5	Coke	0.03

tion) lead to an equation for predicting the penetration of a given size of liquid particle through a packed bed:

$$P_t = \exp\left[\frac{-\pi}{2(j+j^2)(\varepsilon - h_b)}\left(\frac{Z}{d_c}\right)K_p\right] \qquad (14\text{-}231)$$

where

$$K_p = \frac{\rho_p d_p^2 U_g}{9\mu_g d_c} \qquad (14\text{-}232)$$

Values of ρ_p and d_p are droplet density, g/cm^3, and droplet diameter, cm; μ_g is the gas viscosity, P. All other terms were defined previously. Table 14-27 gives values of j calculated from experimental data of Jackson and Calvert. Values of j for most manufactured packing appear to fall in the range from 0.16 to 0.19. The low value of 0.03 for coke may be due to the porosity of the coke itself.

Calvert (R-12) has tested the correlation in cross-flow packed beds, which tend to give better drainage than countercurrent beds, and has found the effect of gas-flow orientation insignificant. However, the onset of reentrainment was somewhat lower in a bed of 2.5-cm (1.0-in) pall rings with gas upflow [6 m/s (20 ft/s)] than with horizontal cross-flow of gas. The onset of reentrainment was independent of liquid loading (all beds were nonirrigated), and entrainment occurred at values somewhat above the flood point for packed beds as predicted by conventional correlations. In beds with more than 3 cm (1.2 in) of water pressure drop, the experimental drop with both vertical and horizontal gas flow was somewhat less than predicted by generalized packed-bed pressure-drop correlations. However, Calvert recommends these correlations for design as conservative.

Calvert's data indicate that packed beds irrigated only with the collected liquid can have collection efficiencies of 80 to 90 percent on mist particles down to 3 μm but have low efficiency on finer mist particles. Frequently, irrigated packed towers and towers with internals will be used with liquid having a wetting capability for the fine mist which must be collected. Tennessee Valley Authority (TVA) experiments with the collection of 1.0-μm mass median phosphoric acid mist in packed towers have shown that the strength of the circulating phosphoric acid is highly important [see Baskerville, *Am. Inst. Chem. Eng. J.*, **37,** 79 (1941); and p. 18–87, 5th ed. of the *Handbook*]. Hesketh (*J. Air Pollut. Control Assoc.*, **24,** 942 (1974)] has reported up to 50 percent improvement in collection efficiency in venturi scrubbers on fine particles with the addition of only 0.10 percent of a low-foaming nonionic surfactant to the scrubbing liquid, and others have experienced similar results in other gas-liquid-contacting devices. Calvert (R-9 and R-10) has reported on the efficiency of various gas-liquid-contacting devices for fine particles. Figure 14-118 gives the particle aerodynamic cut size for a single-sieve-plate gas scrubber as a function of sieve hole size d_h, cm; hole gas velocity u_h, m/s; and froth or foam density on the plate F, g/cm^3. This curve is based on standard air and water properties and wettable (hydrophilic) particles. The cut diameter decreases with an increase in froth density, which must be predicted from correlations for sieve-plate behavior (see Fig. 14-32). Equation (14-231) can be used to calculate generalized design curves for collection in packed columns in the same fashion by finding parameters of packing size, bed length, and gas velocity which give collection efficiencies of 50 percent for various size particles. Figure 14-119 illustrates such a plot for three gas velocities and two sizes of packing.

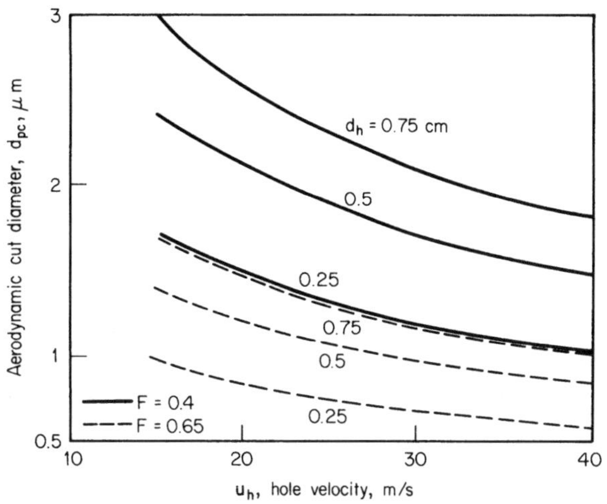

FIG. 14-118 Aerodynamic cut diameter for a single-sieve-plate scrubber as a function of hole size, hole-gas velocity, and froth density, F, g/cm^3. To convert meters per second to feet per second, multiply by 3.281; to convert grams per cubic centimeter to pounds per cubic foot, multiply by 62.43. [*Calvert,* J. Air Pollut. Control Assoc., **24,** 929 (*1974*).]

Wire-Mesh Mist Collectors Knitted mesh of varying density and voidage is widely used for entrainment separators. Its advantage is close to 100 percent removal of drops larger than 5 μm at superficial gas velocities from about 0.2 ms/s (0.6 ft/s) to 5 m/s (16.4 ft/s), depending somewhat on the design of the mesh. Pressure drop is usually no more than 2.5 cm (1 in) of water. A major disadvantage is the ease with which tars and insoluble solids plug the mesh. The separator can be made to fit vessels of any shape and can be made of any material which can be drawn into a wire. Stainless-steel and plastic fibers are most common, but other metals are sometimes used. Generally three basic types of mesh are used: (1) layers with a crimp in the same direction (each layer is actually a nested double layer); (2) layers with a crimp in

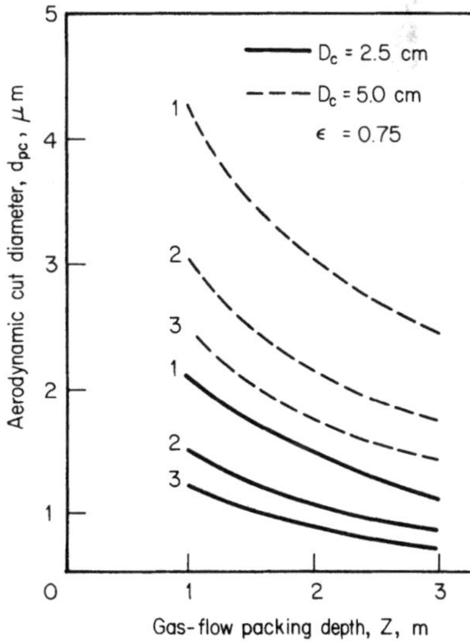

FIG. 14-119 Aerodynamic cut diameter for a typical packed-bed entrainment separator as a function of packing size, bed depth, and three gas velocities: curve 1–1.5 m/s, curve 2–3.0 m/s, and curve 3–4.5 m/s. To convert meters to feet, multiply by 3.281; to convert centimeters to inches, multiply by 0.394. [*Calvert,* J. Air Pollut. Control Assoc., **24,** 929 (*1974*).]

alternate directions, which increases voidage, reduces sheltering and increases target efficiency per layer, and gives a lower pressure drop per unit length; and (3) spiral-wound layers which reduce pressure drop by one-third, but fluid creep may lead to higher entrainment. Some small manufacturers of plastic meshes may offer other weaves claimed to be superior. The filament size can vary from about 0.15 mm (0.006 in) for fine-wire pads to 3.8 mm (0.15 in) for some plastic fibers. Typical pad thickness varies from 100 to 150 mm (4 to 6 in), but occasionally pads up to 300 mm (12 in) thick are used. A typical wire diameter for standard stainless mesh is 0.28 mm (0.011 in), with a finished mesh density of 0.15 g/cm^3 (9.4 lb/ft^3). A lower mesh density may be produced with standard wire to give 10 to 20 percent higher flow rates.

Figure 14-120 presents an early calculated estimate of mesh efficiency as a fraction of mist-particle size. Experiments by Calvert (R-12) confirm the accuracy of the equation of Bradie and Dickson (*Joint Symp. Proc. Inst. Mech. Eng./Yorkshire Br. Inst. Chem. Eng.*, 1969, pp. 24–25) for primary efficiency in mesh separators:

$$\eta = 1 - \exp(-2/3)\pi a l \eta_i) \qquad (14\text{-}232)$$

where η is the overall collection efficiency for a given-size particle; l is the thickness of the mesh, cm, in the direction of gas flow; a is the surface area of the wires per unit volume of mesh pad, cm^2/cm^3; and η_i, the target collection efficiency for cylindrical wire, can be calculated from Fig. 17-39 or the impaction data of Golovin and Putnam [*Ind. Eng. Chem.*, **1,** 264 (1962)]. The factor 2/3, introduced by Carpenter and Othmer [*Am. Inst. Chem. Eng. J.*, **1,** 549 (1955)], corrects for the fact that not all the wires are perpendicular to the gas flow and gives the projected perpendicular area. If the specific mesh surface area a is not available, it can be calculated from the mesh void area ε and the mesh wire diameter d_w in cm, $a = 4(1 - \varepsilon)/d_w$.

York and Poppele (R-17) have stated that factors governing maximum allowable gas velocity through the mesh are (1) gas and liquid density, (2) liquid surface tension, (3) liquid viscosity, (4) specific wire surface area, (5) entering-liquid loading, and (6) suspended-solids content. York (R-18) has proposed application of the Souders-Brown equation [Eq. (14-226)] for correlation of maximum allowable gas velocity with values of K for most cases of 0.1067 m/s to give U in m/s (0.35 for ft/s). When liquid viscosity or inlet loading is high or the liquid is dirty, the value of K must be reduced. Schroeder (M.S. thesis, Newark College of Engineering, 1962) found lower values for K necessary when liquid surface tension is reduced such as by the presence of surfactants in water. Ludwig (*Applied Process Design for Chemical and Petrochemical Plants*, 2d ed., vol. I, Gulf, Houston, 1977, p. 157) recommends reduced K values of (0.061 m/s) under vacuum at an absolute pressure of 6.77 kPa (0.98 lbf/in^2) and K = 0.082 m/s at 54 kPa (7.83 lbf/in^2) absolute. Most manufacturers suggest setting the design velocity at three-fourths of the maximum velocity to allow for surges in gas flow.

York and Poppele (R-17) have suggested that total pressure drop through the mesh is equal to the sum of the mesh dry pressure drop plus an increment due to the presence of liquid. They considered the mesh to be equivalent to numerous small circular channels and used the D'Arcy formula with a modified Reynolds number to correlate friction factor (see Fig. 14-121) for Eq. (14-233) giving dry pressure drop.

$$\Delta P_{dry} = fla\rho_g U_g^2/981\,\varepsilon^3 \qquad (14\text{-}233)$$

where ΔP is in cm of water; f is from Fig. (14-121); ρ_g is the gas density, g/cm^3; U_g is the superficial gas velocity, cm/s; and ε is the mesh porosity or void fraction; l and a are as defined in Eq. (14-232). Figure 14-121 gives data of York and Poppele for mesh crimped in the same and alternating directions and also includes the data of Satsangee, of Schuring, and of Bradie and Dickson.

The incremental pressure drop for wet mesh is not available for all operating conditions or for mesh of different styles. The data of York and Poppele for wet-mesh incremental pressure drop, ΔP_L in cm of water, are shown in Fig. 14-122 or parameters of liquid velocity L/A, defined as liquid volumetric flow rate, cm^3/min per unit of mesh cross-sectional area in cm^2; liquid density ρ_L is in g/cm^3.

York generally recommends the installation of the mesh horizontally with upflow of gas as in Fig. 14-110*f;* Calvert (R-12) tested the mesh horizontally with upflow and vertically with horizontal gas flow. He reports better drainage with the mesh vertical and somewhat higher permissible gas velocities without reentrainment, which is contrary to past practice. With horizontal flow through vertical mesh, he found collection efficiency to follow the predictions of Eq. (14-232) up to 4 m/s (13 ft/s) with air and water. Some reentrainment was encountered at higher velocities, but it did not appear serious until velocities exceeded 6.0 m/s (20 ft/s). With vertical upflow of gas, entrainment was encountered at velocities above and below 4.0 m/s (13 ft/s), depending on inlet liquid quantity (see Fig. 14-123). Figure 14-124 illustrates the onset of entrainment from mesh as a function of liquid loading and gas velocity and the safe operating area recommended by Calvert. Measurements of dry pressure drop by Calvert gave values only about one-third of those predicted from Eq. (14-233). He found the pressure drop to be highly affected by liquid load. The pressure drop of wet mesh could be correlated as a function of $U_g^{1.65}$ and parameters of liquid loading L/A, as shown in Fig. 14-125.

As indicated previously, mesh efficiency drops rapidly as particles decrease in size below 5 μm. An alternative is to use two mesh pads in series. The first mesh is made of fine wires and is operated beyond the

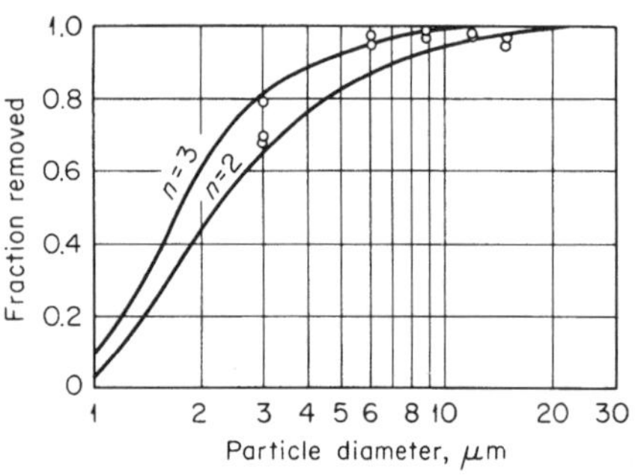

FIG. 14-120 Collection efficiency of wire-mesh separator; 6-in thickness, 98.6 percent free space, 0.006-in-diameter wire used for experiment points. Curves calculated for target area equal to 2 and 3 times the solids volume of packing. To convert inches to millimeters, multiply by 25.4.

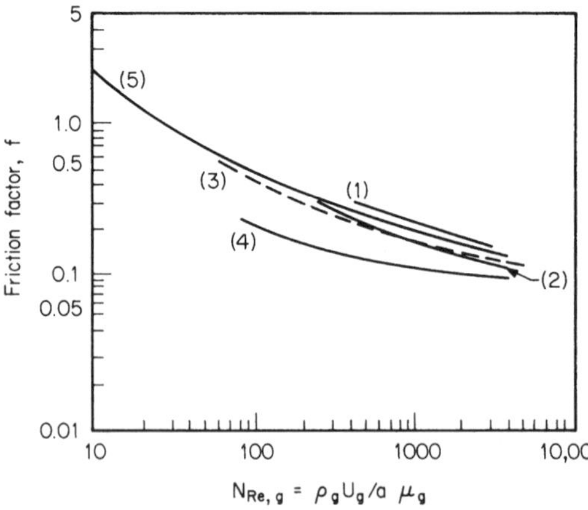

FIG. 14-121 Value of friction factor f for dry knitted mesh for Eq. (14-233). Values of York and Poppele [Chem. Eng. Prog., **50,** *421* (*1954*)] are given in curve 1 for mesh crimped in the alternating direction and curve 2 for mesh crimped in the same direction. Data of Bradie and Dickson (*Joint Symp. Proc. Inst. Mech. Eng./Yorkshire Br. Inst. Chem. Eng., 1969, pp. 24–25*) are given in curve 3 for layered mesh and curve 4 for spiral-wound mesh. Curve 5 is data of Satsangee (M.S. thesis, Brooklyn Polytechnic Institute, 1948) and Schurig (D.Ch.E. dissertation, Brooklyn Polytechnic Institute, 1946). (*From Calvert, Yung, and Leung, NTIS Publ. PB-248050, 1975.*)

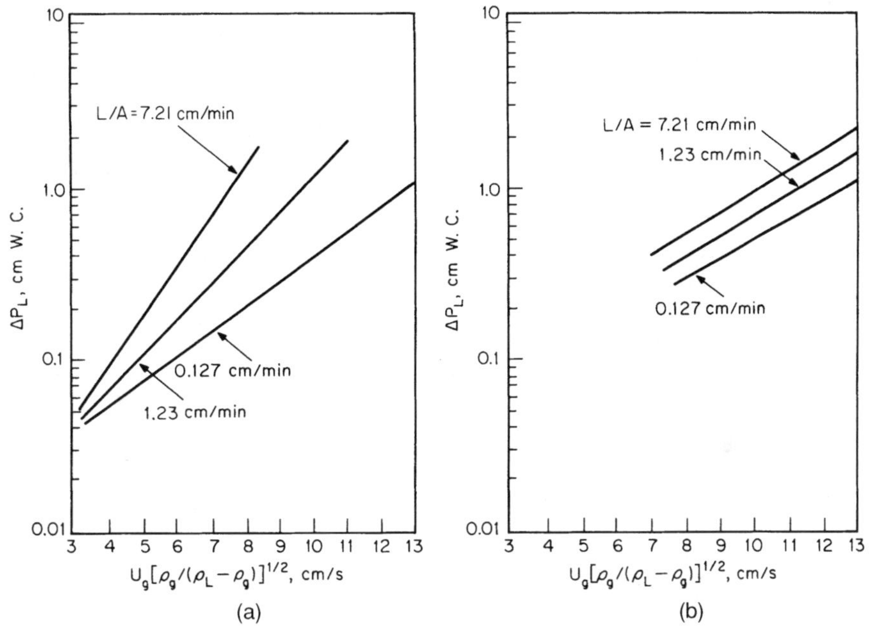

FIG. 14-122 Incremental pressure drop in knitted mesh due to the presence of liquid (*a*) with the mesh crimps in the same direction and (*b*) with crimps in the alternating direction, based on the data of York and Poppele [Chem. Eng. Prog., **50,** 421 (*1954*)]. To convert centimeters per minute to feet per minute, multiply by 0.0328; to convert centimeters per second to feet per second, multiply by 0.0328. (*From Calvert, Yung, and Leung, NTIS Publ. PB-248050, 1975.*)

flood point. It results in droplet coalescence, and the second mesh, using standard wire and operated below flooding, catches entrainment from the first mesh. Coalescence and flooding in the first mesh may be assisted with water sprays or irrigation. Massey [*Chem. Eng. Prog.,* **53**(5), 114 (1959)] and Coykendall et al. [*J. Air Pollut. Control Assoc.,* **18,** 315 (1968)] have discussed such applications. Calvert (R-12) presents data on the particle size of entrained drops from mesh as a function of gas velocity which can be used for sizing the secondary collector. A major disadvantage of this approach is high pressure drop, which can be in the range from 25 cm (10 in) of water to as high as 85 cm (33 in) of water if the mist is mainly submicrometer.

Wet Scrubbers Scrubbers have not been widely used for the collection of purely liquid particulate, probably because they are generally more complex and expensive than impaction devices of the types previously discussed. Further, scrubbers are no more efficient than the former devices for the same energy consumption. However, scrubbers of the types discussed in Sec. 17 and illustrated in Figs. 17-48 to 17-54 can be used to capture liquid particles efficiently. Their use is primarily indicated when it is desired to accomplish simultaneously another task such as gas absorption or the collection of solid and liquid particulate mixtures.

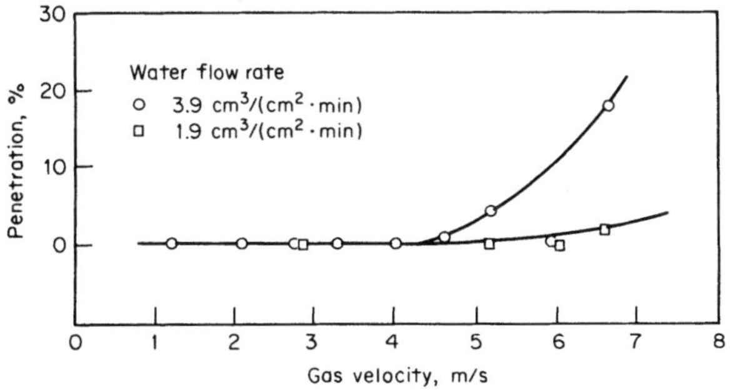

FIG. 14-123 Experimental data of Calvert with air and water in mesh with vertical upflow, showing the effect of liquid loading on efficiency and reentrainment. To convert meters per second to feet per second, multiply by 3.281; to convert cubic centimeters per square centimeter-minute to cubic feet per square foot-minute, multiply by 0.0328. (*Calvert, Yung, and Leung, NTIS Publ. PB-248050, 1975.*)

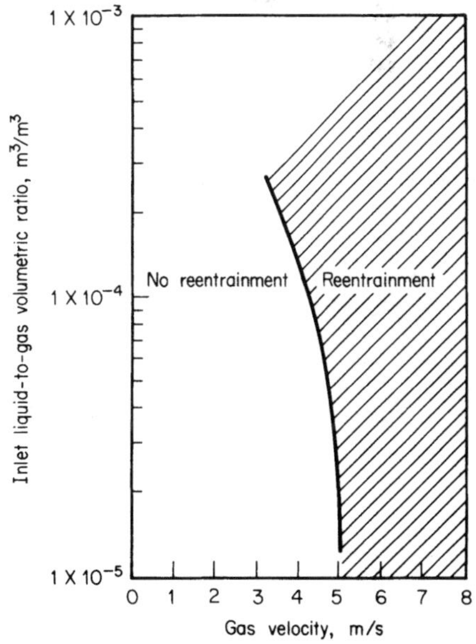

FIG. 14-124 Effect of gas and liquid rates on onset of mesh reentrainment and safe operating regions. To convert meters per second to feet per second, multiply by 3.281. (*Calvert, Yung, and Leung, NTIS Publ. PB-248050, 1975.*)

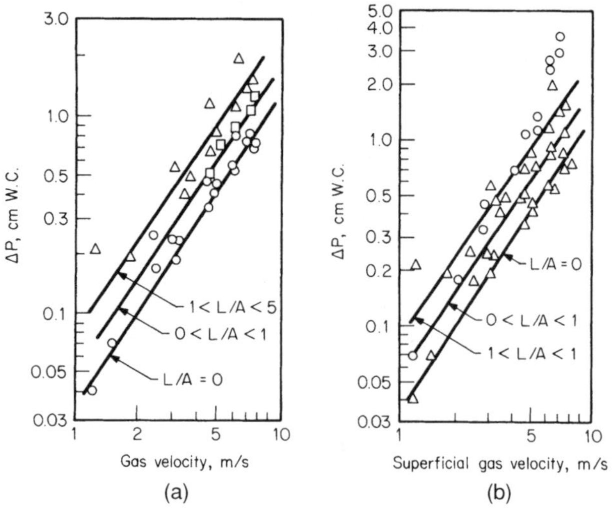

FIG. 14-125 Experimental pressure measured by Calvert as a function of gas velocity and liquid loading for (*a*) horizontal gas flow through vertical mesh and (*b*) gas upflow through horizontal mesh. Mesh thickness was 10 cm with 2.8-mm wire and void fraction of 98.2 percent, crimped in alternating directions. To convert meters per second to feet per second, multiply by 3.281; to convert centimeters to inches, multiply by 0.394. (*Calvert, Yung, and Leung, NTIS Publ. PB-248050, 1975.*)

Table 20-41 [*Chemical Engineers' Handbook*, 5th ed.)], showing the minimum size of particles collectible in different types of scrubbers at reasonably high efficiencies, is a good selection guide. Cyclonic spray towers can effectively remove liquid particles down to around 2 to 3 μm. Figures 20-112 and 20-113 (*Chemical Engineers' Handbook,* 5th ed.), giving target efficiency between spray drop size and particle size as calculated by Stairmand or Johnstone and Roberts, should be considered in selecting spray atomization for the most efficient tower operation. Figure 14-126 gives calculated particle cut size as a function of tower height (or length) for vertical countercurrent spray towers and for horizontal-gas-flow, vertical-liquid-flow cross-current spray towers with parameters for liquid drop size. These curves are based on physical properties of standard air and water and should be used under conditions in which these are reasonable approximations. Lack of uniform liquid distribution or liquid flowing down the walls can affect the performance, requiring empirical correction factors. Calvert (R-10) suggests that a correction factor of 0.2 be used in small-diameter scrubbers to account for the liquid on the walls, i.e., let $Q_L/Q_g = 0.2\ (Q_L/Q_g)_{actual}$. Many more complicated wet scrubbers employ a combination of sprays or liquid atomization, cyclonic action, baffles, and targets. These combinations are not likely to be more efficient than similar devices previously discussed that operate at equivalent pressure drop. The vast majority of wet scrubbers operate at moderate pressure drop [8 to 15 cm (3 to 6 in) of water or 18 to 30 cm (7 to 12 in) of water] and cannot be expected to have high efficiency on particles smaller than 10 μm or 3 to 5 μm respectively. Fine and submicrometer particles can be captured efficiently only in wet scrubbers having high energy input such as venturi scrubbers, two-phase eductor scrubbers, and flux-force-condensation scrubbers.

Venturi Scrubbers One type of venturi scrubber is illustrated in Fig. 17-48. Venturi scrubbers have been used extensively for collecting fine and submicrometer solid particulate, condensing tars and mists, and mixtures of liquids and solids. To a lesser extent, they have also been used for simultaneous gas absorption, although Lundy [*Ind. Eng. Chem.*, **50,** 293 (1958)] indicates that they are generally limited to three transfer units. They have been used to collect submicrometer chemical incinerator fume and mist as well as sulfuric and phosphoric acid mists. The collection efficiency of a venturi scrubber is highly dependent on the throat velocity or pressure drop, the liquid-to-gas ratio, and the chemical nature of wettability of the particulate. Throat velocities may range from 60 to 150 m/s (200 to 500 ft/s). Liquid injection rates are typically 0.67 to 1.4 m^3/1000 m^3 of gas. A liquid rate of 1.0 m^3 per 1000 m^3 of gas is usually close to optimum, but liquid rates as high as 2.7 m^3 (95 ft^3) have been used. Efficiency improves with increased liquid rate but only at the expense of higher pressure drop and energy consumption. Pressure-drop predictions for a given efficiency are hazardous without determining the nature of the particulate and the liquid-to-gas ratio. In general, particles coarser than 1 μm can be collected efficiently with pressure drops of 25 to 50 cm of water. For appreciable collection of submicrometer particles, pressure drops of 75 to 100 cm (30 to 40 in) of water are usually required. When particles are appreciably finer than 0.5 μm, pressure drops of 175 to 250 cm (70 to 100 in) of water have been used.

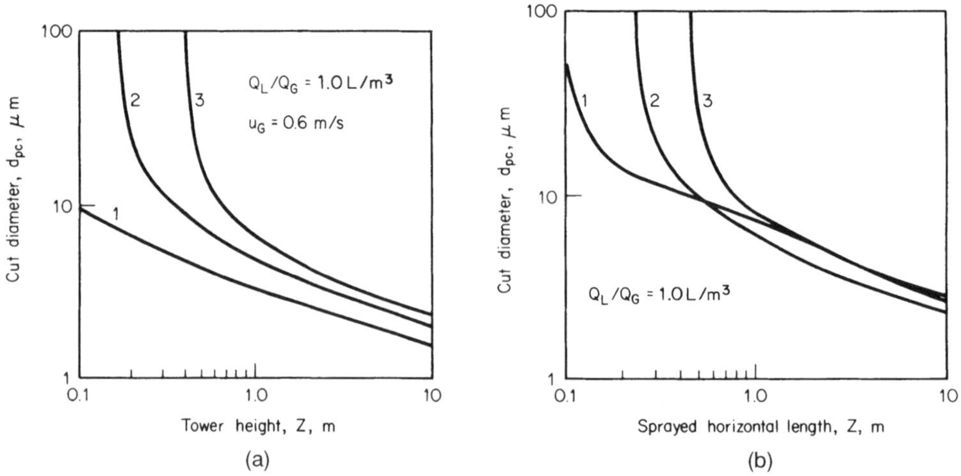

FIG. 14-126 Predicted spray-tower cut diameter as a function of sprayed length and spray droplet size for (*a*) vertical-countercurrent towers and (*b*) horizontal-cross-flow towers per Calvert [J. Air Pollut. Control Assoc., **24,** 929 (*1974*)]. Curve 1 is for 200-μm spray droplets, curve 2 for 500-μm spray, and curve 3 for 1000-μm spray. Q_L/Q_G is the volumetric liquid-to-gas ratio, *L* liquid/m^3 gas, and u_G is the superficial gas velocity in the tower. To convert liters per cubic meter to cubic feet per cubic foot, multiply by 10^{-3}.

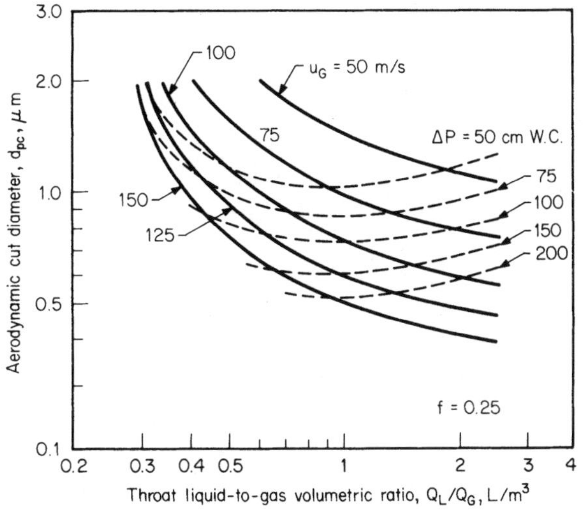

FIG. 14-127 Prediction of venturi-scrubber cut diameter for hydrophobic particles as functions of operating parameters as measured by Calvert [*Calvert, Goldshmid, Leith, and Mehta, NTIS Publ. PB-213016, 213017, 1972; and Calvert,* J. Air Pollut. Control Assoc., **24,** 929 (*1974*).] u_G is the superficial throat velocity, and ΔP is the pressure drop from converging to diverging section. To convert meters per second to feet per second, multiply by 3.281; to convert liters per cubic meter to cubic feet per cubic foot, multiply by 10^{-3}; and to convert centimeters to inches, multiply by 0.394.

One of the problems in predicting efficiency and required pressure drop of a venturi is the chemical nature or wettability of the particulate, which on 0.5-μm-size particles can make up to a threefold difference in required pressure drop for its efficient collection. Calvert (R-9, R-10) has represented this effect by an empirical factor f, which is based on the hydrophobic ($f = 0.25$) or hydrophilic ($f = 0.50$) nature of the particles. Figure 14-127 gives the cut diameter of a venturi scrubber as a function of its operating parameters (throat velocity, pressure drop, and liquid-to-gas ratio) for hydrophobic particles. Figure 14-129 compares cut diameter as a function of pressure drop for an otherwise identically operating venturi on hydrophobic and hydrophilic particles. Calvert (R-9) gives equations which can be used for constructing cut-size curves similar to those of Fig. 14-127 for other values of the empirical factor f. Most real particles are neither completely hydrophobic nor completely hydrophilic but have f values lying between the two extremes. Phosphoric acid mist, on the basis of data of Brink and Contant [*Ind. Eng. Chem.*, 50, 1157 (1958)] appears to have a value of $f = 0.46$. Unfortunately, no chemical-test methods have yet been devised for determining appropriate f values for a particulate in the laboratory.

Pressure drop in a venturi scrubber is controlled by throat velocity. While some venturis have fixed throats, many are designed with variable louvers to change throat dimensions and control performance for changes in gas flow. Pressure-drop equations have been developed by Calvert (R-13, R-14, R-15), Boll [*Ind. Eng. Chem. Fundam.*, **12,** 40 (1973)], and Hesketh [*J. Air Pollut. Control Assoc.*, **24,** 939 (1974)]. Hollands and Goel [*Ind. Eng. Chem. Fundam.*, **14,** 16 (1975)] have developed a generalized pressure-drop equation.

The Hesketh equation is empirical and is based upon a regression analysis of data from a number of industrial venturi scrubbers:

$$\Delta P = U_{gt}^2 \rho_g A_t^{0.155} L^{0.78}/1270 \qquad (14\text{-}234)$$

where ΔP is the pressure drop, in of water; U_{gt} is the gas velocity in the throat, ft/s; ρ_g is the gas density, lb/ft^3; A_t is the throat area, ft^2; and L is the liquid-to-gas ratio, gal/1000 acf.

Calvert (R-15) critiqued the many pressure-drop equations and suggested the following simplified equation as accurate to ±10 percent:

$$\Delta P = \frac{2\rho_\ell U_g^2}{981 g_c}\left(\frac{Q_t}{Q_g}\right)\left[1 - x^2 + \sqrt{(x^4 - x^2)^{0.5}}\right] \qquad (14\text{-}235)$$

where

$$x = (3l_t C_{Di}\rho_g/16 d_l \rho_l) + 1 \qquad (14\text{-}236)$$

ΔP is the pressure drop, cm of water; ρ_ℓ and ρ_g are the density of the scrubbing liquid and gas respectively, g/cm^3; U_g is the velocity of the gas at the throat inlet, cm/s; Q_t/Q_g is the volumetric ratio of liquid to gas at the throat inlet, dimensionless; l_t is the length of the throat, cm; C_{Di} is the drag coefficient, dimensionless, for the mean liquid diameter, evaluated at the throat inlet; and d_l is the Sauter mean diameter, cm, for the atomized liquid. The atomized-liquid mean diameter must be evaluated by the Nukiyama and Tanasawa [*Trans. Soc. Mech Eng.* (*Japan*), 4, 5, 6 (1937–1940)] equation:

$$d_\ell = \frac{0.0585}{U_g}\left(\frac{\sigma_\ell}{\rho_\ell}\right)^{0.5} + 0.0597\left[\frac{\mu_\ell}{(\sigma_\ell \rho_\ell)^{0.5}}\right]^{0.45}\left(\frac{Q_\ell}{Q_g}\right)^{1.5} \qquad (14\text{-}237)$$

where σ_ℓ is the liquid surface tension, dyn/cm; and μ_ℓ is the liquid viscosity; P. The drag coefficient C_{Di} should be evaluated by the Dickinson and Marshall [*Am. Inst. Chem. Eng. J.*, **14,** 541 (1968)] correlation $C_{Di} = 0.22 + (24/N_{Rei})(1 + 0.15\, N_{Rei}^{0.6})$. The Reynolds number, N_{Rei}, is evaluated at the throat inlet considerations as $d_\ell G_g/\mu_g$.

All venturi scrubbers must be followed by an entrainment collector for the liquid spray. These collectors are usually centrifugal and will have an additional pressure drop of several centimeters of water, which must be added to that of the venturi itself.

Other Scrubbers A liquid-ejector venturi (Fig. 17-49), in which high-pressure water from a jet induces the flow of gas, has been used to collect mist particles in the 1- to 2-μm range, but submicrometer particles will generally pass through an eductor. Power costs for liquid pumping are high if appreciable motive force must be imparted to the gas because jet-pump efficiency is usually less than 10 percent. Harris [*Chem. Eng. Prog.*, **42**(4), 55 (1966)] has described their application. Two-phase eductors have been considerably more successful on capture of submicrometer mist particles and could be attractive in situations in which large quantities of waste thermal energy are available. However, the equivalent energy consumption is equal to that required for high-energy venturi scrubbers, and such devices are likely to be no more attractive than venturi scrubbers when the thermal energy is priced at its proper value. Sparks [*J. Air Pollut. Control Assoc.*, **24,** 958 (1974)] has discussed steam ejectors giving 99 percent collection of particles 0.3 to 10 μm. Energy requirements were 311,000 J/m^3(8.25 Btu/scf). Gardenier [*J. Air Pollut. Control Assoc.*, **24,** 954 (1974)] operated a liquid eductor with high-pressure (6900- to 27,600-kPa) (1000- to 4000-lbf/in^2) hot water heated to 200°C (392°F) which flashed into two phases as it issued from the jet. He obtained 95 to 99 percent collection of submicrometer particulate. Figure 14-128 shows the water-to-gas ratio required as a function of particle size to achieve 99 percent collection.

Effect of Gas Saturation in Scrubbing If hot unsaturated gas is introduced into a wet scrubber, spray particles will evaporate to cool and saturate the gas. The evaporating liquid molecules moving away from the target droplets will repel particles which might collide with them. This results in the forces of diffusiophoresis opposing particle collection. Semrau and Witham (Air Pollut. Control Assoc. Prepr. 75-30.1) investigated temperature parameters in wet scrubbing and found a definite decrease in the efficiency of evaporative scrubbers and an enhancement of efficiency when a hot saturated gas is scrubbed with cold water rather than recirculated hot water. Little improvement was experienced in cooling a hot saturated gas below a 50°C dew point.

Energy Requirements for Inertial-Impaction Efficiency Semrau [*J. Air Pollut. Control Assoc.*, **13,** 587 (1963)] proposed a "contacting-power" principle which states that the collecting efficiency of a given size of particle is proportional to the power expended and that the smaller the particle, the greater the power required.

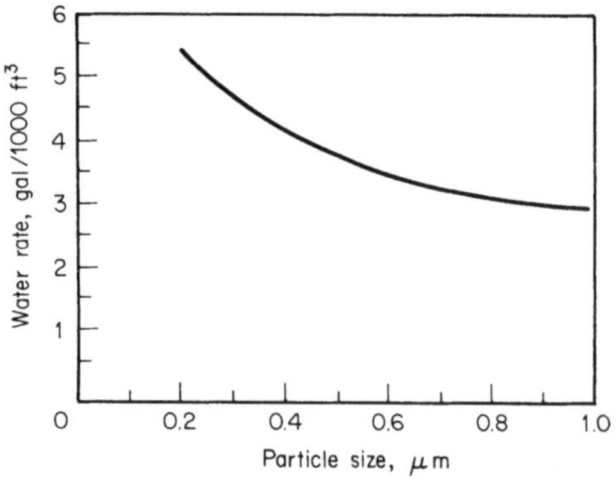

FIG. 14-128 Superheated high-pressure hot-water requirements for 99 percent collection as a function of particle size in a two-phase eductor jet scrubber. To convert gallons per 1000 cubic feet to cubic meters per 1000 cubic meters, multiply by 0.134. [*Gardenier,* J. Air Pollut. Control Assoc., ***24,*** *954 (1974).*]

Mathematically expressed, $N_T = \propto P_T^\gamma$, where N_T is the number of particulate transfer units achieved and P_T is the total energy expended within the collection device, including gas and liquid pressure drop and thermal and mechanical energy added in atomizers. N_T is further defined as $N_T = \ln [1/(1 - \eta)]$, where η is the overall fractional collection efficiency. This was intended as a universal principle, but the constants $\propto$ and γ have been found to be functions of the chemical nature of the system and the design of the control device. Others have pointed out that the principle is applicable only when the primary collection mechanism is impaction and direct interception. Calvert (R-10, R-12) has found that plotting particle cut size versus pressure drop (or power expended) as in Fig. 14-129 is a more suitable way to develop a generalized energy-requirement curve for impaction devices. The various curves fall close together and outline an imaginary curve that indicates the magnitude of pressure drop required as particle size decreases bound by the two limits of hydrophilic and hydrophobic particles. By calculating the required cut size for a given collection efficiency, Fig. 14-129 can also be used as a guide to deciding between different collection devices.

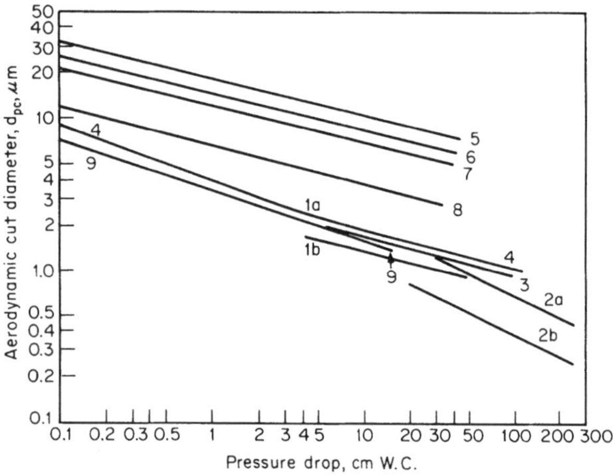

FIG. 14-129 Typical cut diameter as a function of pressure drop for various liquid-particle collectors. Curves 1*a* and *b* are single-sieve plates with froth density of 0.4 g/cm³; 1*a* has sieve holes of 0.5 cm and 1*b* holes of 0.3 cm. Curves 2*a* and *b* are for a venturi scrubber with hydrophobic particles (2*a*) and hydrophilic particles (2*b*). Curve 3 is an impingement plate, and curve 4 is a packed column with 2.5-cm-diameter packing. Curve 5 is a zigzag baffle collector with six baffles at $\theta = 30°$. Curve 7 is for six rows of staggered tubes with 1-cm spacing between adjacent tube walls in a row. Curve 8 is similar, except that tube-wall spacing in the row is 0.3 cm. Curve 9 is for wire-mesh pads. To convert grams per cubic centimeter to pounds per cubic foot, multiply by 62.43; to convert centimeters to inches, multiply by 0.394. [*Calvert,* J. Air Pollut. Control Assoc., ***24,*** *929 (1974); and Calvert, Yung, and Leung, NTIS Publ. PB-248050, 1975.*]

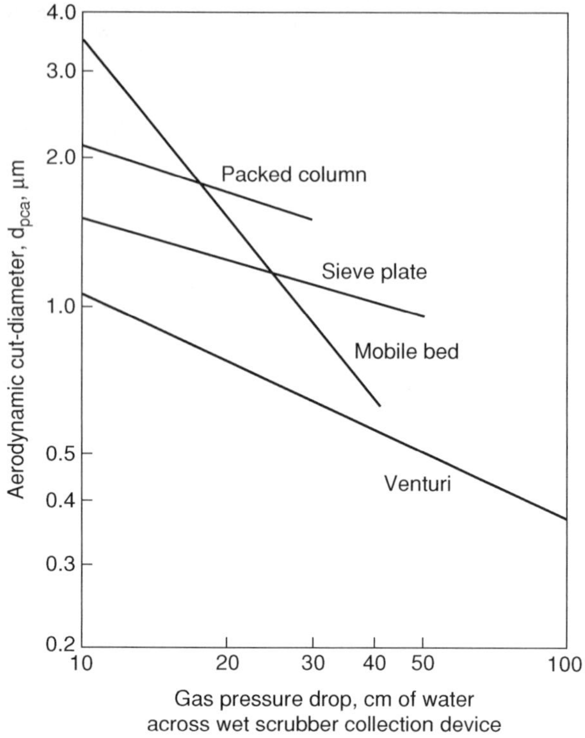

FIG. 14-130 Calvert's refined particle cut-size/power relationship for particle inertial impaction wet collectors. Ref. (R-19) by permission.

Subsequently, Calvert (R-19, p. 228) has combined mathematical modeling with performance tests on a variety of industrial scrubbers and has obtained a refinement of the power-input/cut-size relationship as shown in Fig. 14-130. He considers these relationships sufficiently reliable to use this data as a tool for selection of scrubber type and performance prediction. The power input for this figure is based solely on gas pressure drop across the device.

Collection of Fine Mists Inertial-impaction devices previously discussed give high efficiency on particles above 5 μm in size and often reasonable efficiency on particles down to 3 μm in size at moderate pressure drops. However, this mechanism becomes ineffective for particles smaller than 3 μm because of the particle gaslike mobility. Only impaction devices having extremely high energy input such as venturi scrubbers and a flooded mesh pad (the pad interstices really become miniature venturi scrubbers in parallel and in series) can give high collection efficiency on fine particles, defined as 2.5 or 3 μm and smaller, including the submicrometer range. Fine particles are subjected to brownian motion in gases, and diffusional deposition can be employed for their collection. Diffusional deposition becomes highly efficient as particles become smaller, especially below 0.2 to 0.3 μm. Table 14-28 shows typical displacement velocity of particles. Randomly oriented fiber beds having tortuous and narrow gas passages are suitable devices for utilizing this collection mechanism. (The diffusional collection mechanism is discussed in Sec. 17 under "Mechanisms of Dust Collection.") Other collection mechanisms which are efficient for fine particles are electrostatic forces and flux forces such as thermophoresis and diffusiophoresis. Particle growth and nucleation methods are also applicable. Efficient collection of fine particles

TABLE 14-28 Brownian Movement of Particles*

Particle diameter, μm	Brownian displacement of particle, μm/s
0.1	29.4
0.25	14.2
0.5	8.92
1.0	5.91
2.5	3.58
5.0	2.49
10.0	1.75

*Brink, *Can. J. Chem. Eng.*, **41,** 134 (1963). Based on spherical water particles in air at 21°C and 1 atm.

is important because particles in the range of 2.0 to around 0.2 μm are the ones which penetrate and are deposited in the lung most efficiently. Hence, particles in this range constitute the largest health hazard.

Fiber Mist Eliminators These devices are produced in various configurations. Generally, randomly oriented glass or polypropylene fibers are densely packed between reinforcing screens, producing fiber beds varying in thickness usually from 25 to 75 mm (1 to 3 in), although thicker beds can be produced. Units with efficiencies as high as 99.9 percent on fine particles have been developed (see *Chemical Engineers' Handbook,* 5th ed., p. 18–88). A combination of mechanisms interacts to provide high overall collection efficiency. Particles larger than 2 to 3 μm are collected on the fibers by inertial impaction and direct interception, while small particles are collected by brownian diffusion. When the device is designed to use this latter mechanism as the primary means, efficiency turndown problems are eliminated as collection efficiency by diffusion increases with residence time. Pressure drop through the beds increases with velocity to the first power since the gas flow is laminar. This leads to design capability trade-offs. As pressure drop is reduced and energy is conserved, capital increases because more filtering area is required for the same efficiency.

Three series of fiber mist eliminators are typically available. A spray-catcher series is designed primarily for essentially 100 percent capture of droplets larger than 3 μm. The high-velocity type is designed to give moderately high efficiency on particles down to 1.0 μm as well. Both of these types are usually produced in the form of flat panels of 25- to 50-mm (1- to 2-in) thickness. The high-efficiency type is illustrated in Fig. 14-131. As mist particles are collected, they coalesce into a liquid film which wets the fibers. Liquid is moved horizontally through the bed by the gas drag force and downward by gravity. It drains down the downstream retaining screen to the bottom of the element and is returned to the process through a liquid seal. Table 14-29 gives typical operating characteristics of the three types of collectors. The application of these devices to sulfuric acid plants and other process gases has been discussed by Brink (see *Chemical Engineers' Handbook*, 5th ed., pp. 18–89, 18–90).

Solid particulates are captured as readily as liquids in fiber beds but can rapidly plug the bed if they are insoluble. Fiber beds have frequently been used for mixtures of liquids and soluble solids and with soluble solids in condensing situations. Sufficient solvent (usually water) is atomized into the gas stream entering the collector to irrigate the fiber elements and dissolve the collected particulate. Such fiber beds have been used to collect fine fumes such as ammonium nitrate and ammonium chloride smokes, and oil mists from compressed air.

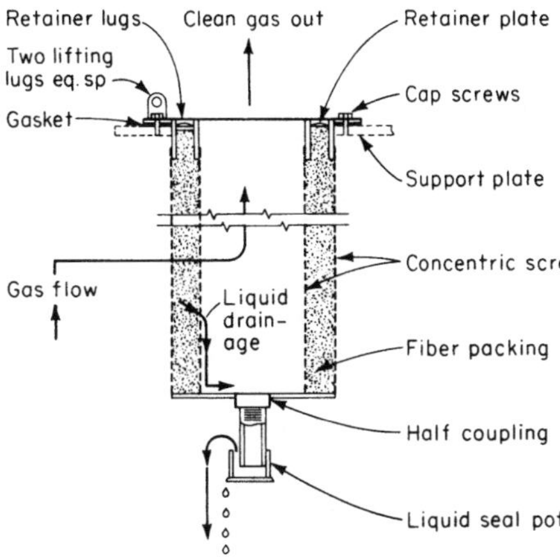

FIG. 14-131 Monsanto high-efficiency fiber-mist-eliminator element. (*Monsanto Company.*)

Electrostatic Precipitators The principles and operation of electrical precipitators are discussed in Sec. 17 under "Gas-Solids Separations." Precipitators are admirably suited to the collection of fine mists and mixtures of mists and solid particulates. Tube-type precipitators have been used for many years for the collection of acid mists and the removal of tar from coke-oven gas. The first practical installation of a precipitator by Cottrell was made on sulfuric acid mist in 1907. Most older installations of precipitators were tube-type rather than plate-type. However, recently two plate-type wet precipitators employing water sprays or overflowing weirs have been introduced by Mikropul Corporation [Bakke, *J. Air Pollut. Control Assoc.*, **25,** 163 (1975)] and by Fluid Ionics. Such precipitators operate on the principle of making all particles conductive when possible, which increases the particle migration velocity and collection efficiency. Under these conditions, particle dielectric strength becomes a much more important variable, and particles with a low dielectric constant such as condensed hydrocarbon mists become much more difficult to collect than water-wettable particles. Bakke (U.S.–U.S.S.R. Joint Work. Group Symp.: Fine Particle Control, San Francisco, 1974) has developed equations for particle charge and relative collection efficiency in wet precipitators that show the effect of dielectric constant. Wet precipitators can also be used to absorb soluble gases simultaneously by adjusting the pH or the chemical composition of the liquid spray. The presence of the electric field appears to enhance absorption. Wet precipitators have found their greatest usefulness to date in handling mixtures of gaseous pollutants and submicrometer particulate (either liquid or solid, or both) such as fumes from aluminum-pot lines, carbon anode baking, fiberglass-fume control, coke-oven and metallurgical operations, chemical incineration, and phosphate-fertilizer operations. Two-stage precipitators are used increasingly for moderate-volume gas streams containing nonconductive liquid mists which will drain from the collecting plates. Their application on hydrocarbon mists has been quite successful, but careful attention must be given to fire and explosion hazards.

Electrically Augmented Collectors A new area for enhancing collection efficiency and lowering cost is the combining of electrostatic forces with devices using other collecting mechanisms such as

TABLE 14-29 Operating Characteristics of Various Types of Fiber Mist Eliminators as Used on Sulfuric Acid Plants*

	High efficiency	High velocity	Spray catcher
Controlling mechanism for mist collection	Brownian movement	Impaction	Impaction
Superficial velocity, m/s	0.075–0.20	2.0–2.5	2.0–2.5
Efficiency on particles greater than 3 μm, %	Essentially 100	Essentially 100	Essentially 100
Efficiency on particles 3 μm and smaller, %	95–99+	90–98	15–30
Pressure drop, cm H_2O	12–38	15–20	1.0–2.5

*Brink, Burggrabe, and Greenwell, *Chem. Eng. Prog.*, **64**(11), 82 (1968). To convert centimeters to inches, multiply by 0.394.

impaction and diffusion. Cooper (Air Pollut. Control Assoc. Prepr. 75-02.1) evaluated the magnitude of forces operating between charged and uncharged particles and concluded that electrostatic attraction is the strongest collecting force operating on particles finer than 2 μm. Nielsen and Hill [*Ind. Eng. Chem. Fundam.,* **15,** 149 (1976)] have quantified these relationships, and a number of practical devices have been demonstrated. Pilat and Meyer (NTIS Publ. PB-252653, 1976) have demonstrated up to 99 percent collection of fine particles in a two-stage spray tower in which the inlet particles and water spray are charged with opposite polarity. The principle has been applied to retrofitting existing spray towers to enhance collection.

Klugman and Sheppard (Air Pollut. Control Assoc. Prepr. 75-30.3) have developed an ionizing wet scrubber in which the charged mist particles are collected in a grounded, irrigated cross-flow bed of Tellerette packing. Particles smaller than 1 μm have been collected with 98 percent efficiency by using two units in series. Dembinsky and Vicard (Air Pollut. Control Assoc. Prepr. 78-17.6) have used an electrically augmented low-pressure [5 to 10 cm (2 to 4 in) of water] venturi scrubber to give 95 to 98 percent collection efficiency on submicrometer particles.

Particle Growth and Nucleation Fine particles may be subjected to conditions favoring the growth of particles either through condensation or through coalescence. Saturation of a hot gas stream with water, followed by condensation on the particles acting as nuclei when the gas is cooled, can increase particle size and ease of collection. Addition of steam can produce the same results. Scrubbing of the humid gas with a cold liquid can bring diffusiophoresis into play. The introduction of cold liquid drops causes a reduction in water-vapor pressure at the surface of the cold drop. The resulting vapor-pressure gradient causes a hydrodynamic flow toward the drop known as Stefan flow which enhances the movement of mist particles toward the spray drop. If the molecular mass of the diffusing vapor is different from the carrier gas, this density difference also produces a driving force, and the sum of these forces is known as diffusiophoresis. A mathematical description of these forces has been presented by Calvert (R-9) and by Sparks and Pilat [*Atmos. Environ.,* **4,** 651 (1970)]. Thermal differences between the carrier gas and the cold scrubbing droplets can further enhance collection through thermophoresis. Calvert and Jhaseri [*J. Air Pollut. Control Assoc.,* **24,** 946 (1974)]; and NTIS Publ. PB-227307, 1973)] have investigated condensation scrubbing in multiple-sieve plate towers.

Submicrometer droplets can be coagulated through brownian diffusion if given ample time. The introduction of particles 50 to 100 times larger in diameter can enhance coagulation, but the addition of a broad range of particle sizes is discouraged. Increasing turbulence will aid coagulation, so fans to stir the gas or narrow, tortuous passages such as those of a packed bed can be beneficial. Sonic energy can also produce coagulation, especially the production of standing waves in the confines of long, narrow tubes. Addition of water and oil mists can sometimes aid sonic coagulation. Sulfuric acid mist [Danser, *Chem. Eng.,* 57(5), 158 (1950)] and carbon black [Stokes, *Chem. Eng. Prog.,* **46,** 423 (1950)] have been successfully agglomerated with sonic energy. Frequently sonic agglomeration has been unsuccessful because of the high energy requirement. Most sonic generators have very poor energy-transformation efficiency. Wegrzyn et al. (U.S. EPA Publ. EPA-600/7-79-004C, 1979, p. 233) have reviewed acoustic agglomerators. Mednikov (U.S.S.R. Akad. Soc. Moscow, 1963) suggested that the incorporation of sonic agglomeration with electrostatic precipitation could greatly reduce precipitator size.

Other Collectors Tarry particulates and other difficult-to-handle liquids have been collected on a dry, expendable phenol formaldehyde-bonded glass-fiber mat (Goldfield, *J. Air Pollut. Control Assoc.,* **20,** 466 (1970)] in roll form which is advanced intermittently into a filter frame. Superficial gas velocities are 2.5 to 3.5 m/s (8.2 to 11.5 ft/s), and pressure drop is typically 41 to 46 cm (16 to 18 in) of water. Collection efficiencies of 99 percent have been obtained on submicrometer particles. Brady [*Chem. Eng. Prog.,* **73**(8), 45 (1977)] has discussed a cleanable modification of this approach in which the gas is passed through a reticulated foam filter that is slowly rotated and solvent-cleaned.

In collecting very fine (mainly submicron) mists of a hazardous nature where one of the collectors previously discussed has been used as the primary one (fiber-mist eliminators of the Brownian diffusion type and electrically augmented collectors are primarily recommended), there is the chance that the effluent concentration may still be too high for atmospheric release when residual concentration must be in the range of 1–2 μm. In such situations, secondary treatment may be needed. Probably removal of the residual mist by adsorption will be in order. See "Adsorption," Sec. 16. Another possibility might be treatment of the remaining gas by membrane separation. A separator having a gas-permeable membrane that is essentially nonliquid-permeable could be useful. However, if the gas-flow volumes are appreciable, the device could be expensive. Most membranes have low capacity (requiring high membrane surface area) to handle high gas-permeation capacity. See "Membrane Separation Processes," Sec. 20.

Continuous Phase Uncertain Some situations exist such as in two-phase gas-liquid flow where the volume of the liquid phase may approach being equal to the volume of the vapor phase, and where it may be difficult to be sure which phase is the continuous phase. Svrcek and Monnery [*Chem. Eng. Prog.,* **89**(10), 53–60 (Oct. 1993)] have discussed the design of two-phase separation in a tank with gas-liquid separation in the middle, mist elimination in the top, and entrained gas-bubble removal from the liquid in the bottom. Monnery and Svrcek [*Chem. Eng. Prog.,* **90**(9), 29–40 (Sept. 1994)] have expanded the separation to include multiphase flow, where the components are a vapor and two immiscible liquids and these are also separated in a tank. A design approach for sizing the gas-liquid disengaging space in the vessel is given using a tangential tank inlet nozzle, followed by a wire mesh mist eliminator in the top of the vessel for final separation of entrained mist from the vapor. Design approaches and equations are also given for sizing the lower portion of the vessel for separation of the two immiscible liquid phases by settling and separation of discontinuous liquid droplets from the continuous liquid phase.

LIQUID-PHASE CONTINUOUS SYSTEMS

Practical separation techniques for gases dispersed in liquids are discussed. Processes and methods for dispersing gas in liquid have been discussed earlier in this section, together with information for predicting the bubble size produced. Gas-in-liquid dispersions are also produced in chemical reactions and electrochemical cells in which a gas is liberated. Such dispersions are likely to be much finer than those produced by the dispersion of a gas. Dispersions may also be unintentionally created in the vaporization of a liquid.

GENERAL REFERENCES: Adamson, *Physical Chemistry of Surfaces,* 4th ed., Wiley, New York, 1982. Akers, *Foams,* Academic, New York, 1976. Bikerman, *Foams,* Springer-Verlag, New York, 1973. Bikerman, et al., *Foams: Theory and Industrial Applications,* Reinhold, New York, 1953. Cheremisinoff, ed., *Encyclopedia of Fluid Mechanics,* vol. 3, Gulf Publishing, Houston, 1986. Kerner, *Foam Control Agents,* Noyes Data Corp, Park Ridge, NJ, 1976. Rubel, *Antifoaming and Defoaming Agents,* Noyes Data Corp., Park Ridge, NJ, 1972. Rosen, *Surfactants and Interfacial Phenomena,* 2d ed., Wiley, New York, 1989. Sonntag and Strenge, *Coagulation and Stability of Disperse Systems,* Halsted-Wiley, New York, 1972. Wilson, ed., *Foams: Physics, Chemistry and Structure,* Springer-Verlag, London, 1989. "Defoamers" and "Foams", *Encyclopedia of Chemical Technology,* 4th ed., vols. 7, 11, Wiley, New York, 1993–1994.

Types of Gas-in-Liquid Dispersions Two types of dispersions exist. In one, gas bubbles produce an unstable dispersion which separates readily under the influence of gravity once the mixture has been removed from the influence of the dispersing force. Gas-liquid contacting means such as bubble towers and gas-dispersing agitators are typical examples of equipment producing such dispersions. More difficulties may result in separation when the gas is dispersed in the form of bubbles only a few micrometers in size. An example is the evolution of gas from a liquid in which it has been dissolved or released through chemical reaction such as electrolysis. Coalescence of the dispersed phase can be helpful in such circumstances.

The second type is a stable dispersion, or foam. Separation can be extremely difficult in some cases. A pure two-component system of gas and liquid cannot produce dispersions of the second type. Stable foams can be produced only when an additional substance is adsorbed

TABLE 14-30 Terminal Velocity of Standard Air Bubbles Rising in Water at 20°C*

Bubble diameter, μm	10	30	50	100	200	300
Terminal velocity, mm/s	0.061	0.488	1.433	5.486	21.95	49.38

*Calculated from Stokes' law. To convert millimeters per second to feet per second, multiply by 0.003281.

at the liquid-surface interface. The substance adsorbed may be in true solution but with a chemical tendency to concentrate in the interface such as that of a surface-active agent, or it may be a finely divided solid which concentrates in the interface because it is only poorly wetted by the liquid. Surfactants and proteins are examples of soluble materials, while dust particles and extraneous dirt including traces of nonmiscible liquids can be examples of poorly wetted materials.

Separation of gases and liquids always involves coalescence, but enhancement of the rate of coalescence may be required only in difficult separations.

Separation of Unstable Systems The buoyancy of bubbles suspended in liquid can frequently be depended upon to cause the bubbles to rise to the surface and separate. This is a special case of gravity settling. The mixture is allowed to stand at rest or is moved along a flow path in laminar flow until the bubbles have surfaced. Table 14-30 shows the calculated rate of rise of air bubbles at atmospheric pressure in water at 20°C (68°F) as a function of diameter. It will be observed that the velocity of rise for 10-μm bubbles is very low, so that long separating times would be required for gas which is more finely dispersed.

For liquids other than water, the rise velocity can be approximated from Table 14-30 by multiplying by the liquid's specific gravity and the reciprocal of its viscosity (in centipoises). For bubbles larger than 100 μm, this procedure is erroneous, but the error is less than 15 percent for bubbles up to 1000 μm. More serious is the underlying assumption of Table 14-30 that the bubbles are rigid spheres. Circulation within the bubble causes notable increases in velocity in the range of 100 μm to 1 mm, and the flattening of bubbles 1 cm and larger appreciably decreases their velocity. However, in this latter size range the velocity is so high as to make separation a trivial problem.

In design of separating chambers, static vessels or continuous-flow tanks may be used. Care must be taken to protect the flow from turbulence, which could cause back mixing of partially separated fluids or which could carry unseparated liquids rapidly to the separated-liquid outlet. Vertical baffles to protect rising bubbles from flow currents are sometimes employed. Unseparated fluids should be distributed to the separating region as uniformly and with as little velocity as possible. When the bubble rise velocity is quite low, shallow tanks or flow channels should be used to minimize the residence time required.

Quite low velocity rise of bubbles due either to small bubble size or to high liquid viscosity can cause difficult situations. With low-viscosity liquids, separation-enhancing possibilities in addition to those previously enumerated are to sparge the liquid with large-diameter gas bubbles or to atomize the mixture as a spray into a tower. Large gas bubbles rising rapidly through the liquid collide with small bubbles and aid their coalescence through capture. Atomizing of the continuous phase reduces the distance that small gas bubbles must travel to reach a gas interface. Evacuation of the spray space can also be beneficial in promoting small-bubble growth and especially in promoting gas evolution when the gas has appreciable liquid solubility. Liquid heating will also reduce solubility.

Surfaces in the settling zone for bubble coalescence such as closely spaced vertical or inclined plates or tubes are beneficial. When clean low-viscosity fluids are involved, passage of the undegassed liquid through a tightly packed pad of mesh or fine fibers at low velocity will result in efficient bubble coalescence. Problems have been experienced in degassing a water-based organic solution that has been passed through an electrolytic cell for chemical reaction in which extremely fine bubbles of hydrogen gas are produced in the liquid within the cell. Near-total removal of hydrogen gas from the liquid is needed for process safety. This is extremely difficult to achieve by gravity settling alone because of the fine bubble size and the need for a coalescing surface. Utilization of a fine fiber media is strongly recommended in such situations. A low-forward liquid flow through the media is desireable to provide time for the bubbles to attach themselves to the fiber media through Brownian diffusion. Spielman and Goren [*Ind. Eng. Chem.*, **62**(10), (1970)] reviewed the literature on coalescence with porous media and reported their own experimental results [*Ind. Eng. Chem. Fundam.*, **11**(1), 73 (1972)] on the coalescence of oil-water liquid emulsions. The principles are applicable to a gas-in-liquid system. Glass-fiber mats composed of 3.5-, 6-, or 12-μm diameter fibers, varying in thickness from 1.3 to 3.3 mm, successfully coalesced and separated 1- to 7-μm oil droplets at superficial bed velocities of 0.02 to 1.5 cm/s (0.00067 to 0.049 ft/s).

In the deaeration of high-viscosity fluids such as polymers, the material is flowed in thin sheets along solid surfaces. Vacuum is applied to increase bubble size and hasten separation. The Versator (Cornell Machine Co.) degasses viscous liquids by spreading them into a thin film by centrifugal action as the liquids flow through an evacuated rotating bowl.

Separation of Foam Foam is a colloidal system containing relatively large volumes of dispersed gas in a relatively small volume of liquid. Foams are thermodynamically unstable with respect to separation into their components of gas and vapor, and appreciable surface energy is released in the bursting of foam bubbles. Foams are dynamic systems in which a third component produces a surface layer that is different in composition from the bulk of the liquid phase. The stabilizing effect of such components (often present only in trace amounts) can produce foams of troubling persistence in many operations. (Foams which have lasted for years when left undisturbed have been produced.) Bendure [TAPPI, 58(2), 83 (1975)], Keszthelyi [*J. Paint Technol.*, **46**(11), 31 (1974)], Ahmad [*Sep. Sci.* **10,** 649 (1975)], and Shedlovsky ("Foams," *Encyclopedia of Chemical Technology,* 2d ed., Wiley, New York, 1966) have presented concise articles on the characteristics and properties of foams in addition to the general references cited at the beginning of this subsection.

Foams can be a severe problem in chemical-processing steps involving gas-liquid interaction such as distillation, absorption, evaporation, chemical reaction, and particle separation and settling. It can also be a major problem in pulp and paper manufacture, oil-well drilling fluids, production of water-based paints, utilization of lubricants and hydraulic fluids, dyeing and sizing of textiles, operation of steam boilers, fermentation operations, polymerization, wet-process phosphoric acid concentration, adhesive production, and foam control in products such as detergents, waxes, printing inks, instant coffee, and glycol antifreeze.

Foams, as freshly generated, are gas emulsions with spherical bubbles separated by liquid films up to a few millimeters in thickness. They age rapidly by liquid drainage and form polyhedrals in which three bubbles intersect at corners with angles of approximately 120°. During drainage, the lamellae become increasingly thinner, especially in the center (only a few micrometers thickness), and more brittle. This feature indicates that with some foams if a foam layer can be tolerated, it may be self-limiting, as fresh foam is added to the bottom of the layer with drained foam collapsing on the top. (A quick-breaking foam may reach its maximum life cycle in 6 s. A moderately stable foam can persist for 140 s.) During drainage, gas from small foam bubbles, which is at a high pressure, will diffuse into large bubbles so that foam micelles increase with time. As drainage proceeds, weak areas in the lamella may develop. However, the presence of a higher concentration of surfactants in the surface produces a lower surface tension. As the lamella starts to fail, exposing bulk liquid with higher surface tension, the surface is renewed and healed. This is known as the *Marangoni effect.* If drainage can occur faster than Marangoni healing, a hole may develop in the lamella. The forces involved are such that collapse will occur in milliseconds without concern for rupture propagation. However, in very stable foams, electrostatic surface forces (zeta potential) prevent complete drainage and collapse. In some cases, stable lamella thicknesses of only two molecules have been measured.

Drainage rate is influenced by surface viscosity, which is very temperature-sensitive. At a critical temperature, which is a function of the system, a temperature change of only a few degrees can change a slow-draining foam to a fast-draining foam. This change in drainage rate can be a factor of 100 or more; thus increasing the temperature of foam can cause its destruction. An increase in temperature may also cause liquid evaporation and lamella thinning. As the lamellae become thinner, they become more brittle and fragile. Thus, mechanical deformation or pressure changes, which cause a change in gas-bubble volume, can also cause rupture.

Bendure indicates 10 ways to increase foam stability: (1) increase bulk liquid viscosity, (2) increase surface viscosity, (3) maintain thick walls (higher liquid-to-gas ratio), (4) reduce liquid surface tension, (5) increase surface elasticity, (6) increase surface concentration, (7) reduce surfactant-adsorption rate, (8) prevent liquid evaporation, (9) avoid mechanical stresses, and (10) eliminate foam inhibitors. Obviously, the reverse of each of these actions, when possible, is a way to control and break foam.

Physical Defoaming Techniques Typical physical defoaming techniques include mechanical methods for producing foam stress, thermal methods involving heating or cooling, and electrical methods. Combinations of these methods may also be employed, or they may be used in conjunction with chemical defoamers. Some methods are only moderately successful when conditions are present to reform the foam such as breaking foam on the surface of boiling liquids. In some cases it may be desirable to draw the foam off and treat it separately. Foam can always be stopped by removing the energy source creating it, but this is often impractical.

Thermal Methods Heating is often a suitable means of destroying foam. As indicated previously, raising the foam above a critical temperature (which must be determined experimentally) can greatly decrease the surface viscosity of the film and change the foam from a slow-draining to a fast-draining foam. Coupling such heating with a mechanical force such as a revolving paddle to cause foam deformation is frequently successful. Other effects of heating are expansion of the gas in the foam bubbles, which increases strain on the lamella walls as well as requiring their movement and flexing. Evaporation of solvent may occur causing thinning of the walls. At sufficiently high temperatures, desorption or decomposition of stabilizing substances may occur. Placing a high-temperature bank of steam coils at the maximum foam level is one control method. As the foam approaches or touches the coil, it collapses. The designer should consider the fact that the coil will frequently become coated with solute.

Application of radiant heat to a foam surface is also practiced. Depending on the situation, the radiant source may be electric lamps, Glowbar units, or gas-fired radiant burners. Hot gases from burners will enhance film drying of the foam. Heat may also be applied by jetting or spraying hot water on the foam. This is a combination of methods since the jetting produces mechanical shear, and the water itself provides dilution and change in foam-film composition. Newer approaches might include foam heating with the application of focused microwaves. This could be coupled with continuous or intermittent pressure fluctuations to stress lamella walls as the foam ages.

Cooling can also destroy foam if it is carried to the point of freezing since the formation of solvent crystals destroys the foam structure. Less drastic cooling such as spraying a hot foam with cold water may be effective. Cooling will reduce the gas pressure in the foam bubbles and may cause them to shrink. This is coupled with the effects of shear and dilution mentioned earlier. In general, moderate cooling will be less effective than heating since the surface viscosity is being modified in the direction of a more stable foam.

Mechanical Methods Static or rotating breaker bars or slowly revolving paddles are sometimes successful. Their application in conjunction with other methods is frequently better. As indicated in the theory of foams, they will work better if installed at a level at which the foam has had some time to age and drain. A rotating breaker works by deforming the foam, which causes rupture of the lamella walls. Rapidly moving slingers will throw the foam against the vessel wall and may cause impact on other foam outside the envelope of the slinger. In some instances, stationary bars or closely spaced plates will limit the rise of foam. The action here is primarily one of providing surface for coalescence of the foam. Wettability of the surface, whether moving or stationary, is frequently important. Usually a surface not wetted by the liquid is superior, just as is frequently the case of porous media for foam coalescence. However, in both cases there are exceptions for which wettable surfaces are preferred. Shkodin [*Kolloidn. Zh.,* **14,** 213 (1952)] found molasses foam to be destroyed by contact with a wax-coated rod and unaffected by a clean glass rod.

Goldberg and Rubin [*Ind. Eng. Chem. Process Des. Dev.,* **6** 195 (1967)] showed in tests with a disk spinning vertically to the foam layer that most mechanical procedures, whether centrifugation, mixing, or blowing through nozzles, consist basically of the application of shear stress. Subjecting foam to an air-jet impact can also provide a source of drying and evaporation from the film, especially if the air is heated. Other effective means of destroying bubbles are to lower a frame of metal points periodically into the foam or to shower the foam with falling solid particles.

Pressure and Acoustic Vibrations These methods for rupturing foam are really special forms of mechanical treatment. Change in pressure in the vessel containing the foam stresses the lamella walls by expanding or contracting the gas inside the foam bubbles. Oscillation of the vessel pressure subjects the foam to repeated film flexing. Parlow [*Zucker,* 3, 468 (1950)] controlled foam in sugar-sirup evaporators with high-frequency air pulses. It is by no means certain that high-frequency pulsing is necessary in all cases. Lower frequency and higher amplitude could be equally beneficial. Acoustic vibration is a similar phenomenon causing localized pressure oscillation by using sound waves. Impulses at 6 kHz have been found to break froth from coal flotation [Sun, *Min. Eng.,* **3,** 865 (1958)]. Sonntag and Strenge (*Coagulation and Stability of Disperse Systems,* Halsted-Wiley, New York, 1972, p. 121) report foam suppression with high-intensity sound waves (11 kHz, 150 dB) but indicate that the procedure is too expensive for large-scale application. The Sontrifuge (Teknika Inc., a subsidiary of Chemineer, Inc.) is a commercially available low-speed centrifuge employing sonic energy to break the foam. Walsh [*Chem. Process.,* **29,** 91 (1966)], Carlson [*Pap. Trade J.,* **151,** 38 (1967)], and Thorhildsen and Rich [*TAPPI,* **49,** 95A (1966)] have described the unit.

Electrical Methods As colloids, most foams typically have electrical double layers of charged ions which contribute to foam stability. Accordingly, foams can be broken by the influence of an external electric field. While few commercial applications have been developed, Sonntag and Strenge (op. cit., p. 114) indicate that foams can be broken by passage through devices much like electrostatic precipitators for dusts. Devices similar to two-stage precipitators having closely spaced plates of opposite polarity should be especially useful. Sonntag and Strenge, in experiments with liquid-liquid emulsions, indicate that the colloid structure can be broken at a field strength of the order of 8 to 9×10^5 V/cm.

Chemical Defoaming Techniques Sonntag and Strenge (op. cit., p. 111) indicate two chemical methods for foam breaking. One method is causing the stabilizing substances to be desorbed from the interface, such as by displacement with other more surface-active but nonstabilizing compounds. Heat may also cause desorption. The second method is to carry on chemical changes in the adsorption layer, leading to a new structure. Some defoamers may act purely by mechanical means but will be discussed in this subsection since their action is generally considered to be chemical in nature. Often chemical defoamers act in more than one way.

Chemical Defoamers The addition of chemical foam breakers is the most elegant way to break a foam. Effective defoamers cause very rapid disintegration of the foam and frequently need be present only in parts per million. The great diversity of compounds used for defoamers and the many different systems in which they are applied make a brief and orderly discussion of their selection difficult. Compounds needed to break aqueous foams may be different from those needed for aqueous-free systems. The majority of defoamers are insoluble or nonmiscible in the foam continuous phase, but some work best because of their ready solubility. Lichtman (*Defoamers,* 3d ed., Wiley, New York, 1979) has presented a concise summary of the application and use of defoamers. Rubel (*Antifoaming and Defoaming*

TABLE 14-31 Major Types and Applications of Defoamers

Classification	Examples	Applications
Silicones	Dimethyl silicone, trialkyl and tetraalkyl silanes	Lubricating oils; distillation; fermentation; jam and wine making; food processing
Aliphatic acids or esters	Mostly high-molecular-weight compounds; diethyl phthalate; lauric acid	Papermaking; wood-pulp suspensions; water-based paints; food processing
Alcohols	Moderate- to high-molecular-weight monohydric and polyhydric alcohols; octyl alcohol; C-12 to C-20 alcohols; lauryl alcohol	Distillation; fermentation; papermaking; glues and adhesives
Sulfates or sulfonates	Alkali metal salts of sulfated alcohols, sulfonic acid salts; alkyl-aryl sulfonates; sodium lauryl sulfate	Nonaqueous systems; mixed aqueous and nonaqueous systems; oil-well drilling muds; spent H_3SO_4 recovery; deep-fat frying
Amines or amides	Alkyl amines (undecyloctyl and diamyl methyl amine); polyamides (acyl derivatives of piperazine)	Boiler foam; sewage foam; fermentation; dye baths
Halogenated compounds	Fluochloro hydrocarbons with 5 to 50 C atoms; chlorinated hydrocarbons	Lubrication-oil and grease distillation; vegetable-protein glues
Natural products	Vegetable oils; waxes, mineral oils plus their sulfated derivatives (including those of animal oils and fats)	Sugar extraction; glue manufacture; cutting oils
Fatty-acid soaps	Alkali, alkaline earth, and other metal soaps; sodium stearate; aluminum stearate	Gear oils; paper stock; paper sizing; glue solutions
Inorganic compounds	Monosodium phosphate mixed with boric acid and ethyl carbonate, disodium phosphate; sodium aluminate, bentonite and other solids	Distillation; instant coffee; boiler feedwater; sugar extraction
Phosphates	Alkyl-alkalene diphosphates; tributyl phosphate in isopropanol	Petroleum-oil systems; foam control in soap solutions
Hydrophobic silica	Finely divided silica in polydimethyl siloxane	Aqueous foaming systems
Sulfides or thio derivatives	Metallic derivatives of thio ethers and disulfides, usually mixed with organic phosphite esters; long-chain alkyl thienyl ketones	Lubricating oils; boiler water

Agents, Noyes Data Corp., Park Ridge, N.J., 1972) has reviewed the extensive patent literature on defoamers. Defoamers are also discussed extensively in the general references at the beginning of this subsection.

One useful method of aqueous defoaming is to add a nonfoam stabilizing surfactant which is more surface-active than the stabilizing substance in the foam. Thus a foam stabilized with an ionic surfactant can be broken by the addition of a very surface-active but nonstabilizing silicone oil. The silicone displaces the foam stabilizer from the interface by virtue of its insolubility. However, it does not stabilize the foam because its foam films have poor elasticity and rupture easily.

A major requirement for a defoamer is cost-effectiveness. Accordingly, some useful characteristics are low volatility (to prevent stripping from the system before it is dispersed and does its work), ease of dispersion and strong spreading power, and surface attraction-orientation. Chemical defoamers must also be selected in regard to their possible effect on product quality and their environmental and health suitability. For instance, silicone antifoam agents are effective in textile jet dyeing but reduce the fire retardancy of the fabric. Mineral-oil defoamers in sugar evaporation have been replaced by specifically approved materials. The tendency is no longer to use a single defoamer compound but to use a formulation specially tailored for the application comprising carriers, secondary antifoam agents, emulsifiers, and stabilizing agents in addition to the primary defoamer. Carriers, usually hydrocarbon oils or water, serve as the vehicle to support the release and spread of the primary defoamer. Secondary defoamers may provide a synergistic effect for the primary defoamer or modify its properties such as spreadability or solubility. Emulsifiers may enhance the speed of dispersion, while stabilizing agents may enhance defoamer stability or shelf life.

Hydrophobic silica defoamers work on a basis which may not be chemical at all. They are basically finely divided solid silica particles dispersed in a hydrocarbon or silicone oil which serves as a spreading vehicle. Kulkarni [*Ind. Eng. Chem. Fundam.*, **16**, 472 (1977)] theorizes that this mixture defoams by the penetration of the silica particle into the bubble and the rupture of the wall. Table 14-31 lists major types of defoamers and typical applications.

Other Chemical Methods These methods rely chiefly on destroying the foam stabilizer or neutralizing its effect through methods other than displacement and are applicable when the process will permit changing the chemical environment. Forms stabilized with alkali esters can be broken by acidification since the equivalent free acids do not stabilize foam. Foams containing sulfated and sulfonated ionic detergents can be broken with the addition of fatty-acid soaps and calcium salts. Several theories have been proposed. One suggests that the surfactant is tied up in the foam as double calcium salts of both the sulfonate and the soap. Another suggests that calcium soaps oriented in the film render it inelastic.

Ionic surfactants adsorb at the foam interface and orient with the charged group immersed in the lamellae and their uncharged tails pointed into the gas stream. As the film drains, the charged groups, which repel each other, tend to be moved more closely together. The repulsive force between like charges hinders drainage and stabilizes the film. Addition of a salt or an electrolyte to the foam screens the repulsive effect, permits additional drainage, and can reduce foam stability.

Foam Prevention Chemical prevention of foam differs from defoaming only in that compounds or mixtures are added to a stream prior to processing to prevent the formation of foam either during processing or during customer use. Such additives, sometimes distinguished as antifoam agents, are usually in the same chemical class of materials as defoamers. However, they are usually specifically formulated for the application. Typical examples of products formulated with antifoam agents are laundry detergents (to control excess foaming), automotive antifreeze, instant coffee, and jet-aircraft fuel. Foaming in some chemical processes such as distillation or evaporation may be due to trace impurities such as surface-active agents. An alternative to antifoam agents is their removal before processing such as by treatment with activated carbon [Pool, *Chem. Process.*, **21**(9), 56 (1958)].

Automatic Foam Control In processing materials when foam can accumulate, it is often desirable to measure the height of the foam layer continuously and to dispense defoamer automatically as required to control the foam. Other corrective action can also be taken automatically. Methods of sensing the foam level have included electrodes in which the electrical circuit is completed when the foam touches the electrode [Nelson, *Ind. Eng. Chem.*, **48**, 2183 (1956); and Browne, U.S. Patent 2,981,693, 1961], floats designed to rise in a foam layer (Carter, U.S. Patent 3,154,577, 1964), and change in power input required to turn a foam-breaking impeller as the foam level rises (Yamashita, U.S. Patent 3,317,435, 1967). Timers to control the duration of defoamer addition have also been used. Browne has suggested automatic addition of defoamer through a porous wick when the foam level reaches the level of the wick. Foam control has also been discussed by Kroll [*Ind. Eng. Chem.*, **48**, 2190 (1956)].

Made in United States
Orlando, FL
04 November 2026

90206128R00072